Recent Topics in Electromagnetic Compatibility

Edited by Ahmed Kishk

Published in London, United Kingdom

IntechOpen

Supporting open minds since 2005

Recent Topics in Electromagnetic Compatibility
http://dx.doi.org/10.5772/intechopen.95160
Edited by Ahmed Kishk

Contributors
Ting-Wei Wang, Ting-Tse Lin, Donglin Meng, Sassia Hedia, Bessem Zitouna, Lionel Pichon, Pai-Yen Chen, Mohamed Farhat, Danilo Erricolo, Hakan Bagci, Muhammad Amin, Zhilu Ye, Roberto Linares y Miranda, Luis E. Martínez Santos, Fermin P. Espino-Cortés, Yuhua Guo, Jin Shichao, Fridolin Heidler, Kataka Frank Banaseka, Kofi Sarpong Adu-Manu, Godfred Yaw Koi-Akrofi, Selasie Aformaley Brown, Ion Razvan Radulescu, Bogdana Mitu, Cristian Morari, Manuel Sierra Castañer, Rubén Tena Sánchez, Lars Jacob Foged, Walid Labiedh, Mohamed Tlig, Jaleleddine Ben Hadj Slama, Debasish Nath, Antonio Carlos Siqueira S. Lima, Marco Aurélio O. Schroeder, Rodolfo Antônio R. Moura, Pedro H.N. Vieira, Bystrík Dolník, Shiva Hayati Raad, Lilioara Surdu, Emilia Visileanu, Yuhong Guo

First published in London, United Kingdom, 2022 by IntechOpen
IntechOpen is the global imprint of INTECHOPEN LIMITED, registered in England and Wales, registration number: 11086078, 5 Princes Gate Court, London, SW7 2QJ, United Kingdom
Printed in Croatia

British Library Cataloguing-in-Publication Data
A catalogue record for this book is available from the British Library

Additional hard and PDF copies can be obtained from orders@intechopen.com

Recent Topics in Electromagnetic Compatibility
Edited by Ahmed Kishk
p. cm.
Print ISBN 978-1-83969-668-8
Online ISBN 978-1-83969-669-5
eBook (PDF) ISBN 978-1-83969-670-1

For EU product safety concerns:
IN TECH d.o.o., Prolaz Marije Krucifikse Kozulić 3, 51000 Rijeka, Croatia, info@intechopen.com or visit our website at intechopen.com.

Meet the editor

Ahmed A. Kishk is a professor and Tier 1 Canada Research Chair in Advanced Antenna Systems at Concordia University, Montréal, Québec, Canada. His research interests include millimeter-wave antenna arrays and passive devices. He is the recipient of several international awards. He was a technical program committee member for several international conferences. He was an Institute of Electrical and Electronics Engineers (IEEE) Antennas and Propagation Society (AP-S) AdCom member from 2013 to 2015 and the 2017 AP-S president. Dr. Kishk is a life fellow of IEEE and a fellow of the Electromagnetic Academy and Applied Computational Electromagnetics Society (ACES). He is a member of several IEEE societies such as AP-S, Microwave Theory and Techniques Society (MTT-S), Electromagnetic Compatibility (EMC) Society, Signal Processing Society (SPS), Communications Society (ComSoc), Engineering in Medicine and Biology Society (EMBS), Aerospace and Electronic Systems Society (AESS), Vehicular Technology Society (VTS), and Photonics Society. He is also a senior member of the International Union of Radio Science, Commission B, and Applied Computational Electromagnetics Society.

Contents

Preface

Electromagnetic compatibility (EMC) and electromagnetic interference (EMI) are of interest in almost all aspects of modern daily life. The increased use of and dependence on wireless communication, remote control of critical devices, and machine-to-machine communications make the world more connected and thus more vulnerable to interference and attacks. Therefore, new applications require different treatments and materials that suit their respective environments. For example, flexible and smart reconfigurable material could be of use in biomedical applications. In avionic applications, lightweight, aerodynamic material is of interest. This book addresses recent developments in EMC and EMI in medicine and health care, automotive and avionic industries, the Internet of Things (IoT), industry, and mobile communication. It also presents and examines efficient design tools and testing methods. The book consists of five sections containing fifteen chapters.

The first section covers the electromagnetic spectrum of corona discharge (CD), which concentrates its energy at a very narrow signal below 70 MHz to a level that does not affect radio communication systems. Also addressed are clinical issues and EMC requirements for implantable medical devices such as cardiac pacemakers and vagus nerve stimulators. A life cycle assessment of flexible electromagnetic shields assesses flexible EM shields of fabrics with inserted conductive yarns with and without magnetron plasma coating. Results show that plasma-coated fabric has a substantial impact on the environment. In addition, the section discusses the design, construction, and validation of a high-performance open-area test site (OATS) based on long experience that helps calibrate laboratories and write EMC standards.

The second section analyzes the possibility of cost and space reduction for EMC systems based on multiprobe nearfield measurement systems in combination with OTA (over the air measurements), reference-less systems, spherical nearfield transformation, phase reconstruction, modal filtering, source reconstruction, and software-defined radio receivers. A chapter on sources reconstruction using electromagnetic time reversal (EMTR) identifies the transient disturbance sources in power electronics. This is followed by a study about the characterization of the electromagnetic nearfield radiated emissions using a time-domain analysis to provide an equivalent model consisting of a set of electromagnetic dipole parameters.

The third section discusses analysis and design of absorbers for EMC applications. It introduces analysis and modeling of ferrite-based absorbers for low-frequency applications (below 1 GHz). Different kinds of pyramidal absorbers are presented. Analytical and numerical approaches for predicting their performance and the combination of the ferrite tile and pyramidal dielectric absorbers, including the influence of carbon loading and the use of a matching layer on their performance, are also discussed. This section introduces graphene material in addition to the millimeter-wave application using metamaterial. Artificial surfaces and media for EMC and EMI shielding are discussed for recent advances in lightweight, low-profile electromagnetic absorbing media, such as metamaterials, metasurfaces, and nanomaterial-based solutions, which may suppress unwanted RF and microwave noises.

The fourth section focuses on EMI issues. EMI pre-compliance measurements reveal sources of interference focusing on the EMC of a prototype of electrical equipment such as LED boards and audio equipment. Electromagnetic emissions due to power converters and methods to measure and reduce the noise fields are presented through simulations and computations. Engineers and researchers working in the development of electrical equipment and the general public interested in EMC issues will find this information helpful.

Finally, the book presents several computational methods on different aspects of EMC. It introduces the Traveling Current Source (TCS) model fundamentals and current reflections at the ground and the upper end of the return stroke channel. It also includes a generic model of the radiation from rectangular film capacitors as a power electronics component, which are sources of electromagnetic radiation. An efficient simulation approach based on hybrid time and frequency-domain algorithms is presented. The computational complexity of the proposed algorithm is analyzed, demonstrating the effectiveness and efficiency of the proposed approach of modeling grounding systems for application in EMC to illustrate the impact of nearby grounding grids. A two-dimensional finite element method of electromagnetic wave propagation through the soil is also formulated. In addition, the boundary value problem is employed to solve the Helmholtz time-harmonic electromagnetic model.

Ahmed Kishk
Department of Electrical and Computer Engineering,
Concordia University,
Montreal, Canada

Section 1

Some Issues in EMC

Chapter 1

Electromagnetic Spectrum of the Corona Discharge and Their Fundamental Frequency

Luis E. Martínez Santos, Roberto Linares y Miranda and Fermín P. Espino-Cortés

Abstract

Historically, the Electromagnetic Compatibility (EMC) began with the disturbances at the radio navigation systems generated by the electrical power distribution lines; hence it was referred to as Radio Interference (RI). This disturbance is an Electromagnetic Interference (EMI). Although this type of EMI has been studied since the first decades of the past century, it still maintains a continued interest of the researchers, especially with the Corona Discharge (CD), generated by High Voltage Direct Current (HVDC) systems. Because of its design criterion and the concern that this phenomenon may affect the new radio communication systems in the very high frequency (VHF), ultra high frequency (UHF), and microwave bands, interest in their studies continues. In this chapter, an analysis of the electromagnetic spectrum of the CD is presented. The CD is generated at a short transmission line located within a semi-anechoic chamber. To be sure of the phenomenon, the CD is identified by its current pulse, which is well studied. The instruments used are an oscilloscope of 2 GHz and 2 GS/s, a spectrum analyzer, and an EMI test receiver. The results show that the CD concentrates its energy at frequencies below 70 MHz. In the UHF band, only narrowband signals very separated were found, with levels that cannot affect radio communication systems.

Keywords: corona discharge, electromagnetic spectrum, radio interferences

1. Introduction

The issue of electromagnetic coexistence of the Corona Discharge (CD) generated by High Voltage (HV) systems, with broadcasting, especially with the Amplitude Modulated (AM), can be said to be thoroughly studied. However, disturbances persist because the CD is an inherent part of HV systems. Now, broader spectrum must be analyzed, due to possible effects on all electronic systems, including the radio communication by their high sensitivity at the VHF, UHF, and microwave band, that may be in the presence of said systems, especially at HVDC systems [1–3]. In addition, this issue is a design requirement for high voltage transmission lines [4–6]. These problems are subjects of Electromagnetic Compatibility (EMC), which were born from the beginning of AM radio [7]. Nowadays, radio communication technologies have changed, as well as services and their applications. Thus, Electromagnetic Interferences (EMI) have increased because there is a higher density of high sensitivity systems [8]. The EMI is identified in the high voltage power distribution area as Radio Interference (RI).

The HV power and radio communication systems are every day in human life. Therefore, all of them should be coexist, and they must comply with the conformity assessment of the EMC. However, the rules are not yet very clear with respect to CD at frequencies greater than 30 MHz [9]. In the frequency range of 30 MHz–1 GHz, there are several investigations about RI, but some of them are associated with CD since the detection, in general, corresponds to partial discharges (PD) because the measurements are made in uncontrolled electromagnetic environments [8, 10]. Within the literature on Direct Current Corona Discharge (DC-CD), there are no reports of the frequency spectrum; they only exist for AC [11]. In general, the measurements reported are related to the electromagnetic radiation of the DC-CD. They have been carried out as a function of time, whose data are processed by software to determine the electromagnetic spectrum of the phenomenon [12, 13]. In the analytical part, the developments that have served as a reference for both AC and DC corona discharge are reported in [14, 15].

With respect to the standards, these must be developed in accordance with technological advances. However, they are always delayed, e.g., with the change that is taking place in the transmission of high voltage from AC to DC and with the increase of radio communication services. The CISPR-18 standard [9] in its recent version covers the range of 150 kHz–3 GHz, but it does not define the limits at frequencies greater than 30 MHz due to the great uncertainty that exists. This fact indicates that there are challenges for researchers on the subject since now it is had much more sensitive and complex electronic systems, such as electric vehicles and smart grids, among other systems. Therefore, it is first necessary to know the levels of electromagnetic emissions of CD in the frequency range specified by the CISPR in order to define the interference levels that exceed the immunity level of high-sensitivity systems.

This chapter presents an experimental analysis of the electromagnetic spectrum of CD, covering the frequency range from 150 kHz to 1 GHz. The aim is to respond to the problems that are constantly manifested with regard to interference from radio communication systems. Despite, it is known that the CD effect is very strong in the AM frequency band, be it analog or digital, but in the frequency bands above 30 MHz, many questions remain. The experimentation is carried out within an electromagnetically controlled environment that is a semi-anechoic chamber, and a short transmission line is used as the source. Measurements were carried out in the time domain and the frequency domain, using three antennas to cover the specified frequency range. Filters based on the Wavelet Transform (WT) were applied to the results obtained to smooth or minimize the noise in order to better appreciate the levels of the radiated signal from the CD, according to the procedure of [14]. WT was applied to the time domain response to obtain a spectrogram and better observe the distribution of energy levels. The results present information that can be compared with the sensitivity levels of electronic systems, including radio communication systems, in order to define whether these can be interfered with in a corona discharge environment.

2. Experimental setup

The measurements for analysis of CD electromagnetic spectrum on-situ or at HV laboratories are not reliable because the environment is contaminated by various RF signals and reflection multiples. Thus, it is necessary for these measurements to use a controlled environment, such as an anechoic or semi-anechoic

Antenna	Mark	Model	Wide band (MHz)
Loop-antenna	ETS Lindgren (EMCO)	6507	0.001–30
Biconical antenna	ETS Lindgren (EMCO)	3110B	30–300
SAS 521-1 hybrid biconical/log periodic antenna	A. H. Systems, Inc.	SAS-521-2	25–2000

Table 1.
Antenna's characteristics.

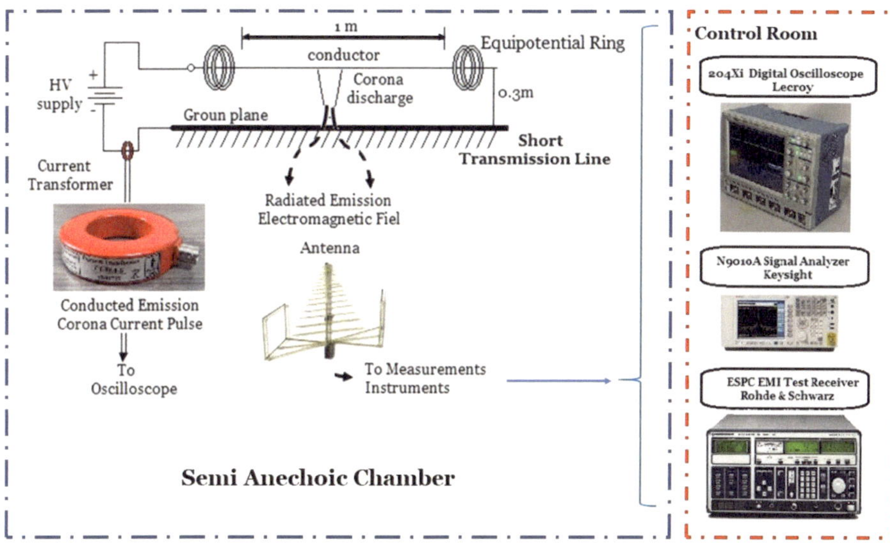

(a)

(b)

Figure 1.
Experimental setup scheme and short transmission lines. (a) Experimental setup. (b) Short transmission lines [16].

chamber. In our case, we used a semi-anechoic chamber of EMC. As a CD source, a short transmission line was used, which was supplied with a positive DC high power supply (Bertan 205B Series). The measurement of the conducted corona current pulse is carried out with a high-frequency current transformer, which has an operating range of 4.8 kHz–400 MHz (Bergoz CT-E5.0). To cover the frequency range of the radiated emission, three antennas were used, which characteristics are shown in **Table 1**. As measurements instruments are used: oscilloscope (204Xi Lecroy, 2 GHz, 10 GS/s), signal analyzer (N9010A EXA X Keysight, 10 MHz–7GHz), and EMI test receiver (ESPC EMI Test Receiver Rohde & Schwarz, 9KHz–1GHz).

The schemes of the experimental setup and a picture of the short transmission line short are shown in **Figure 1**.

The scheme of the experimental setup is shown in **Figure 1a**, and a picture of the short transmission line short is shown in **Figure 1b**.

The short transmission line was made with a magnetic wire 28-gauge AWG of long 1 m and a ground plane with 0.3 m of separation. The line is fixed on a wood table with a dimension of 0.8 × 1.2 × 0.8 m^3, high, long, and wide, respectively. The current transformer (CT) is placed at the retorn plane cable, and the antennas were fixed at 1.5 m of the corona source. This kit was placed inside the semi-anechoic chamber.

3. Procedure

Measurements were carried out in a semi-anechoic chamber located at a latitude of 2250 m and pressure of 585 mmHg (Mexico City). The temperature inside the chamber was 25°C with a relative humidity of 50%. To start the measurements, the electromagnetic environment was first analyzed in the range of measurement frequencies under test in order to ensure that there would not be a presence of spurious signals.

CD onset corona voltage is well identified for the different configurations of needle plane, corona cage, and others [17, 18]. A good reference for the presence of the phenomenon is the corona current pulse, which is detected in the conducted form [18]. The detection of such as current may be carried out between the power supply and configuration under test or in return plane. In the first case, a coupling circuit is needed, which is a complex method of measurement. The measurement of corona current at the return plane is simpler because measurements with a current probe are enough as a CT.

The rise time of the corona current pulse gives information of the phenomenon, and the fall time is a function of the parasitic electrical components of the configuration. Therefore, to have a better shaped pulse, it is necessary to optimize the circuit of the discharge configuration. Knowing the rise time of the corona pulse, the fundamental frequency can be determined, both for conducted and radiated radio electric emissions [19].

To carry out the electromagnetic spectrum analysis of the corona discharge, we did the following:

1. Equipment installation: Inside the semi-anechoic chamber, the short transmission line was installed, as well as the antennas were fixed at 1.5 m in front of the line.

2. Measurement of corona current conducted: The corona current was measured in the return plane with the CT and the oscilloscope. With the presence of the pulse, we ensure that the corona phenomenon is present. This was done at every measurement.

3. Measurement radiation of the corona discharge: Due to the dimension of the short transmission line and the semi-anechoic chamber dimensions, we could not perform measurements in the far field. Therefore, the measurements were

carried out in the near field. The results are presented in dBμV; from these, the current can be calculated to predict the far field.

4. Measurement of frequency spectrum: This measurement is carried out in accordance with the frequency ranges, where are more activities of EMI: 0.15–30 MHz, 30–300 MHz, and 470–950 MHz, as is specified in the CISPR-18-1 [20]. For the first range, the loop antenna is used; for the second range, the biconical antenna is used, and for the third range, the hybrid biconical/log periodic antenna is used. These antennas used the signal analyzer and EMI test receiver for covering the frequency band of 150 kHz until 1 GHz. The signal analyzer has several functions; in this case, it is used as a spectrum analyzer.

5. Measurement of corona current radiated in the time domain: The corona current in the time domain was determined by the antennas specified in **Table 1** and the oscilloscope.

6. Antennas polarization: For the case of the loop antenna, the three components of the field were measured (x, y, z), and the total field was obtained. For the other antennas, the two polarizations considered were vertical and horizontal.

7. Presentation of the results: All results were smoothed by filters using the wavelet transform to estimate the emissions radiated by corona discharge in time and frequency domains simultaneously.

4. Results

4.1 Conducted emission

The CD analyzed in the laboratories is identified in different ways. As already mentioned, one of the ways is by means of the conducted current. Therefore, we connected the CT to the return plane, and we observed in the oscilloscope the pulse. With this, we identified the presence of the discharge for an onset voltage of 12 kV ± 100 V. The current of corona pulse obtained of conducted way is shown in **Figure 2**, where **Figure 2a** corresponds to the pulse shape, and **Figure 2b** is the spectrogram obtained by WT.

In accordance with the spectrogram of **Figure 2**, the energies are concentrated at the range of 7 MHz and 16 MHz. In this last range is where there is the greatest amount of energy.

4.2 Radiated emission

The electromagnetic radiation of the CD generated by the short transmission line under test was measured in the time and frequency domain covering the frequency ranges in accordance with CISPR 16-1-1 (**Table 2**) [21]: band B, C, and D. For the time domain an oscilloscope was used, and for the frequency domain a signal analyzer in spectrum analyzer mode and an EMI test receiver were used. In both cases, the antenna factors were considered.

In order to cover the frequency intervals, three antennas were used, which are specified in **Table 2**. Whose frequency bands are the basis for displaying the results.

The results of the antennas are presented in dBμV vs. frequency, which can be converted to current, to predict the far field using the expression of a short dipole in the first instance [22].

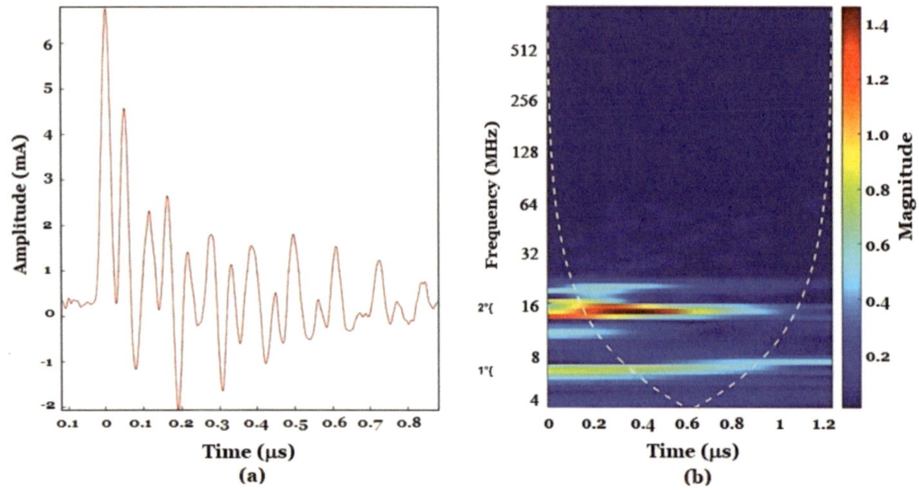

Figure 2.
Corona discharge pulse conducted. (a) pulse shape; (b) WT spectrogram.

Antenna	Frequency bands
Loop antenna	Band B (300 kHz–30 MHz)
Biconical antenna	Band C (30–300 MHz)
SAS 521-1 hybrid biconical/log periodic antenna	Band D (300–1000 MHz)

Table 2.
Antennas and frequency band accordance with CSPR 16-1-1 [21].

4.2.1 Frequency band B (0.3–30 MHz)

In this frequency range, the loop antenna was used. The result corresponds to the total field formed by the three components (x, y, z). For the temporal part, an oscilloscope was used, whose data was processed using the WT to obtain its spectrograms; these results are shown in **Figure 3**. In the frequential part, we made two measurements: one with the spectrum analyzer and the other which EMI test receiver. The results are expressed in dBμV for both instruments, as shown in **Figure 4**, where graph (a) corresponds to the spectrum analyzer and graph (b) corresponds to EMI test receiver.

The results show that most amount of energy is concentrated in the 8 MHz, 11 MHz, 18 MHz and 25 MHz range; this agrees both in the temporal part and in the frequency part. At frequencies below 15 MHz, the EMI test receiver exhibits a higher signal level because it responds to the envelope and has a higher fall time constant on the input filter. An appropriate resolution bandwidth (RBW) in the spectrum analyzer allows a sweep so that the peaks of the RI are seen with greater definition, which makes the frequencies higher than 15 MHz the graphs of both instruments coincide.

4.2.2 Frequency band C (30–300 MHz)

The antenna used in this frequency range is biconical, which is recommended for this application by international standards. This antenna is basically a dipole with conical shape elements and has linear polarization. The results of the radiated emissions in the 30–300 MHz frequency range are shown for vertical and horizontal polarization. For the temporal response, the pulse with its spectrogram is shown in

DOI: http://dx.doi.org/10.5772/intechopen.101550

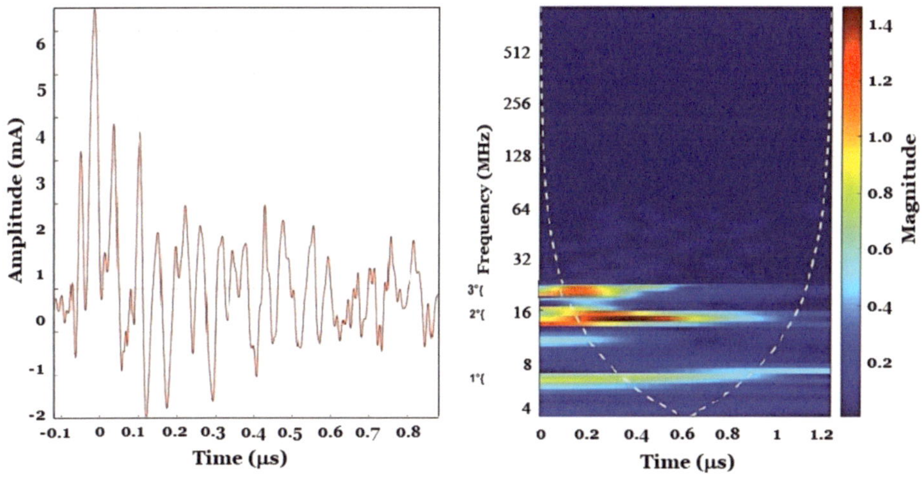

Figure 3.
Temporal answer of the corona discharge at B band (0.3–30 MHz) measured with a loop antenna.

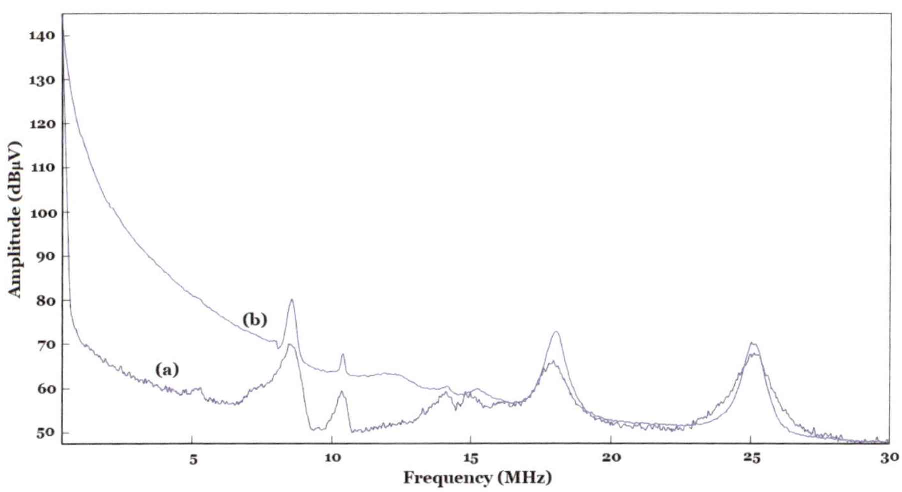

Figure 4.
Frequency spectrum of the corona discharge at B band (0.3–30 MHz) measured with a loop antenna: (a) spectrum analyzer, (b) EMI test receiver.

Figure 5. The frequency response measured with a spectrum analyzer and with the EMI test receiver are shown in **Figure 6**.

In this frequency band, we have the following:

In the spectrogram, it can be observed that the energy concentration for vertical polarization covers the range from 10 to 65 MHz. In this case, the measurement frequency range is not limited, and it is understood that the antenna responds outside its operating band.

With the frequency behavior, the measurement with the spectrum analyzer in vertical polarization shows that there are signal levels in almost the entire band, concentrating in the 30–60 MHz range. In EMI test receiver also the energy of RI is concentrating at 30 MHz and 60 MHz, but there is no signal at higher frequencies. For horizontal polarization, both measuring instruments present the same behavior, concentrating the greatest number of electromagnetic emissions in the 30–60 MHz interval; the energy is most high, around 40 MHz.

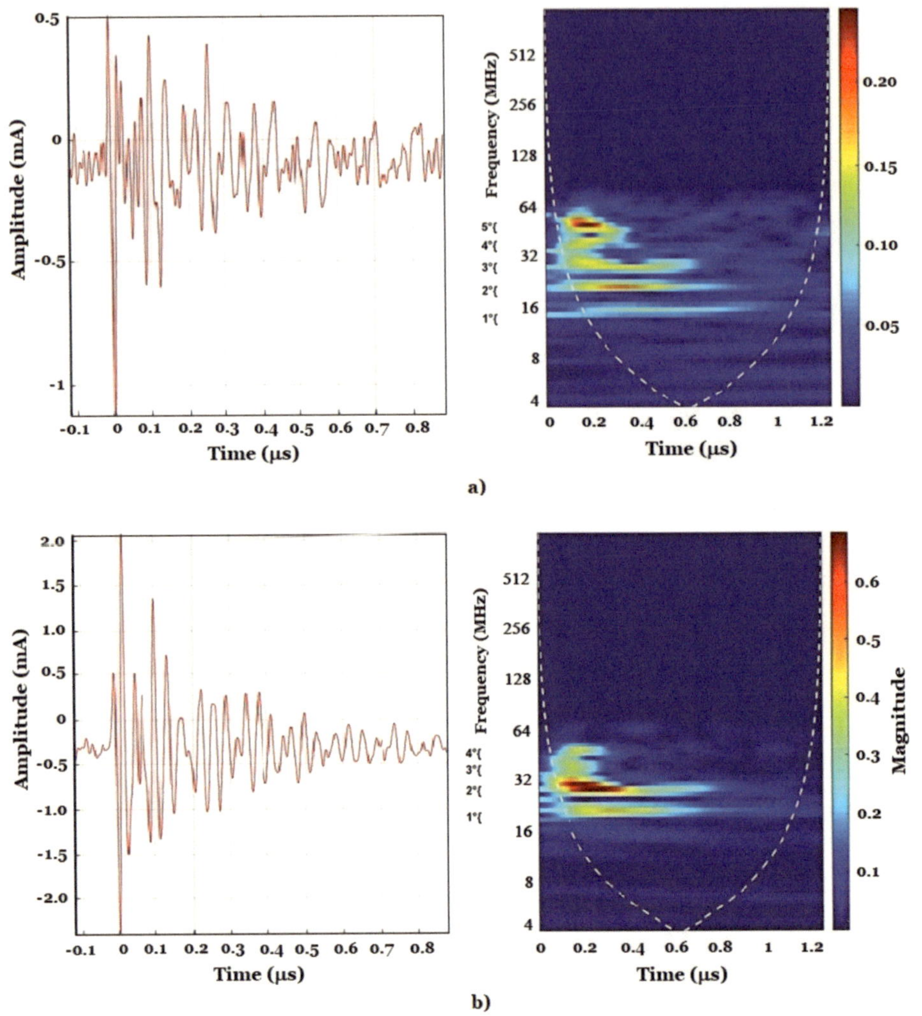

Figure 5.
Temporal answer of the corona discharge at C band (30–300 MHz) measured with a biconical antenna. (a) Vertical polarization. (b) Horizontal polarization.

According to the above, the highest amount of electromagnetic energy radiated by a corona discharge generated by a short transmission line is in the range of 30–60 MHz for both polarizations. In vertical polarization, some significant emissions at frequencies greater than 100 MHz, such as in 120 MHz, are only identified with the spectrum analyzer due to their higher resolution.

The noise floor of each instrument is an important part. In the case of the spectrum analyzer, the noise is greater due to the used span. In EMI test receiver, the noise is less and can not identify narrowband signals, because detect only the envelope and the filters are not able to resolve those signals. The signal levels in the high activity band of RI are almost equal, considering the background noise in both instruments.

4.2.3 Frequency band D (300–1000 MHz)

This frequency band is important because there are lots of portable radio communication services. For example, some of the services are the trunk radio, which

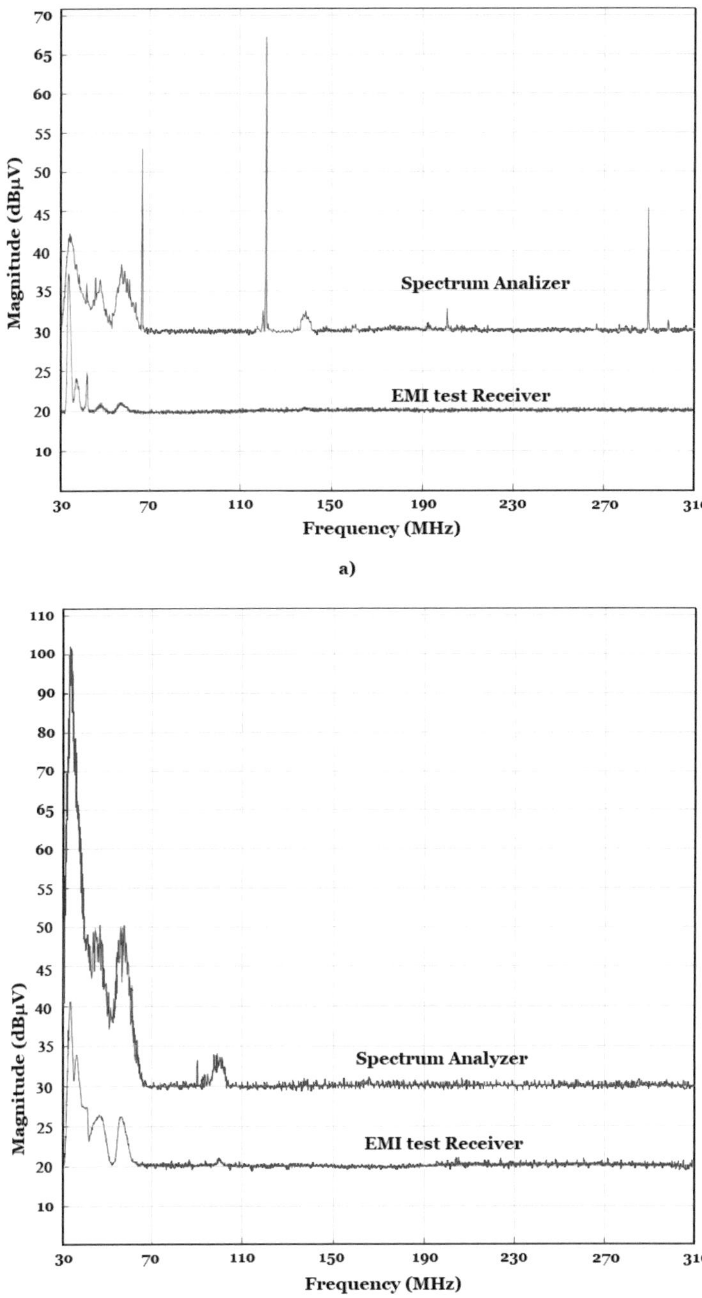

Figure 6.
Frequency spectrum of the corona discharge at C band (30–300 MHz) measured with a biconical antenna. (a) Vertical polarization. (b) Horizontal polarization.

is a typical means of communication used by those who maintain the high-voltage transmission lines, the open digital television, the first cell phone band that one continues to apply, systems 5G and others. Therefore, it is necessary to analyze the levels of electromagnetic emissions from the corona phenomenon. In this case, a hybrid biconical/log periodic antenna was used to cover the D band frequency range. The results in the time domain for vertical and horizontal polarization are presented in **Figure 7**, including the spectrogram. For the frequency domain part at both polarities with the spectrum analyzer and the EMI test receiver, the spectra are shown in **Figure 8**.

From the temporal results in this band, it is observed that there is noise only for vertical polarization. Then, it is difficult to identify if this corresponds to the corona discharge. As the measurement is made with an oscilloscope, the time base that is set is where the maximum amount of energy of the phenomenon occurs. What we see corresponds to the distribution of the noise signal at frequencies related to the time base of the oscilloscope. However, for horizontal polarization, the energy is concentrated in the frequency bands where the corona is identified.

Regarding the frequency part, in this case, the measurement with the spectrum analyzer and the antenna at vertical polarization, the spectrum shows a signal at the frequency of 336 MHz and noise. The same signal of 336 MHz is present with the horizontal polarization and one at 960 MHz. These signals are not seen with the EMI test receiver because they are narrowband. Therefore, in both polarizations, only the background noise is seen.

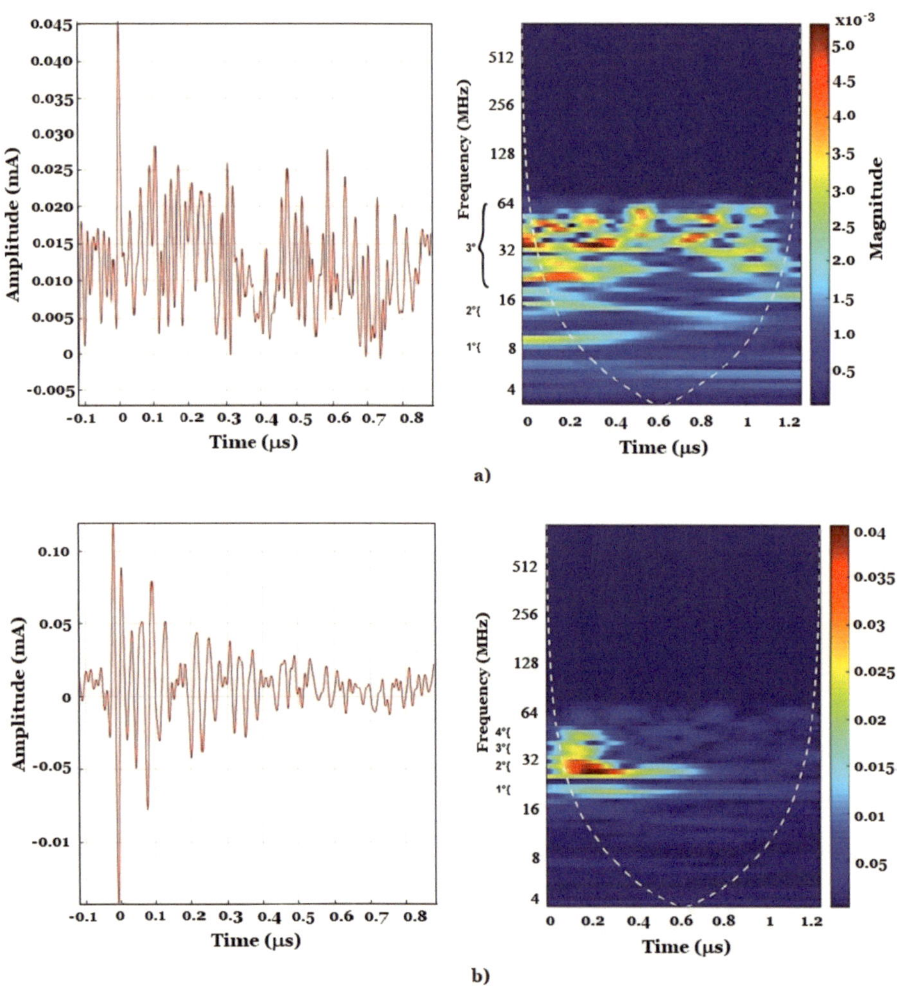

Figure 7.
Temporal answer of the corona discharge at D band (300–1000 MHz) measured with a hybrid biconical/log periodic antenna. (a) Vertical polarization. (b) Horizontal polarization.

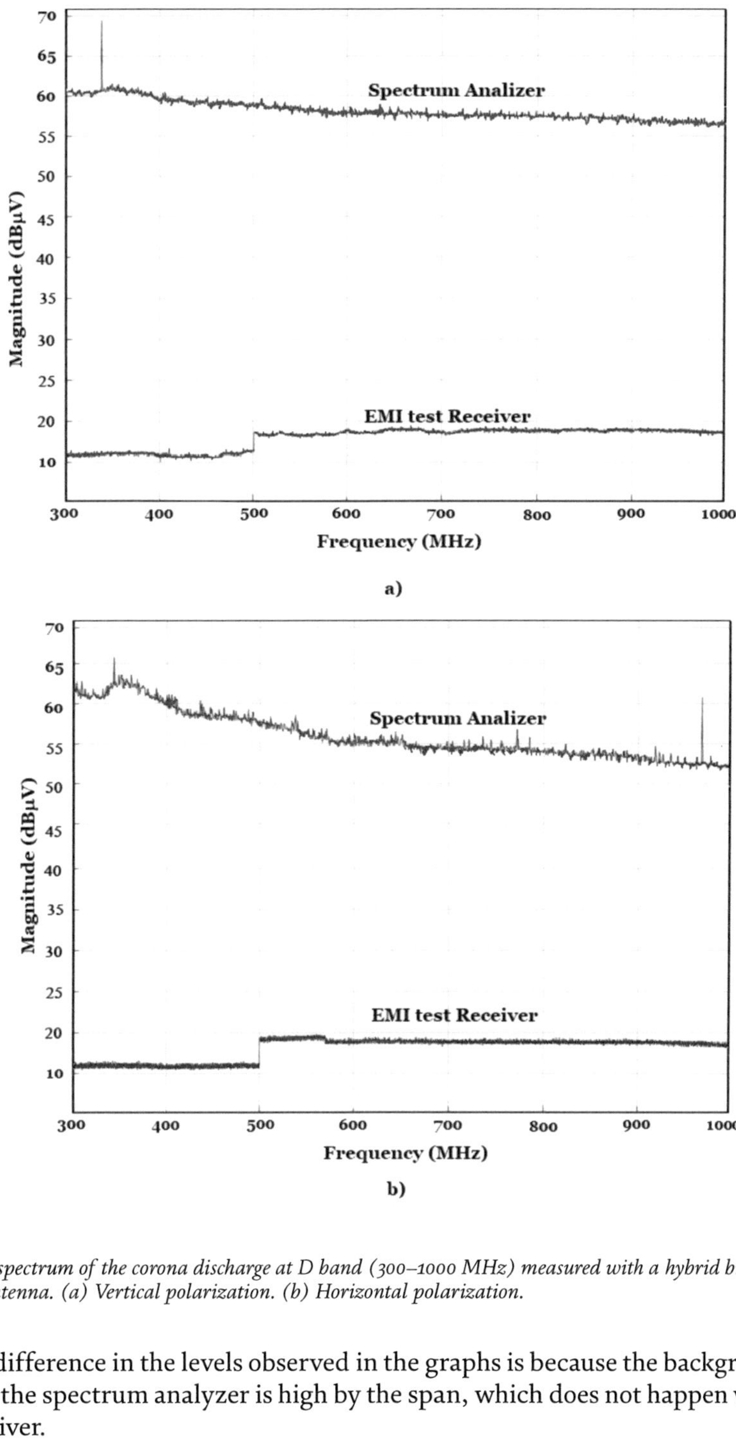

Figure 8.
Frequency spectrum of the corona discharge at D band (300–1000 MHz) measured with a hybrid biconical/log periodic antenna. (a) Vertical polarization. (b) Horizontal polarization.

The difference in the levels observed in the graphs is because the background noise of the spectrum analyzer is high by the span, which does not happen with EMI test receiver.

5. Discussion

The measurements of the electromagnetic radiation generated by corona discharge at media electromagnetically controlled (semi-anechoic chamber) give us information secure in time and frequency domains about the energy amount emitted by this phenomenon to the environment. In such as environment, only the signal referring to the corona was present. We ensure the presence of the corona by measuring the conducted current, which is a well-known parameter. This current was monitored in all measurements. The emissions radiated in dBμV are indicated, and those are compared to observe the frequency behavior, where are shown the band of most activity of RI.

In frequency band B, we measure one decade from 0.3 MHz to 30 MHz. The antenna used is a loop, so its three components (x, y, z) were measured. In the time domain, the current pulse was obtained, and the wavelet transform was applied in order to obtain information on the energy distribution. In the frequency domain, measurements were made with a spectrum analyzer and an EMI test receiver with the peak detector. In this case, the information coincides with the frequency. At frequencies lower than 6 MHz, what is observed is the background noise in both instruments. The spectrum analyzer presented less noise since the span and RBW were conditioned for this decade, which implies less bandwidth in both parameters. In this regard, the frequency peaks are matching.

For the C frequency band that corresponds to 30–300 MHz, a biconical antenna was used. The answer of the time domain that corresponds to the current pulse was processed by the transform wavelet, which is shown the energy distribution. For this case, the major amount of energy is below that of the 70 MHz. The vertical polarization presents more activity of emission than polarization horizontal, which is also shown with both the spectrum analyzer and the EMI test receiver. The answers at both instruments are similar, where the greatest amount of energy is found. In vertical polarization, the spectrum obtained with the spectrum analyzer shows two peaks at 120 MHz and 260 MHz, which are not shown in the measurements with EMI test receiver. For the horizontal polarization, the spectrums of both instruments have the same performance.

Measurements in the D frequency band (300 MHz–1GHz) with vertical polarization, only a signal in the 335 MHz frequency was obtained, with a low amplitude level; the rest is the background noise. This is corroborated with the EMI test receiver and with temporal measurements, where the distribution in the spectrogram is observed. In the case of horizontal polarization, in the temporal part, a more defined signal can be seen with respect to the corona, keeping the energy concentration at 16 MHz and 32 MHz, as observed in the spectrogram. In the response of the spectrum analyzer, there are two signals, one at 335 MHz and the other at 960 MHz; the first is also presented with vertical polarization, the rest is background noise. Measurements with the EMI test receiver in this frequency range do not present information; only the background noise is seen.

As an important part of this work analysis, we can say that the measurements of the electromagnetic radiation of corona discharges are concentrated below 70 MHz, which has already been reported. However, it does not present a spectrogram like we do, which gives information on the distribution of radiation both in time and in frequency. It was also determined that, for frequencies greater than 30 MHz, the most appropriate is to use a spectrum analyzer since the bandwidth resolution can be adjusted to obtain more frequency components. With the EMI test receiver, the envelope of the signal is obtained by its filter; it has a large recovery constant.

The measurements of the corona discharge in a semi-anechoic chamber and safely verify the presence of the phenomenon gives the certainty that the emissions only correspond to this discharge and that it can also be ensured that there are no interfering signals in the band from 300 MHz to 1 GHz. Therefore, radio communication systems that are highly sensitive cannot be affected by this phenomenon. This leads us to ensure that the measurements that have been reported regarding this problem both on-site and the development of antennas have a complete spectrum of partial discharges where corona discharge may be included.

For measurements of radiated emissions in open spaces, the fact of identifying the corona with an ultraviolet camera does not indicate that only the corona discharge is present, it may be the main discharge with frequencies of partial discharges, but as the corona has been verified, it does not emit high-frequency signals (UHF).

In HV systems greater than those used (>12 kV) in this work, of course, the corona discharge has levels of radiated emissions higher than those obtained. In the bands with higher activity (<60 MHz), their presence is significant, but at high frequencies (UHF and microwaves), this phenomenon is not a problem for radio-communication systems.

6. Conclusion

The energy of conducted emissions of the CD that can be affirmed is located in the frequency range of 8–32 MHz. The fundamental frequency is in 8 MHZ, which indicates that the frequency fundamental of the corona discharge is independent of the source configuration.

A summary is presented in **Table 3** for the analysis of the spectrogram of radiated CD emissions.

The temporal and frequency responses of the radiated emissions of the CD present the energy concentration in the same frequency range, which indicates that the process used in the measurements is adequate.

In relation to the radiated emissions in the frequency range from 0.3 to 30 MHz, the three measurement processes agree, which indicates the RBW selected in both instruments is adequate (10 kHz). In the case of the 30–300 MHz frequency band, in both instruments, RBW of 120 kHz was set as specified in CSPR-1-1. However, there are differences according to the results. For this case, the EMI test receiver can be reliable up to 100 MHz. The opposite for the spectrum analyzer, which has a better resolution because narrowband emissions can be identified at 120 and 260 MHz. In the band from 300 MHz to 1 GHz, having the same RBW (120 kHz) on both instruments, the EMI test receiver is unresponsive; it only presents background noise. Instead, the spectrum analyzer makes it possible to determine narrowband emissions. Finally, the results show that in the 100 MHz–1 GHz band, the radiated emissions from corona discharge are not a threat to radio communication systems.

Antenna	Polatization	Frequency range MHz
Loop	x, y, z	8, 16, 16–32
Biconical	Vertical	16, 16–64
Biconical	Horizontal	16–64
Hybrid biconical/log periodic antenna	Vertical	8, 16, 24–64
Hybrid biconical/log periodic antenna	Horizontal	16–64

Table 3.
Concentration of energy in the spectrograms.

Acknowledgements

The authors kindly acknowledge the grant awarded by the National Polytechnic Institute (IPN) and the National Council of Science and Technology (CONAYT) Mexico.

Conflict of interest

The authors declare no conflict of interest.

Author details

Luis E. Martínez Santos, Roberto Linares y Miranda* and Fermín P. Espino-Cortés*
Laboratorio de Compatibilidad Electromagnética, ESIME-Zac, Instituto Politécnico Nacional, Mexico

*Address all correspondence to: rlinaresy@ipn.mx; fespinoc@ipn.mx

References

[1] Lia X, Cuib X, Luc T, Wangd D, Wange Z, He Z. A platform for multiple DC corona effects measurements and analysis. In: Proceeding IEEE Instrumentation and Measurement Technology Conference (I2MTC); 11-14 May 2015, July 2015. Pisa, Italy: IEEE; 2015. pp. 181-186. DOI: 10.1109/I2MTC.2015.7151262

[2] Radio Interference. 2010. Available from: https://www.sciencedirect.com/topics/engineering/radio-interference [Accessed: January 20, 2021]

[3] Riba J-R, Gómez-Pau Á, Moreno-Eguilaz M. Experimental study of visual corona under aeronautic pressure conditions using low-cost imaging sensors. Sensors. 2020;**20**(2): 411. DOI: 10.3390/s20020411

[4] Riba J-R, Morosini A, Capelli F. Comparative study of AC and positive and negative DC visual corona for sphere-plane gaps in atmospheric air. Energies. 2018;**11**:2671. DOI: 10.3390/en11102671

[5] Tejada-Martínez C, Espino-Cortes FP, Ilhan S, Ozdemir A. optimization of radio interference levels for 500 and 600 kV bipolar HVDC transmission lines. Energies. 2019;**12**:3187. DOI: 10.3390/en12163187

[6] Phillips DB, Olsen RG, Pedrow PD. Corona onset as a design optimization criterion for high voltage hardware. IEEE Transactions on Dielectrics and Electrical Insulation. 2000;**7**(6):744-751. DOI: 10.1109/94.891984

[7] Tanner RL. Radio Interference from Corona Discharge. Technical Report No 37. SRI Project No 591. Air Force Contract No AF 19(604)-266. 1953

[8] Moongilan D. Corona noise considerations for smart grid wireless communication and control network planning. In: Proceedings IEEE International Symposium on Electromagnetic Compatibility (EMC); 6-10 August 2012; Pittsburgh, PA. USA: IEEE; 2012. pp. 357-362. DOI: 10.1109/ISEMC.2012.6351822

[9] CISPR TR 18-2. Radio interference characteristics of overhead power lines and high-voltage equipment—Part 2: Methods of measurement and procedure for determining limits. Edition 3.0. 2017

[10] Xia Y, Song X, Jia Z, Wang X, Qi J, Xu Z. Research on antenna for detecting the corona discharge of transmission line. In: Proceeding the 14th IET International Conference on AC and DC Power Transmission (ACDC2018). United Kingdom; 19th September 2018. DOI: 10.1049/joe.2018.8392

[11] Kysela A. Electromagnetic emissions of AC high-voltage corona. In: Intensive Programme, Renewable Energy Sources. Czech Republic: Železná Ruda-Špičák, University of West Bohemia; 2011. pp. 114-118

[12] Madhar SA, Mráz P, Mor AR, Ross R. Study of corona configurations under DC Conditions and recommendations for an identification test plan. International Journal of Electrical Power & Energy Systems. 2020;**118**:1-10. DOI: 10.1016/j.ijepes.2020.105820

[13] Korzhov AV, Okrainskaya IS, Sidorov AI, Kufel VD. A study of electromagnetic radiation of corona discharge near 500-kV electric installations. Power Technology and Engineering. 2004;**38**(1):57-60. DOI: 10.1023/B:HYCO.0000029636.31951.91

[14] Robert GO, Bradley OS. Predicting VHF/UHF electromagnetic noise from corona on power-line conductors. IEEE Transactions on Electromagnetic Compatibility. 1988;**30**(1):13-22. DOI: 10.1109/15.19883

[15] Chartier VL, Sheridan R, DiPlacido JN, Loftness MO. Electromagnetic interference measurements at 900 MHz on 230-kV and 500-kV transmission lines. IEEE Transactions on Power Delivery. 1986;**PWRD-1**(2):140-149. DOI: 10.1109/TPWRD.1986.4307944

[16] Abdel-Salam M, Abdel-Sattar S. Calculation of corona V-I characteristics of monopolar bundles using the charge simulation method. IEEE Transactions on Electrical Insulation. 1989;**24**(4): 669-679. DOI: 10.1109/14.34202

[17] Wang D, Lin D, Yao C. Statistical study on space charge effects and stage characteristics of needle-plate corona discharge under DC voltage. Energies. 2019;**12**:2732. DOI: 10.3390/en12142732

[18] Huang C, Yin H, Xu P, Zhang B, He J, Liu J. Prediction of radio interference from HVDC transmission lines based on corona discharge characteristics. High Voltage. 2020;**5**(6):679-687. DOI: 10.1049/hve.2019.0050

[19] Salazar-Hernández JC, Linares y Miranda R, Espino-Cortes FP, Ozdemir A. Far-field prediction for radio-interference produced by fundamental frequency of a DC corona discharge at a rod-plane configuration. IEEE Electromagnetic Compatibility Magazine. 2020;**9**(Quarter 3):37-43. DOI: 10.1109/MEMC.2020.9241549

[20] CISPR TR 18-1. Radio interference characteristics of overhead power lines and high-voltage equipment—Part 1: Description of phenomena. Edition 3.0. 2017

[21] CISPR 16-1-1. Specification for radio disturbance and immunity measuring apparatus and methods—Part 1-1: Radio disturbance and immunity measuring apparatus—Measuring apparatus. ED. 5.0 B. 2019

[22] Shixiu C, Youlin S, Hengkun M. Characteristics of electromagnetic wave radiated from corona discharge. In: Proceeding of the IEEE EMC International Symposium on Electromagnetic Compatibility (Cat. No. 01CH37161); 13-17 August 2001; Montreal, QC. Canada: IEEE; 2001. pp. 1279-1282. DOI: 10.1109/ISEMC.2001.950630

Chapter 2

Electromagnetic Compatibility Issues in Medical Devices

Ting-Wei Wang and Ting-Tse Lin

Abstract

Electromagnetic compatibility (EMC) in biomedical applications is a significant issue related to the user's life safety, especially in implantable medical devices. Cardiovascular diseases and neurodegenerative disorders are the main chronic disease worldwide that rely on implantable treatment devices such as cardiac pacemakers and vagus nerve stimulators. Both devices must have high EMC to avoid electromagnetic interference-induced health risks, even death during the treatment. Thus, it is important to understand how EMI can affect implantable devices and proactively protect devices from electromagnetic interference, providing reliable and safe implantable device therapy. To this end, this chapter comprehensively introduces the clinical issues and provides EMC requirements for the implantable device such as a cardiac pacemaker and vagus nerve stimulator. The significance of this chapter is to present the EMC important issues in medical engineering that can help to evolve reliable and secure implantable device development in the future.

Keywords: Cardiac pacemaker, Cardiovascular disease, Electromagnetic compatibility, Implantable device, Medical device, Neuroscience, Vagus nerve stimulator

1. Introduction - Physiological background for implantable medical device

1.1 Cardiac pacemaker

Cardiovascular diseases (CVDs) are the major cause of death globally, which takes an estimated 17.9 million lives per year based on the World Health Organization (WHO) statistics. Especially, arrhythmia has a strong clinical correlation with sudden cardiac death (SCD) [1]. The irregular heart rhythm called arrhythmia can mainly be divided into two main types: tachycardia arrhythmia and bradycardia arrhythmia. As shown in **Figure 1**, the tachycardia and bradycardia arrhythmia represent the heart beats too fast and slow, respectively. The physical treatment for tachycardia and bradycardia syndrome required a regulator to suppress the abnormal heart rhythm. The implantable cardiac pacemaker is commonly applied in the cardiac modulation that generates the electrical stimulation pulse to regulate the heart's sinoatrial node, thus obtaining the normal rhythm.

As shown in **Figure 2**, the cardiac pacemaker would be activated once the heart detector measures the abnormal cardiac rhythm. Thus, the stable operation of a cardiac pacemaker is important for adverse patients, thus providing prompt treatment in arrhythmia.

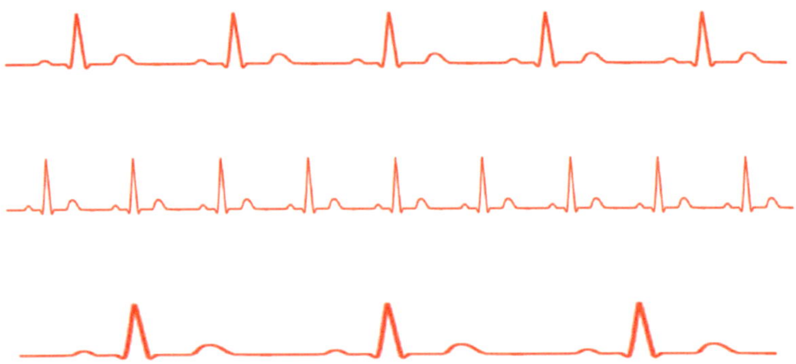

Figure 1.
Heart rhythm is divided into normal rhythm, tachycardia arrhythmia, and bradycardia arrhythmia.

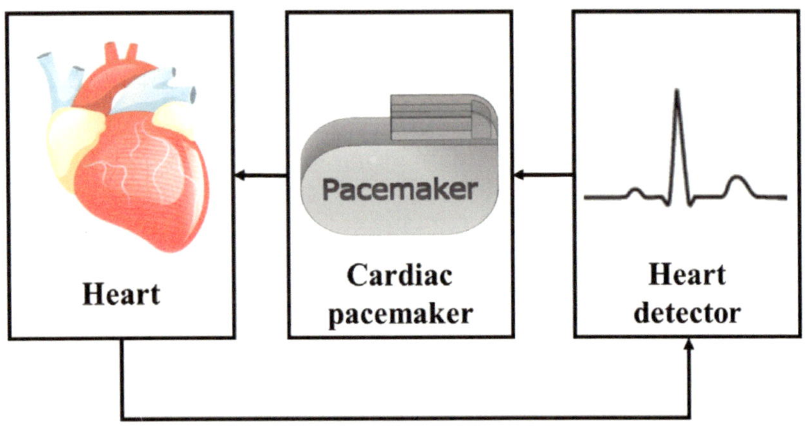

Figure 2.
Schematic of the detection and treatment system for cardiac arrhythmia.

1.2 Vagus nerve stimulator

Epilepsy is a neurological disorder that can induce the brain activities abnormal, causing seizures and further loss of awareness suddenly [2–4]. According to WHO statistics, epilepsy is a chronic non-communicable brain disease that affects humans of all ages, around 50 million globally, becoming one of the most common neurological diseases worldwide. The risk in the death probability of epilepsy patients is up to three times higher than healthy people. In epilepsy diagnosis, Electroencephalogram (EEG) is the most common non-invasive approach to record the brain's electrical activity and identify the measured signals, whether it is epilepsy or not [5]. In addition, the imaging-based diagnosis of computed tomography (CT) [6], positron emission tomography (PET) [7], and MRI [8] can help further examining brain-tumor-induced epilepsy. In epilepsy treatment, an vagus nerve stimulator is the common physical approach implanted near the left chest area, as shown in **Figure 3**. The electrode is attached around the vagus nerve in the neck to generates the electrical pulses to the brain via the vagus nerve. The delay time more than 10 minutes from seizure onset would increase mortality, which emphasizes the importance of timely treatment and time and medical emergency [9]. Therefore, high reliability is crucial for vagus nerve stimulation (VNS) during incidental neurological disease. Moreover, VNS is also widely applied in significant disease treatment, including cardiac function [10], depression [11], anxiety [12], Parkinson's disease [13], and Alzheimer's disease [14].

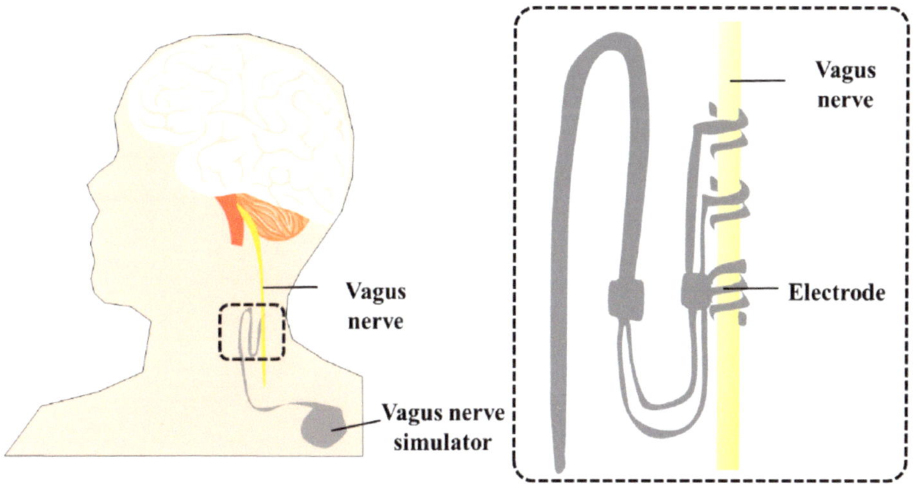

Figure 3.
Schematic of vagus nerve stimulation and its installations.

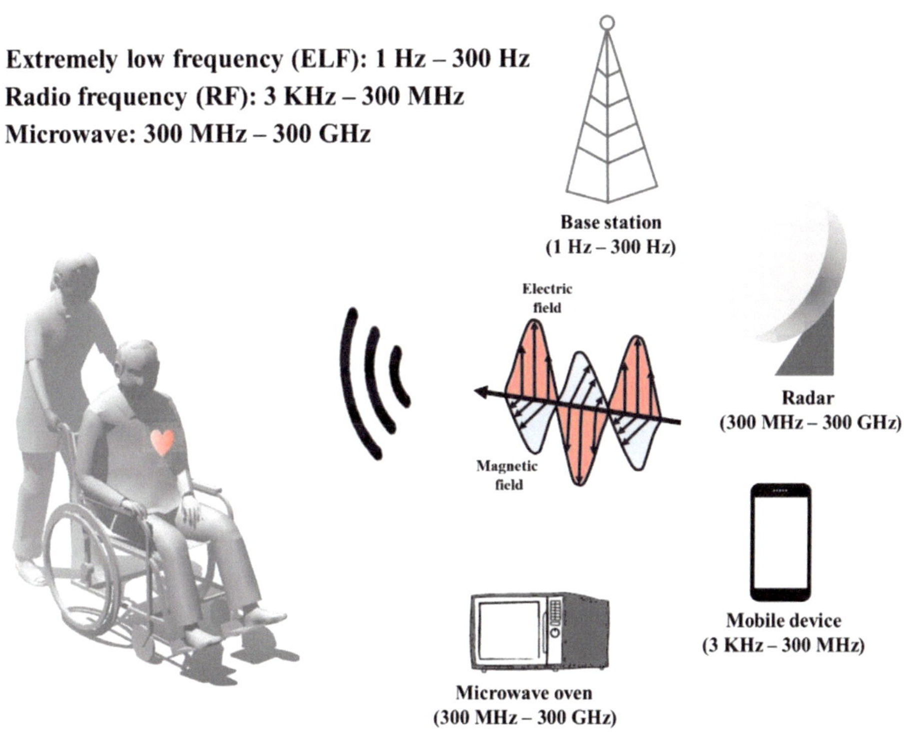

Figure 4.
EMI in medical devices from external sources with time-varying electrical and magnetic fields such as base station, radar, mobile device, and microwave oven [16].

2. Electromagnetic interference for implantable medical device

The electromagnetic interference (EMI) in the implantable medical device can be produced by the external source with the combined electric and magnetic fields [10, 15], as shown in **Figure 4**. EMI is due to radiation that can be through the air from many possible sources (**Table 1**) in our daily life [17–20], including the

Home environment
Mobile phone (RF)
Microwave oven (Microwave)
Remote controller (RF)
Refrigerator (ELF)
Electric razor (ELF)
Outdoor environment
Radar (Microwave)
Base station (Microwave)
High voltage power lines (ELF)
Medical environment
Magnetic resonance image machine (RF)
Radio-based therapy (RF)
Ionizing-based radiation therapy (X-ray)
Defibrillation (ELF)
Lithotripsy (ELF)
Industrial environment
Transformer (ELF)
High voltage power lines (ELF)
Electric motor (ELF)
Radiofrequency identification (RF)

Table 1.
EMI sources from home, outdoor, medical, and industrial environments.

common consumer device such as mobile phones, radio frequency identification (RFID) based systems, and microwaves. Moreover, the medical procedure-induced EMI is a critical concern. For example, dental equipment and magnetic resonance imaging (MRI) can generate EMI. In particular, the MRI equipment can cause a strong EMI that is very hard to guard against. The MRI machine can produce an intense magnetic field of about two or three teslas that are dangerous to any electronic device. The electromagnetic susceptibility (EMS) is frequently used to define immunity for EMI, which implies the degree of electronic system malfunctions under varying levels of EMI. Therefore, electromagnetic compatibility (EMC) in implantable medical devices such as cardiac pacemakers and vagus nerve stimulators is important to sustain the stable and normal function to treat accidental cardiac issues because humans are always surrounded by electrical equipment [21, 22].

3. Electromagnetic compatibility for implantable medical device

3.1 EMI shield

The EMI shield is designed to decrease the electromagnetic (EM) wave transmission using a shield to increase the reflection or absorption of EM wave incident at the interfaces between different mediums. As shown in **Figure 5**, the electric and magnetic fields of EM waves are perpendicular to each component and the EM

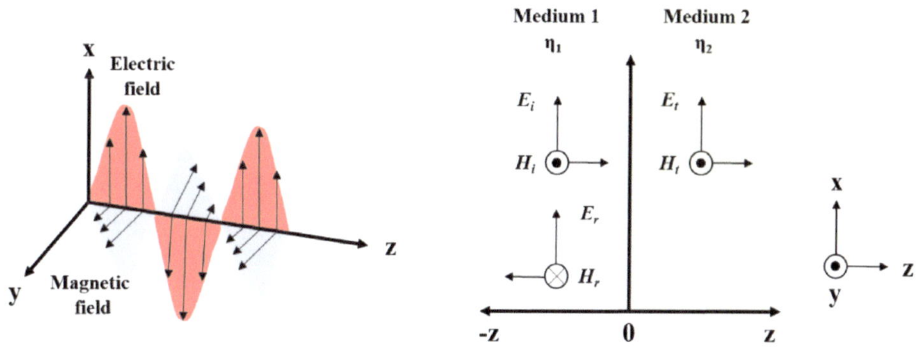

Figure 5.
Time-varying EM waves and the propagation, transmission, and reflection of the EM waves in different mediums.

propagation direction, which can be expressed as phasor form [23], according to Eqs. (1) and (2). Where γ, α, β are the propagation, attenuation, phase constants of the medium, respectively; E_0 and H_0 are the amplitude of the electric and magnetic fields.

$$\mathbf{E} = \hat{a}_y E_0 e^{-\gamma z} = \hat{a}_y E_0 e^{-\alpha z} e^{-j\beta z} \tag{1}$$

$$\mathbf{H} = \hat{a}_y H_0 e^{-\gamma z} = \hat{a}_y H_0 e^{-\alpha z} e^{-j\beta z} \tag{2}$$

The EM wave propagates at the interface of two different mediums that induce reflection due to impedance mismatching. The reflection coefficient (R_{12}) and transmission coefficient (T_{12}) at the interface between two mediums can be determined, according to Eqs. (3) and (4). E_i, E_r, η_1, η_2 represent incident, reflected electric fields, impedances in Medium 1 and Medium 2, respectively. The impedance in the medium can be defined by a ratio of electric field and magnetic field, which is related to the permittivity (ε), permeability (μ), and conductivity (σ), according to Eq. (5).

$$R_{12} = \frac{E_r}{E_i} = \frac{\eta_2 - \eta_1}{\eta_2 + \eta_1} \tag{3}$$

$$\mathbf{T}_{12} = \frac{E_t}{E_i} = \frac{2\eta_2}{\eta_2 + \eta_1} \tag{4}$$

$$\eta = \frac{|\mathbf{E}|}{|\mathbf{H}|} = \sqrt{\frac{j\omega\mu}{\sigma + j\omega\varepsilon}} \tag{5}$$

The time-domain electric field can be rewritten as Eq. (6), according to Eq. (1).

$$\mathbf{E} = \hat{a}_y E_0 e^{-\gamma z} = \hat{a}_y E_0 e^{-\alpha z} \cos(\omega t - \beta z) \tag{6}$$

When the EM wave propagates in the conductive shield (loss medium) at a time of zero, the expression between the distance and amplitude of the electric field can be obtained, according to (7).

$$\mathbf{E} = \hat{a}_y E_0 e^{-\alpha z} \cos(\beta z) \tag{7}$$

Figure 6 demonstrates the EM wave propagation in the conductive shield. The skin depth (δ) can be defined as that penetration distance at which the intensity of

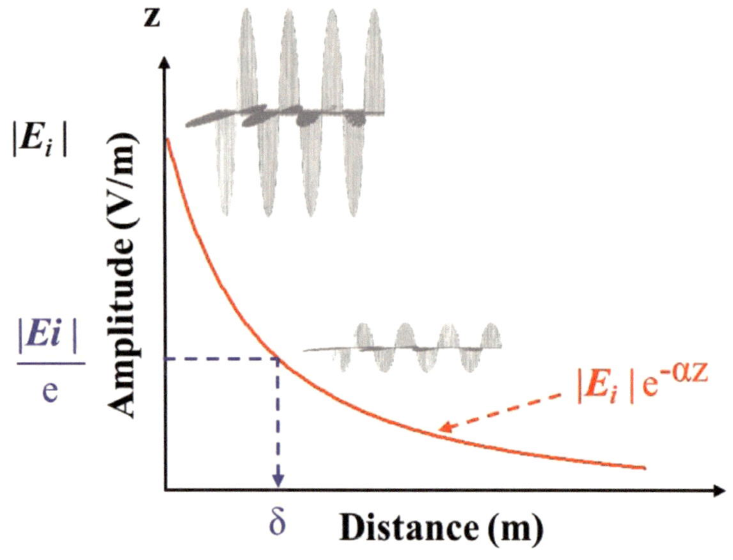

Figure 6.
Schematic of skin depth of EM waves using the shield.

the electric field attenuates to 1/e of the original incident wave intensity. According to the Eqs. (8) and (9), the skin depth (δ) is the reciprocal of attenuation constant (α), which is related to the EM operating frequency, permeability (μ), and conductivity (σ) in the medium [10].

$$\frac{E_0}{e} = E_0 e^{-\alpha\delta} \tag{8}$$

$$\delta = \frac{1}{\alpha} = \frac{1}{\sqrt{\pi f \mu \sigma}} \tag{9}$$

Thus, conductivity and permeability in shielding design play an important role in EM wave absorption enhancement, thus increasing the overall EMI shielding effectiveness.

However, the implantable medical device is not a fully closed system that must require the openings of the shield to interact with external equipment such as body sensing devices for signal transmit or receive. In some cases, the external controlled magnetic fields or electrical signals can be utilized to externally modulate the stimulation protocol of implantable medical devices according to patients' clinical requirements. So, the selective filtering of EMI waves is important for implantable medical devices to classify the noise and external signals. Thus, the EMI filter was provided in the following subsection to promote the EMC applications in implantable medical devices.

3.2 EMI filter

The filter implementation is also a strategy for EMI elimination in medical devices [24–27]. For example, the typical ranges of P, R, T waves in ECG are 20 to 40 Hz, 18 to 50 Hz, and 0 to 10 Hz [28], as shown in **Table 2**.

Filtering can be divided into active and passive modes. Active filters consist of several operational amplifiers and passive elements such as capacitors and resistors. [29–32]. The active filters are applied in wide applications owing to excellent filter performance. However, the active filters need a power source to sustain the

	Frequency (Hz)	Signal amplitude (mV)
T-waves	0–10	3.5
R-waves	18–50	30
P-waves	20–40	4.5
Muscle signals	30–200	3.5

Table 2.
Significant frequency bandwidth of ECG waveform [28].

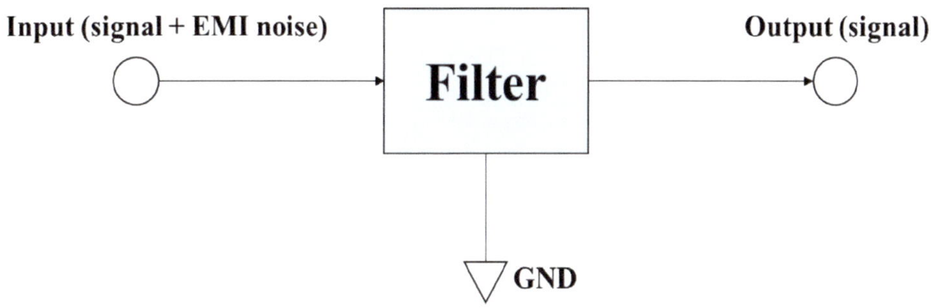

Figure 7.
Filter implementation for removing the time-varying electric and magnetic fields.

operations. Moreover, the upper frequency of active filters may be limited. Thus, the active filter is not suitable for EMI filtering in implantable medical devices. The active filter can perform programmatical filtering for the received signals, thus separating the signal from noise. However, programmatical computation requires a high-cost and complex circuit with larger power consumption to sustain the processing functions. Because the implantable devices aim to sustain life, such devices are not expected to remove or insert frequently because of extremely high costs for device failure.

Moreover, it is not easy to replace it if the devices fail due to the high risk of surgery. The concern regarding the surgery risk, which makes battery life issues more important. Owing to the battery requirements of implantable medical devices, the minimization of filters' power is crucial to prolong the implantable device lifespan.

The capacitor-based passive filter is frequently used for most high-frequency noise in the surrounding ambient for the filter design regarding implantable medical devices. Capacitors can filter EMI noise utilizing absorption and smoothing of electromagnetic noise. The high-frequency noise attenuates as quickly as charging and discharging the capacitor-based filter. Absorbing such EMI noise to the ground will neutralize or prevent specific frequencies from passing through the circuit, as shown in **Figure 7**.

The discoidal capacitor and feedthrough capacitor array were commonly utilized in the practical applications of medical devices, which deliver high-density performance with low-volume packaging [33–35], as shown in **Figure 8**.

The circular-shaped discoidal capacitor is one of the most common constructions for feed-through-style EMI filters. Circular capacitors outside and inside diameters serve as connection points for the case and the lead and serve as the capacitor poles. Moreover, several discoidal capacitors can be assembled to integrate as a capacitor array on a single piece of ceramic [36]. Such assembly offers the highest filter performance within the limited physical dimension. Thus, the feedthrough

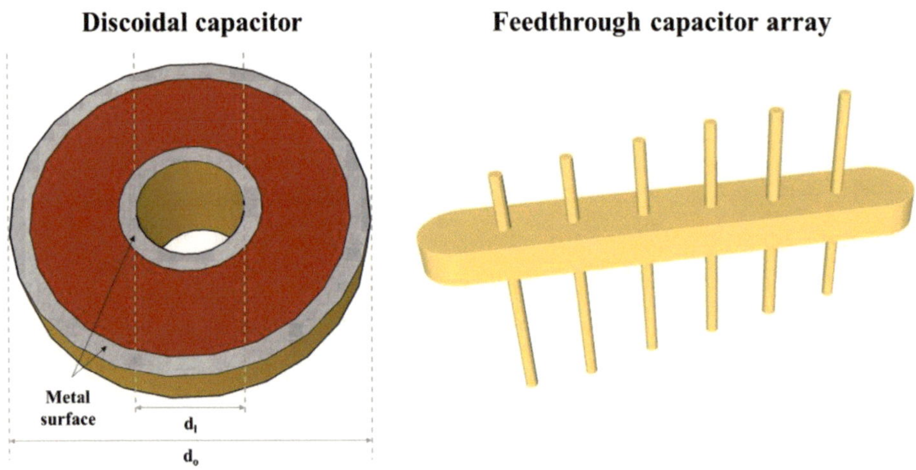

Figure 8.
The feedthrough capacitor array and discoidal capacitor for the implantable medical device.

capacitor array provides the merits of the miniature dimension and lightweight within a high-density implantable device [37].

However, a feedthrough filter will have a double impact on battery life. First, a minimal amount of current always flows between the plates of the charging capacitor. Since one capacitor is processing the signal and the other capacitor is grounded, the leakage current will drain the battery over time. A strong dielectric with an appropriate thickness can resist this current flow, thereby significantly reducing battery consumption. Besides, filter design in implantable medical devices is to minimize the loss of expected signals. The filter's insertion loss implies how much a signal will be lost or reduced for each frequency. An excellent filter requires a lower insertion loss for signal frequencies and a higher insertion loss for noise frequencies. Some energy in the expected signal will attenuate in internal resistance and inductance of the filter, which implies that the implantable battery needs optimize in the power design. Thus, an optimized design can suppress the energy loss. The less power dissipated in the battery, which extends battery life and improves effectiveness.

4. Conclusion

This chapter has demonstrated the EMC methodology for implantable medical devices. A common effective EMI removal approach was provided by EMI shielding and filtering. Such EMC design in implantable medical devices can resist EMI day-to-day exposure to ensure the stable and reliable operation in implantable medical devices such as cardiac pacemakers and vagus nerve stimulators.

Author details

Ting-Wei Wang[1,2] and Ting-Tse Lin[3,4,5]*

1 Department of Electrical Engineering, California Institute of Technology, Pasadena, CA, USA

2 Department of Medical Engineering, California Institute of Technology, Pasadena, CA, USA

3 Division of Cardiology, Department of Internal Medicine, National Taiwan University Hospital Hsinchu Branch, Hsinchu, Taiwan

4 Cardiovascular Center, National Taiwan University Hospital Hsinchu Branch, Hsinchu, Taiwan

5 College of Medicine, National Taiwan University, Taipei, Taiwan

*Address all correspondence to: ttlin111@gmail.com

References

[1] N. T. Srinivasan and R. J. Schilling, "Sudden Cardiac Death and Arrhythmias," (in eng), *Arrhythmia & Electrophysiology Review,* vol. 7, no. 2, pp. 111-117, 2018.

[2] O. Devinsky *et al.*, "Epilepsy," *Nature Reviews Disease Primers,* vol. 4, no. 1, p. 18024, 2018/05/03 2018.

[3] P. J. Karoly *et al.*, "Cycles in epilepsy," *Nature Reviews Neurology,* vol. 17, no. 5, pp. 267-284, 2021/05/01 2021.

[4] C. E. Stafstrom and L. Carmant, "Seizures and epilepsy: an overview for neuroscientists," (in eng), *Cold Spring Harbor Perspectives in Medicine,* vol. 5, no. 6, p. a022426, 2015.

[5] A. Sohrabpour, Z. Cai, S. Ye, B. Brinkmann, G. Worrell, and B. He, "Noninvasive electromagnetic source imaging of spatiotemporally distributed epileptogenic brain sources," *Nature Communications,* vol. 11, no. 1, p. 1946, 2020/04/23 2020.

[6] J. S. Chopra, I. M. S. Sawhney, N. Suresh, S. Prabhakar, U. K. Dhand, and S. Suri, "Vanishing CT lesions in epilepsy," *Journal of the Neurological Sciences,* vol. 107, no. 1, pp. 40-49, 1992/01/01/ 1992.

[7] I. Sarikaya, "PET studies in epilepsy," (in eng), *American Journal of Nuclear Medicine and Molecular Imaging,* vol. 5, no. 5, pp. 416-430, 2015.

[8] H. Wood, "Ultra-high-field MRI improves detection of epileptic lesions," *Nature Reviews Neurology,* vol. 16, no. 12, pp. 654-654, 2020/12/01 2020.

[9] A. Verrotti and C. Mazzocchetti, "Timely treatment of refractory convulsive status epilepticus," *Nature Reviews Neurology,* vol. 14, no. 5, pp. 256-258, 2018/05/01 2018.

[10] T. W. Wang, Y. L. Sung, and S. F. Lin, "Cardiac Influence of Repetitive Transcranial Magnetic Stimulation in Small Animals," *IEEE Journal of Electromagnetics, RF and Microwaves in Medicine and Biology,* vol. 4, no. 4, pp. 279-285, 2020.

[11] R. L. Johnson and C. G. Wilson, "A review of vagus nerve stimulation as a therapeutic intervention," (in eng), *Journal of Inflammation Research,* vol. 11, pp. 203-213, 2018.

[12] M. S. George *et al.*, "A pilot study of vagus nerve stimulation (VNS) for treatment-resistant anxiety disorders," (in eng), *Brain Stimul,* vol. 1, no. 2, pp. 112-21, Apr 2008.

[13] B. Mondal *et al.*, "Non-invasive vagus nerve stimulation improves clinical and molecular biomarkers of Parkinson's disease in patients with freezing of gait," *NPJ Parkinson's Disease,* vol. 7, no. 1, p. 46, 2021/05/27 2021.

[14] P. Boon, I. Moors, V. De Herdt, and K. Vonck, "Vagus nerve stimulation and cognition," *Seizure,* vol. 15, no. 4, pp. 259-263, 2006/06/01/ 2006.

[15] R. Beinart and S. Nazarian, "Effects of external electrical and magnetic fields on pacemakers and defibrillators: from engineering principles to clinical practice," (in eng), *Circulation,* vol. 128, no. 25, pp. 2799-2809, 2013.

[16] J. B. Marion, "14 - ELECTROMAGNETIC RADIATION," in *Physics in the Modern World (Second Edition)*, J. B. Marion, Ed.: Academic Press, 1981, pp. 377-397.

[17] *L. Santini*, G. B. Forleo, and M. Santini, "Implantable devices in the electromagnetic environment," *Journal of Arrhythmia,* vol. 29, no. 6, pp. 325-333, 2013/12/01/ 2013.

[18] M. Tiikkaja, *Environmental electromagnetic fields: interference with cardiac pacemakers and implantable cardioverter-defibrillators*. Työterveyslaitos, 2014.

[19] N. Mizutani, I. Kato, and T. Kobayashi, "A survey of the effect of electromagnetic interference on currently marketed pacemakers," *Journal of Artificial Organs,* vol. 3, no. 1, pp. 47-52, 2000/03/01 2000.

[20] J. L. N. Violette, D. R. J. White, and M. F. Violette, "Sources of Electromagnetic Interference," in *Electromagnetic Compatibility Handbook*, J. L. N. Violette, D. R. J. White, and M. F. Violette, Eds. Dordrecht: Springer Netherlands, 1987, pp. 13-62.

[21] O. Erdogan, "Electromagnetic interference on pacemakers," (in eng), *Indian Pacing and Electrophysiology Journal,* vol. 2, no. 3, pp. 74-78, 2002.

[22] H. W. Roberts, "The effect of electrical dental equipment on a vagus nerve stimulator's function," *The Journal of the American Dental Association,* vol. 133, no. 12, pp. 1657-1664, 2002/12/01/ 2002.

[23] K. L. Kaiser, *Electromagnetic shielding*. CRC Press, 2005.

[24] T.-W. Wang, Y.-L. Sung, H.-W. Chu, and S.-F. Lin, "IPG-based field potential measurement of cultured cardiomyocytes for optogenetic applications," *Biosensors and Bioelectronics,* vol. 179, p. 113060, 2021/05/01/ 2021.

[25] T.-W. Wang *et al.*, "Bio-Impedance Measurement Optimization for High-Resolution Carotid Pulse Sensing," *Sensors,* vol. 21, no. 5, 2021.

[26] T. W. Wang and S. F. Lin, "Negative Impedance Capacitive Electrode for ECG Sensing Through Fabric Layer," *IEEE Transactions on Instrumentation and Measurement,* vol. 70, pp. 1-8, 2021.

[27] T.-W. Wang and S.-F. Lin, "Non-contact capacitive sensing for ECG recording in small animals," *Measurement Science and Technology,* vol. 31, no. 12, p. 125703, 2020/10/17 2020.

[28] M. W. Sweesy, J. L. Holland, and K. W. Smith, "Electromagnetic interference in cardiac rhythm management devices," (in eng), *AACN Clin Issues,* vol. 15, no. 3, pp. 391-403, Jul-Sep 2004.

[29] T.-W. Wang and S.-F. Lin, "Wearable Piezoelectric-Based System for Continuous Beat-to-Beat Blood Pressure Measurement," *Sensors,* vol. 20, no. 3, 2020.

[30] T. W. Wang, W. X. Chen, H. W. Chu, and S. F. Lin, "Single-Channel Bioimpedance Measurement for Wearable Continuous Blood Pressure Monitoring," *IEEE Transactions on Instrumentation and Measurement,* vol. 70, pp. 1-9, 2021.

[31] T. W. Wang *et al.*, "Single-Channel Impedance Plethysmography Neck Patch Device for Unobtrusive Wearable Cardiovascular Monitoring," *IEEE Access,* vol. 8, pp. 184909-184919, 2020.

[32] T. W. Wang, H. Zhang, and S. F. Lin, "Influence of Capacitive Coupling on High-Fidelity Non-Contact ECG Measurement," *IEEE Sensors Journal,* vol. 20, no. 16, pp. 9265-9273, 2020.

[33] R. A. Stevenson and D. N. Pruett, "Feedthrough filter capacitor assembly for human implant," ed: Google Patents, 1994.

[34] J. H. Coleman, "Discoidal monolithic ceramic capacitor," ed: Google Patents, 1981.

[35] H. V. Trinh, D. F. Devoe, A. D. Devoe, M. L. Trinh, and M. Petkova,

"MLC Discoidal Capacitors for EMI-RFI Filters Employing Non-Overlapping Electrodes Yield Substantial Performance Improvements," ed: CARTS USA Palm Springs, CA, USA, 2005.

[36] S. Grob, P. A. Tass, and C. Hauptmann, "Capacitive Feedthroughs for Medical Implants," (in eng), *Frontiers in Neuroscience,* vol. 10, pp. 404-404, 2016.

[37] P. J. Barry and D. J. Boettger, "Feedthrough assembly for an implantable device," ed: Google Patents, 2013.

Chapter 3

Life Cycle Assessment of Flexible Electromagnetic Shields

Ion Răzvan Rădulescu, Lilioara Surdu, Emilia Visileanu, Bogdana Mitu and Cristian Morari

Abstract

Nowadays, fiber based flexible electromagnetic shields have widespread applications in ensuring Electromagnetic Compatibility (EMC). Shielding is a solution of EMC, and the main methods to estimate shielding effectiveness are represented by the circuit method and the impedance method. Magnetron sputtering of metallic layers represents a novel technique to impart electric conductive properties to fabrics. Coating of fabrics represents a second main option to manufacture textile shields beside the insertion of conductive yarns in the fabric structure. Life Cycle Assessment (LCA) is often used to assess a comparatively modern with a classical manufacturing process in order to prove its eco-friendly character. This chapter comparatively assesses flexible EM shields manufactured of fabrics with inserted conductive yarns with and without magnetron plasma coating. The copper plasma coating of cotton fabrics with inserted silver yarns increases shielding effectiveness (EMSE) by 8–10 dB. In order to keep for the LCA study the same functional unit of 50 dB at 100 MHz for one sqm of fabric, the fabric structure is modeled with a reduced distance between the inserted conductive yarns. Results of the LCA study show a substantial impact on the environment for the plasma coated fabric upon using a laboratory scale deposition set-up.

Keywords: LCA, EMC, textiles, magnetron plasma, copper coating, silver yarns, woven fabrics

1. Introduction

This book chapter tackles a highly interdisciplinary theme: a Life Cycle Assessment (LCA) study of plasma coated textiles for electromagnetic interference (EMI) shielding. Thus, the introduction presents some indicative aspects for all these particular domains.

1.1 EMI shielding textiles

The textile materials destined for Electromagnetic Interference (EMI) shielding include some additional properties when compared to metallic shields, according to the latest research studies [1–5]:

- A significant electromagnetic shielding effectiveness (EMSE - expressed in dB), as the main functionality of the textile shield, with values in the range of 30–80 dB;
- Lightweight, flexibility, adaptability to the shape and mechanical resistance;
- Preservation of specific textile properties like drape, bending and resistance to repeated washing;
- Sustainable and eco-friendly character by the application of the new technologies;
- Cost-effectiveness based on these new manufacturing technologies.

The additional functionalities offer added value to the textile substrates. Starting from the research phase, the textile shields may be used as promising EM shields in numerous applications, such as civil and military EMI protection, medical devices, or the buildings' EMI shielding. All these applications tackle the attenuation of EMI, needed for the proper functioning of electronic equipment (by electromagnetic compatibility principles), and the protection of living beings against non-ionizing radiation (medicine).

As known from the literature, there is a strong correlation between the electrical conductivity of the textile shield [S/m] and their electromagnetic shielding effectiveness [dB] [6]. Eddy currents induced by the incident EM field generate an opposed EM field, with an attenuating result [7]. Some of the recent advances in manufacturing textile shields with additional properties include integrating various electric conductive raw materials into the fabric structure [1–5].

An eco-efficient method to manufacture a hydrophobic textile substrate with excellent EM shielding properties by carbon nanotubes and graphene dispersion is presented in [1]. Various solutions of manufacturing textile shields by preserving the specific properties of textiles, such as good air permeability, good bending, flexibility, and stability in a hot and wet environment, were achieved by polymerization of Pyrrole and subsequent coating with Nickel [2]. Additional resistance to washing and end-applications of the electric conductive fabric as patch antenna are included in [2]. Carbon nanotubes with nanometer deposition of copper layers were integrated into fabrics for EM shielding [3]. Additional mechanical resistance properties were proved on the achieved fabrics [3]. The special properties of intrinsic conductive polymers (Pyrrole) are presented within the review paper [4]. Another contribution to EM shields was achieved by sintering silver on the fabrics, however, with a significant decrease of EMSE after repeated washing cycles [5]. Hence, textile EM shields combine the EMI shielding properties with the specific advantages of textile fabrics.

1.2 Methods to estimate shielding effectiveness

The main methods to estimate shielding effectiveness based on material properties are represented by the impedance and circuit methods [6]. Several physical premises apply for each of the methods. Both methods consider an infinite perpendicular plane shield to the incident EM field. While the impedance method is valid for the electromagnetic far field, the circuit method is valid for the near field.

For woven fabrics with inserted metallic yarns in warp and weft direction, the resulting conductive grid structure may be modeled by the impedance method with correction factors [8]. Adaptation of this method for textiles may be found in [9].

Another method for modeling conductive grid structures by balancing EMSE of the grid with EMSE of the sheet, with regard to the electric thin materials, was provided by [10, 11], with specific adjustment for textiles [9, 12]. Further contributions to model EMSE of woven fabrics with metallic yarns were provided by analogy with an RLC circuit [13] and by analogy with small aperture antennas [14].

The main material constants used for modeling EMSE of textiles are electric conductivity and magnetic permeability of the metallic yarns and the composite fabrics, the optical diameter of the yarn/fabric thickness, distance between metallic yarns in the fabric structure. The skin depth of yarns/fabrics has a great significance both in the circuit method and in the impedance method with correction factors [9, 15].

1.3 Magnetron plasma sputtering of textiles

Textile shields are usually manufactured by two main methods: inserting metallic yarns into the fabric structure and coating with metallic layers [16]. While the insertion of metallic yarns by weaving, knitting, and nonwoven making already represents the classical method, the coating of nanometer scale metallic layers by magnetron plasma represents a modern, promising technique. The main advantage of the plasma coating is the enhancement of EMSE and preservation of flexibility of the fabrics by the nanometer thickness [17].

The copper coating onto the textile fabrics was performed at INFLPR into a dedicated stainless steel spherical vacuum chamber (K.J. Lesker, UK), pumped out by an assembly of afore pump and turbomolecular pump (Pfeiffer, DE), which allowed obtaining of a base pressure down to 3×10^{-5} mbar. A constant argon flow (purity 6.0) of 50 sccm was continuously introduced into the chamber by means of a Bronkhorst mass flow controller, which allowed to establish the processing pressure around 5×10^{-3} mbar. The chamber is provisioned with a rectangular magnetron sputtering gun from K.J. Lesker, accommodating the high purity copper target. The discharge was ignited by a radio frequency generator (13.56 MHz) provisioned with an automatic matching box for adapting the impedance and the deposition time was set to ensure coating thicknesses of 1200 nm on each side of the textile fabric. Enhanced deposition uniformity was achieved by rotating the samples during the deposition process (200 rotations/ min). **Figure 1** presents a sketch of the experimental set-up of the magnetron plasma equipment of INFLPR.

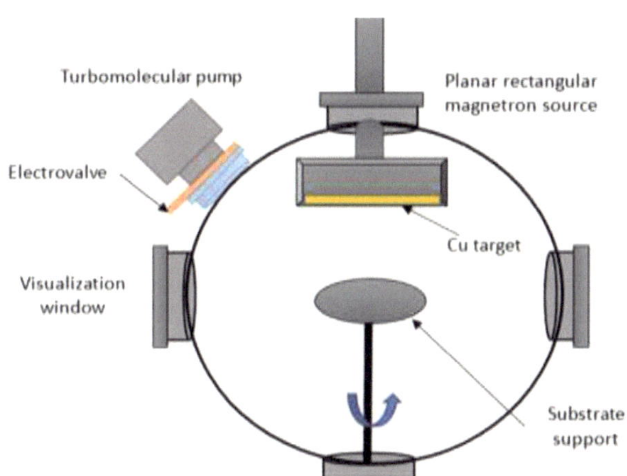

Figure 1.
Sketch of the experimental set-up in INFLPR utilized for copper coating of textiles.

1.4 Life cycle assessment of textiles

Life Cycle Assessment (LCA) studies are conducted according to the standard ISO 14040 [18]. The standard specifies four phases of an LCA study: i) the Goal and scope, ii) the Life Cycle Inventory (LCI), iii) the Life Cycle Impact Assessment (LCIA), and iv) the Interpretation (including the Sensitivity Analysis). **Figure 2** presents graphically this approach, as well as some of the main applications:

The Goal and scope phase defines the aim of the study as well as its limitations. LCA studies may be conducted for various reasons, such as benchmarking for different products, identifying Key Environment Performance Indicators (KEPI), motivating green acquisitions, or comparative assessment between a modern and a classical technology [19]. The latter application is used mainly for research to prove the environmentally friendly character of modern technology [20].

The Life Cycle Inventory (LCI) phase means collecting for the product or process the inputs and outputs into the system, such as electric energy, heat, raw materials, auxiliary substances, transport, etc. The Life Cycle Impact Assessment (LCIA) phase is usually supported by software applications. The inputs and outputs into the system are discretized in elemental units upon the environment, included within a data basis, and computed by the software application on impact categories. Impact categories characterize various aspects of the impact on the environment and are grouped in specific methods. The EcoIndicator 99E method includes impact categories such as carcinogens, ozone layer, fossil fuels, ecotoxicity, etc. It is based on the calculation of the damages to Human Health, Ecosystem Quality, and Resources [21]. More examples of methods include ReCiPe (which includes global warming, terrestrial acidification, water consumption, etc.) [22] and IMPACT2002+ refereeing human health, ecosystem quality, climate change, and resources in a combined midpoint/damage structure [23].

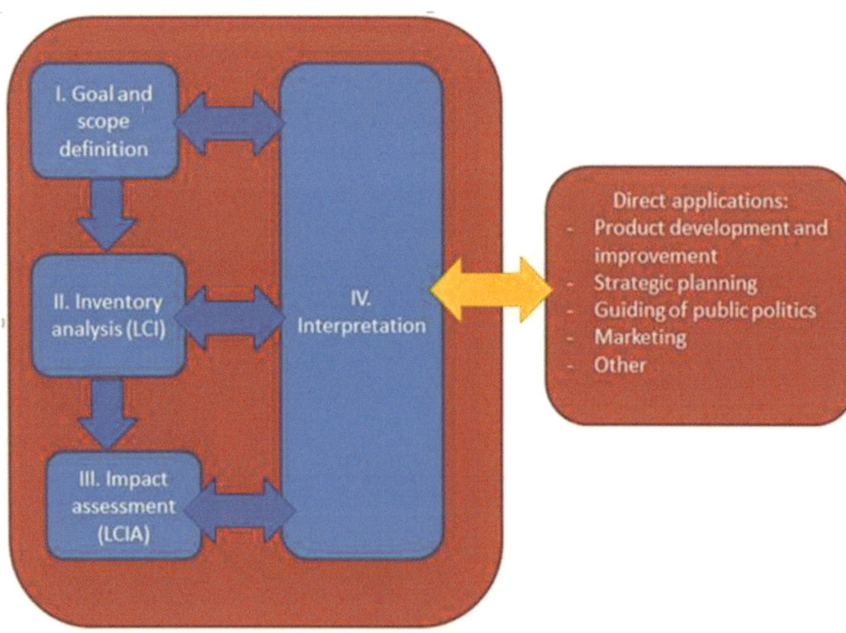

Figure 2.
The four phases of LCA accordingly to the Standard 14040 [18].

The SimaPro7 software and the EcoInvent 3.0 data basis were used in our study [24, 25]. The following types of LCIA diagrams may be generated by the software, according to the following principles:

- Characterization: Impact on the environment is computed by characterization factor of each specific substance;
- Normalization: Impact categories are related to a common reference;
- Weighting: Impact categories are weighted according to their specific relevance;
- Single score: Presents the impact categories on a single column and is useful for comparative assessment.

The functional unit in a comparative LCA study represents the common reference for the two products or processes [19]. Some of the recent advances in LCA for textile materials include the following literature studies [26–30]. LCA studies were used to foster decisions for implementing new technologies on the management level of SMEs [26]. The research was accomplished on LCA studies for textile raw materials, namely cotton, polyester, nylon, acryl, and elastane, within a benchmarking study [27]. The research was directed towards LCA for various treatment processes on textiles, such as the fireproof treatment of fabrics and its environmental impact reduction by eco-path disposal treatment [28]. Three different recycling PES trousers (chemical/ mechanical and energy recovery) were analyzed in [29].

Moreover, attention was focused on LCA for smart and e-textiles with conductive fibers made of conjugated polymers, carbon nanotubes, graphene, polymer blend, or nanocomposite [30]. Eco-design in smart textiles plays an important role: a comparative LCA study for eco-designed and original smart textiles products was achieved in [31]. End-of-life and recycling management of the textile chain was analyzed by LCA in countries such as Finland [32] and Denmark [33]. A review of the overall impact of nanomaterials on the environment was performed in [34].

1.5 Aim of this book chapter

The aim of this chapter is to comparatively assess the impact on the environment of two types of electromagnetic shields:

- Woven fabric with inserted silver yarns in warp and weft (**Figure 3**);
- Woven fabric with inserted silver yarns in warp and weft and magnetron plasma copper coating (**Figure 4**).

EMSE was experimentally determined according to the ASTM ES-07 (TEM cell) standard and represented in **Figure 5**.

Since magnetron sputtering copper coating enhances the shielding effectiveness of the woven fabrics with 8–10 dB on the frequency domain of 0.1–1000 MHz, the woven fabric with inserted metallic yarns was modeled considering a smaller distance of the grid in order to reach the same EMSE. The increase of EMSE in smaller grid distances is given by relations of the EMC literature [6]. This approach was considered in order to have the same LCA functional unit for both textile shields, namely: **one sqm of shielding fabric with 50 dB at 100 MHz.**

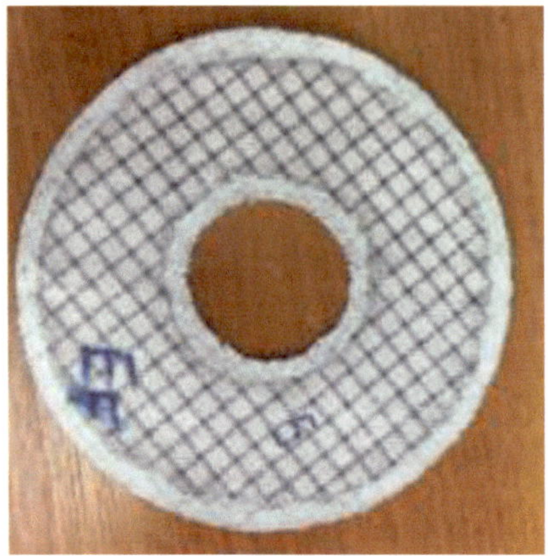

Figure 3.
Woven fabric with inserted silver yarns in warp and weft.

Figure 4.
Same woven fabric with plasma copper coating.

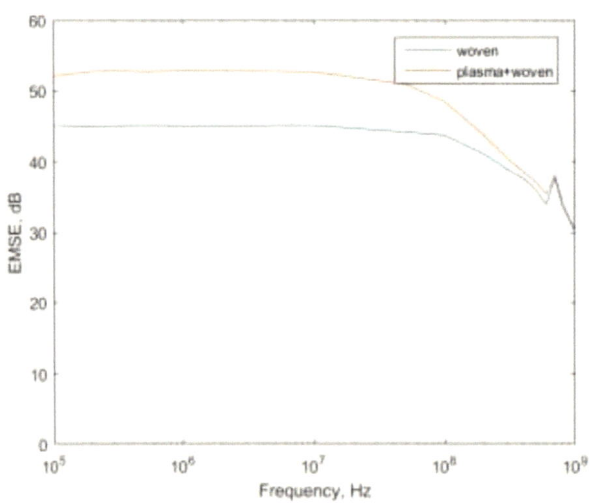

Figure 5.
Experimental EMSE values for the woven fabric with conductive yarns (blue) and the plasma coated woven fabric with conductive yarns (red).

Yarn code	M1	M2
Yarn type	100% Cotton, spun yarn	Silver coated PA yarn (Shieldtex 117/17 dtex)
Linear density [Nm, dtex]	Nm 50/2	140x2 dtex
Linear electrical resistance [Ω/m]	—	220
Linear electrical conductivity [S/m]	—	121842
Relative Magnetic Permeability [1]	1	1
Optical diameter [μm]	293	218

Table 1.
The yarn's properties.

2. Comparative LCA on plasma coated woven fabrics with conductive yarns destined to EMI shielding

2.1 The design of the textile shields

The woven fabric with inserted metallic yarns was manufactured at SC Majutex SRL, a woven fabric producer of technical textiles (www.majutex.ro). **Table 1** presents the physical-mechanical and electric properties of the cotton and silver yarns used for the fabric weaving, while **Table 2** presents the physical-mechanical and electric properties of the resulted fabric.

The woven fabric was manufactured by an industrial weaving process on SOMET weaving machines with a width of 1.90 m.

2.2 The modeling of the woven fabric with conductive yarns

The following relations of electric conductive grids from the EMC literature was applied to model the woven fabric with silver yarns by reducing the distance of the grid in order to increase EMSE [14]:

$$EMSE = 20\log\left(\frac{\lambda/2}{W\sqrt{\pi}}\right) \quad (1)$$

Yarn type	Woven fabric system	M1 + M2
Float repeat (cotton yarns: silver yarns) SR6431:2012	Warp	6:2
	Weft	5:2
Fabric Density [no.yarns/10 cm] EN1049–2:2000	Warp	168
	Weft	150
Distance between Conductive Yarns *W* [mm]	Warp	5
	Weft	5
Fabric Thickness, [mm] ISO5084:2001	—	0.490
Specific Mass, [g/m^2] EN12127:2003	—	118

Table 2.
The woven fabric properties.

$$W = \frac{\lambda}{2\sqrt{\pi}} \times 10^{-\frac{EMSE}{20}} \quad (2)$$

With W = distance between metallic yarns in a woven fabric structure.λ = wavelength of the incident EM field.

According to the model (**Figure 6**), the distance of the electric conductive grid has to shift from W = 5 mm to W = 3 mm, in order to increase the EMSE of the woven fabric by 8 dB at 100 MHz. Since the functional unit for the comparative LCA study was set to one sqm of shielding fabric with 50 dB at 100 MHz; it resulted in two variants of textile shields:

1. Woven fabric with inserted metallic yarns with the distance of 3 mm

2. Plasma coated woven fabric with inserted metallic yarns with the distance of 5 mm

The next step was to collect the LC Inventory data for these two EM shields.

2.3 The LC inventory data

LCI includes all the inputs and outputs into the system, such as raw materials, electricity consumption, auxiliary materials, etc. **Table 3** presents the mass per sqm for the two textile shields: sample 1 is manufactured, and sample 2 is modeled. Sample 1 was subsequently coated by magnetron plasma.

The LCI of silver yarns was computed by considering the ratio of Silver and PA6.6 mass of the Statex Silver yarn [35] and mean data for the spinning process [27]. The LCI of copper coated fabrics was computed by considering the mass

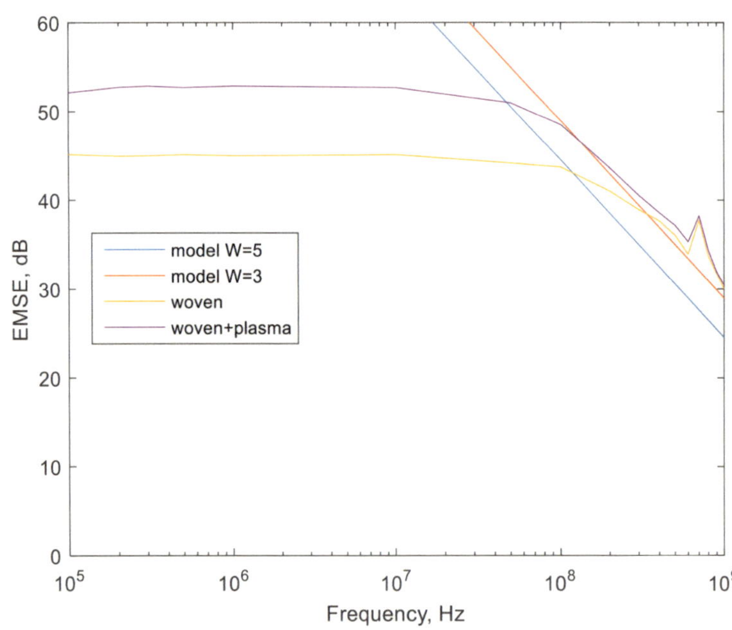

Figure 6.
Modeled and experimental EMSE.

Samples		Cotton yarns [g/sqm]	Silver coated PA yarns [g/sqm]	Silver content out of yarn [g/sqm]
Sample 1 W = 5 mm	Warp	51,84	6,08	3,73
	Weft	46,22	6,51	4,00
	Total	**98,06**	**12,59**	**7,73**
Sample 2 W = 3 mm	Warp	41,47	9,73	5,98
	Weft	38,88	9,13	5,61
	Total	**80,35**	**18,86**	**11,59**

Table 3.
Specific mass for the woven structure with silver yarns.

Input/output into the system	Value	Unit
Functional unit: 1 sqm of shielding fabric with 50 dB at 100 MHz		
Cotton yarn (product)	98.06	g
Silver Statex yarn (product)	12.59	g
Argon, liquid, at the plant	100	g
Copper, primary, at the refinery	50	g
Electricity, low voltage, at grid/RO	0.46	kWh
Weaving Majutex (process)	110	g

Table 4.
LC inventory – Sample 1 (plasma coated woven fabric).

Input/output into the system	Value	Unit
Functional unit: 1 sqm of shielding fabric with 50 dB at 100 MHz		
Cotton yarn (product)	80.35	g
Silver Statex yarn (product)	18.86	g
Weaving Majutex (process)	110	g

Table 5.
LC inventory – Sample 2 (modeled woven fabric).

of copper, the energy consumption of the magnetron sputtering equipment for laboratory scale deposition, and the release of Argon into the air (data provided by INFLPR). LCI data is included within **Tables 4** and **5**.

The main challenge of the comparative LCA study is the additional Silver coated yarn needed for the grid of W = 3 mm related to the plasma copper coating of 1200 nm on both sides of the fabric with the grid of W = 5 mm. The manufacturing process of the PA6.6 coated silver yarns was estimated with mean values according to the spinning processes of various raw materials based yarns [27]. Single main input/output factors into the system were considered within the indicative LCA study. Limitations applied for transport, heat, etc., of the industrial weaving process at SC Majutex SRL. INFLPR provided the magnetron sputtering inputs/outputs into the system.

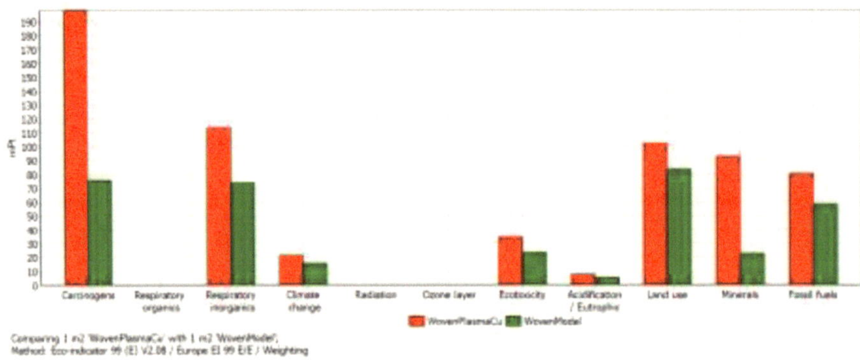

Figure 7.
Comparative weighting diagram: Sample1 (red) and Sample2 (green).

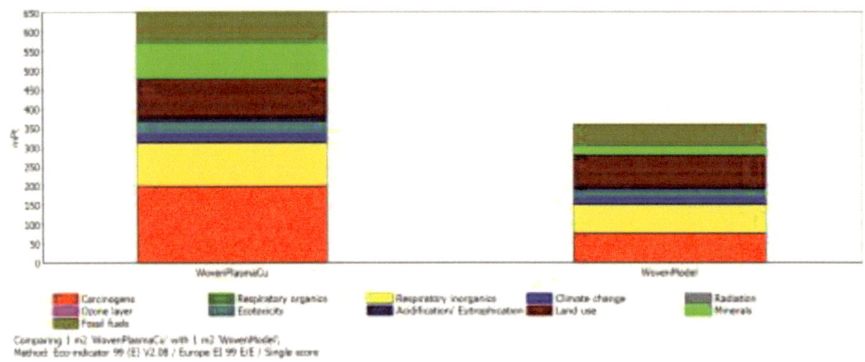

Figure 8.
Comparative single score diagram: Sample1 (left) and Sample2 (right).

2.4 Life Cycle Impact Assessment diagrams

By introducing the LCI data into the SimaPro software, the following LCIA diagrams resulted:

Figure 7 presents the comparative impact of the two textile shields on impact categories of the method EcoIndicator 99E. The main negative impact on the environment is given by the carcinogens followed by Respiratory inorganics, which could be explained by releasing Argon into the air within the plasma process. Another wide difference between the two shields is at the impact category Minerals, while Fossil fuels for electric energy are quite balanced.

The Single score diagram (**Figure 8**) presents the total comparative impact on the environment for the two textile shields by a single view. One of the processes is industry (weaving of metallic yarns – Sample 2), and the other process is of laboratory (coating with metallic layers – Sample 1), which is why manufacturing one sqm of shielding fabric has a significant impact on the environment in the laboratory process. The corresponding LCIA data of the two processes offers significant differences when related to the functional unit.

2.5 Scale-up of plasma equipment

One of the LCA study consequences is the need to scale up the plasma equipment for industrial use. A business plan considering the investment for the equipment and the Return-On-Investment and Break-even point analysis has been done and will be published in the near future. The scale-up magnetron plasma equipment represents the key expertise of INFLPR. Textile shields properties are key expertise of INCDTP and the experimental set-up for EMSE measurement expertise of ICPE-CA.

3. Conclusion

This chapter has included theoretical aspects and applications within a highly interdisciplinary domain: electromagnetic compatibility, plasma physics, textiles, and Life Cycle Assessment. Textile EM shields have been manufactured by two methods – insertion of metallic yarns and plasma coating of metallic layers and their shielding effectiveness has been experimentally determined. In order to assess the environmentally friendly character of plasma coating, a comparative LCA study has been conducted between the woven fabric with conductive yarns and the plasma coated woven fabric. One sqm of shielding fabric with 50 dB at 100 MHz has been considered as a functional unit. Since plasma coating enhances EMSE with 8–10 dB on the frequency domain of 0.1–1000 MHz for the woven fabric with metallic yarns, the grid of the woven fabric with metallic yarns has been reduced in order to reach the same EMSE. Modeling according to relations from the EMC literature has been applied for this purpose. The main challenge of the comparative LCA study has been the relation between additional metallic yarns for the model and the plasma coating of the woven fabric. Since the plasma coating is a laboratory process, the impact on the environment was higher. A business plan has been achieved to implement a scale up plasma equipment for industrial use. Further research is needed to evaluate the eco-friendly character of textile electromagnetic shields.

Acknowledgements

This research was funded by ERA-NET Manunet project Contract no. 28/2018 TexEMFiRe—“Manufacturing textiles with electromagnetic shielding and fire-retardant properties by plasma based methods.”

Author details

Ion Răzvan Rădulescu[1*], Lilioara Surdu[1], Emilia Visileanu[1], Bogdana Mitu[2] and Cristian Morari[3]

1 INCDTP, Bucharest, Romania

2 INFLPR – Măgurele, Bucharest, Romania

3 ICPE-CA, Bucharest, Romania

*Address all correspondence to: razvan.radulescu@incdtp.ro

References

[1] Dai M, Zhai Y, Zhang Y, A green approach to preparing hydrophobic, electrically conductive textiles based on waterborne polyurethane for electromagnetic interference shielding with low reflectivity, Chemical Engineering Journal, 2020, https://doi.org/10.1016/j.cej.2020.127749

[2] Liu. Q, Yi C, Chen J, Xia M, Lu Y, Wang Y, Liu X, Li M, Liu K, Wang D, Flexible, breathable, and highly environmental-stable Ni/PPy/PET conductive fabrics for efficient electromagnetic interference shielding and wearable textile antennas, Composites Part B 215, 2021, 108752, https://doi.org/10.1016/j.compositesb.2021.108752

[3] Xu C, Zhao J, Chao Z, Wang J, Wang W, Zhang X, Li Q, Developing thermal regulating and electromagnetic shielding textiles using ultra-thin carbon nanotube films, Composites Communications 21, 2020, 100409, https://doi.org/10.1016/j.coco.2020.100409

[4] Pakdel E, Wang J, Kashi S, Sun L, Wang X, Advances in photocatalytic self-cleaning, superhydrophobic and electromagnetic interference shielding textile treatments, Advances in Colloid and Interface Science 277, 2020, 102116, https://doi.org/10.1016/j.cis.2020.102116

[5] Kardarian K, Busani T, Osorio I, Domingos H, Igreja R, Franco R, Cortez J, Sintering of nanoscale Silver coated textiles, a new approach to attain conductive fabrics for electromagnetic shielding, Materials Chemistry, and Physics 147, 2014, 815e822, http://dx.doi.org/10.1016/j.matchemphys.2014.06.025

[6] Schwab A, Kuerner W, Electromagnetische Vertraeglichkeit, Springer, 2011, DOI:10.1007/978-3-642-16610-5

[7] Kaden H., Wirbelstroeme und Schirmung in der Nachrichtentechnik, Springer, 1959

[8] Keiser B, Principles of electromagnetic compatibility, CRC Press, 1981

[9] Radulescu I R, Surdu L, Visileanu E, Morari C, Costea M, (29.01.2021). Modeling and Validating Analytic Relations for Electromagnetic Shielding Effectiveness of Fabrics with Conductive Yarns, IntechOpen, DOI: 10.5772/intechopen.95524.

[10] Henn A R, Crib R M, Modeling the shielding effectiveness of metallized fabrics. In Proceedings of the International Symposium on Textiles for Functional Applications Electromagnetic Compatibility, Anaheim, CA, USA, August 1992.

[11] Neelakanta P S, Handbook of Electromagnetic Materials: Monolithic and Composite Versions and Their Applications. CRC Press: Boca Raton, USA, 1995; p. 447-490, ISBN 0-8493-2500-5.

[12] Neruda M., and Vojtech L., Electromagnetic Shielding Effectiveness of Woven Fabrics with High Electrical Conductivity: Complete Derivation and Verification of Analytical Model, In: Materials, 2018, 11, 1657, http://doi.org/10.3390/ma11091657

[13] Rybicki T, EMI Shielding and reflection from textile mesh grids compared with analytic models. IEEE Trans. Electromagn. Compat. 2019, 61, 372-380. doi: 10.1109/TEMC.2018.2830968

[14] Vojtech, L, Neruda M, Modelling of Surface and Bulk Resistance for Wearable Textile Antenna Design, PRZEGLĄD ELEKTROTECHNICZNY, ISSN 0033-2097, R. 89 NR 2b/2013

[15] Rădulescu I R., Surdu L, Visileanu E., Costea M., Pătru I, Voicu V., Modelling and testing the electromagnetic near field shielding effectiveness achieved by woven fabrics with conductive yarns, In: Industria Textila, 2018, 69, 3, 169-176, http://doi.org/10.35530/IT.069.03.1508

[16] Ziaja J, Jaroszewski M., (2011). EMI Shielding using Composite Materials with Plasma Layers, Electromagnetic Waves, Vitaliy Zhurbenko, IntechOpen, DOI: 10.5772/16553.

[17] Rădulescu I.R., Surdu L., Mitu B., Morari C., Costea M., Golovanov N., Conductive textile structures and their contribution to electromagnetic shielding effectiveness, In: Industria Textila, 2020, 71, 5, 432-437, http://doi.org/10.35530/IT.071.05.1783

[18] Standard ISO 14040:2006, Environmental management — Life cycle assessment — Principles and framework, https://www.iso.org/standard/37456.html

[19] ILCD Handbook, European Commission - Joint Research Centre - Institute for Environment and Sustainability: International Reference Life Cycle Data System (ILCD) Handbook – General guide for Life Cycle Assessment - Detailed guidance. First edition March 2010.

[20] Surdu L, Barbu I, Rădulescu I R, Life cycle assessment for medical textiles treated with plasma, Industria Textila 6/2015, 360-364

[21] Goedkoop M, The Eco-indicator 99 Methodology, Journal of Life Cycle Assessment, Japan, Vol.3 No.1 January 2007, pages 32-38

[22] Dekker E, Zijp M, van de Kamp M, Temme E, van Zelm R, A taste of the new ReCiPe for life cycle assessment: consequences of the updated impact assessment method on food product LCAs, The International Journal of Life Cycle Assessment (2020) 25:2315-2324, https://doi.org/10.1007/s11367-019-01653-3

[23] Jolliet O, Margni M, Charles R, Humbert S, Payet J, Rebitzer G, Rosenbaum R, IMPACT 2002+: A new life cycle impact assessment methodology, The Intl. Journal of Life Cycle Assessment volume 8, Article number: 324 (2003)

[24] SimaPro user manual: https://pre-sustainability.com/legacy/download/ SimaPro8Tutorial.pdf, Accessed 2021-06-02

[25] SimaPro database manual, https://simapro.com/wp-content/uploads/2020/10/DatabaseManualMethods.pdf, Accessed 2021-06-02

[26] Kloepffer W. (editor), Background and Future Prospects in Life Cycle Assessment, Springer, Dordrecht, 2014, ISSN 2214-3513

[27] Van der Velden N., Patel M, Vogtlaender J, (2014), LCA benchmarking study on textiles made of cotton, polyester, nylon, acryl or elastane, Int. J. Life Cycle Assess, 19, 2014:331-356, DOI 10.1007/s11367-013-0626-9

[28] Pesnel S., Perwuelz A., LCA: a decision-making tool for recycling process in textile industry, 2011, https://docplayer.net/24614249-Lca-a-decision-making-tool-for-recycling-processes-in-textile-industry.html

[29] Yasin S., Behary N., Perwuelz A., Life cycle assessment of flame retardant cotton textiles with optimized end-of-life phase, Journal of Cleaner Production 172, 2018, 1080-1088, https://doi.org/10.1016/j.jclepro.2017.10.198

[30] Lund A., Van der Velden N., Persson N-K., Hamedi M, Mueller C, Electrically

conducting fibres for e-textiles: An open playground for conjugated polymers and carbon nanomaterials, Materials Science & Engineering R 126, 2018, 1-29, https://doi.org/10.1016/j.mser.2018.03.001

[31] Van der Velden N, Kuusuk K, Koehler A., Life cycle assessment and eco-design of smart textiles: The importance of material selection demonstrated through e-textile product redesign, Materials and Design, 84, 2015, 313-324, http://dx.doi.org/10.1016/j.matdes.2015.06.129

[32] Dahlbo H., Aalto K., Eskelinen H, Salmenpera H,, Increasing textile circulation—Consequences and requirements, Sustainable Production and Consumption, 2017, 44-57, http://dx.doi.org/10.1016/j.spc.2016.06.005

[33] Koligkioni A., Parajuly K, Sorensen B, Cimpan C, Environmental assessment of end-of-life textiles in Denmark, Procedia CIRP 69, 2018, 962 – 967, doi:10.1016/j.procir.2017.11.090

[34] Miseljic M, Olsen S., Life-cycle assessment of engineered nanomaterials: a literature review of assessment status, J Nanopart Res, 2014, 16:2427 DOI 10.1007/s11051-014-2427-x

[35] Shieldex manufacturer of Silver coated PA yarns: Available from: https://www.shieldex.de/, Accessed 2021-06-02

Chapter 4

Design, Construction and Validation of a High-Performance OATS

Donglin Meng

Abstract

The state of the art on the open-area test site (OATS) has been introduced. Key technologies on the design and validation of a high-performance OATS have been provided. Some famous OATS in the world regarding their structure, the dimensions of the ground plane (GP), the location of the control room, and performance are listed in a table. A case study is provided on NIM's high-performance OATS. Many details are open for the first time, which show the fine design. A measurement uncertainty example has been provided in measuring the free-space antenna factor of biconical antennas. These results are based on the author's many years of experience, with lots of valuable data and photos. It is intended for calibration laboratories, for EMC antenna users, for writing EMC standards, as well as for the assessors in EMC.

Keywords: antenna factor, antenna calibration, open-area test site, calculable dipole antennas, site validation, site insertion loss

1. Introduction

An open-area test site (OATS) is one of the key facilities in EMC. An OATS is a basic facility to measure the antenna factor [1] and the "benchmark" for semi-anechoic chambers over (30–1000) MHz [2].

When we make radiated disturbance measurements, both antennas and anechoic chambers are required. An OATS is the best choice for measuring the antenna factor over 30 MHz to 1000 MHz. We use antennas to "capture" the invisible electric fields (E-fields) or magnetic fields (M-fields) by converting the fields into voltage signals, which can be measured by an EMI receiver. However, the converting capabilities are different for different antennas. Thus, we need a parameter to "eliminate" this difference; thus, the antenna factor is defined as the ratio between the magnitude of E-field and the voltage induced on the load (usually 50 Ω) connected into the feed of an antenna [1]. One of the major activities for a calibration laboratory is to measure their value. In order to get a very accurate value for the antenna factor over 30 to 1000 MHz, OATS is the best choice since it can provide the most accurate value for broadband EMC antennas over this frequency band.

An OATS is the "benchmark" for semi-anechoic chambers over (30–1000) MHz. An EMC chamber can isolate the radiated emission by an EUT from others by a shielding enclosure outside the antenna and EUT. Usually, absorbers are lined inside to reduce the reflections along the four sides of an EUT and from the ceiling. However, it is very difficult to reduce totally due to the limits of technology and cost. Therefore, a quality verification process called validation is needed. The basic idea of this validation procedure is to measure the site attenuation (SA) between a pair of broadband antennas in the EMC chamber over (30–1000) MHz and compared with some standard value, which is defined as antenna pair reference site attenuation, A_{apr} [3]. If the difference is less than some value (e.g., 4 dB or less), then the EMC chamber can be regarded as a qualified "compatible test site" (COMTS). Thus the accuracy of A_{apr} is significant for the suppliers and end-users of the EMC chamber. Up to now, the OATS is the only way to measure the A_{apr} accurately. "Normalized site attenuation" (NSA) can be used to validate the performance of EMC chambers [3], too; however, OATS is also required to measure the free-space antenna factor and the mutual coupling, etc., are very difficult to get [4, 5].

Due to the significance of OATS in the EMC area, it is worthwhile reviewing the technologies of design, construction and validation of a high-performance OATS for the following:

- the antenna users
- calibration Labs
- potential builders of high-performance OATS
- certificate examinants
- revisions of international standards CISPR 16–1-6, CISPR 16–1-5, ANSI C63.5, and so on.

The start of the art on the OATS is introduced in Section 2; a detailed case study is provided on the design, construction, and validation in Section 3. An uncertainty example in measuring A_{apr} is provided in Section 4.

2. The state of the art on OATS in the world

The theoretical model of an OATS is listed in **Figure 1**. The electromagnetic fields (EM fields) radiated by a transmit antenna (Tx) are reflected only by the ground plane (GP), and the GP is infinitely large, consisting of a perfect electric conductor. Then, there are no other reflections. The receive antenna (Rx) received the EM fields radiated directly from the Tx and the reflections from the ground plane. Any other reflections are regarded as uncertainty sources. This means the better the ground plane size, the better agreement with an ideal OATS. Standard CISPR 16–1-5 states the minimum should be 30 m by 20 m. The most common ground plane size is 60 m by 30 m. The flatness of the metal ground plane is also important, especially above 700 MHz [6].

There are basically two types of structures on the large ground plane, as shown in **Figure 2**. In the solid type, the metal ground plane is placed directly on the surface of the concrete layer. This structure makes sure the metal ground plane can hold heavy EUT. In order to adjust the flatness of the ground plane, some space is designed between the metal ground plane and the concrete layer, which is called the floating type.

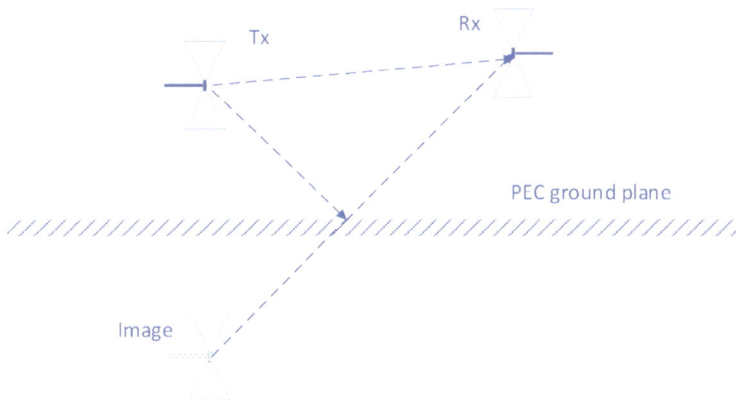

Figure 1.
The theoretical model of an OATS—Half-space model.

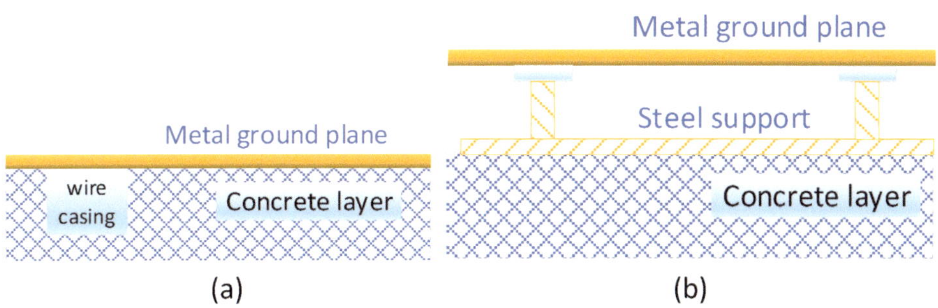

Figure 2.
The structures of the ground plane. (a) Solid type and (b) floating type.

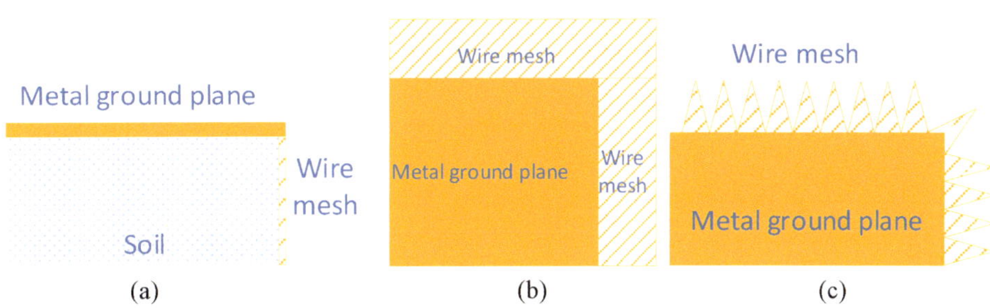

Figure 3.
The design of stainless wire mesh (showing the corner of an OATS). (a) Inverted L-shape (vertical cut), (b) belt (bird view), and (c) triangular (bird view).

The impedance transition between the edge of the ground plane and the soil is also important. Usually, stainless wire mesh is designed for this purpose. **Figure 3** shows some typical designs for the stainless wire mesh. In (a), the wire mesh is mounted vertically down to the soil. In (b), a wire-mesh belt is mounted around the edge of the metal ground plane. In (c), the belt is replaced by some triangular mire mesh. Simulation shows design (c) agrees best with the theoretical mode in **Figure 1**, under the same ground plane size and wire mesh area.

Figure 4 shows the surface current induced. A pair of dipole antennas are resonated at 30 MHz and VP; the dipole antennas are separated by 20 m (R = 20 m)

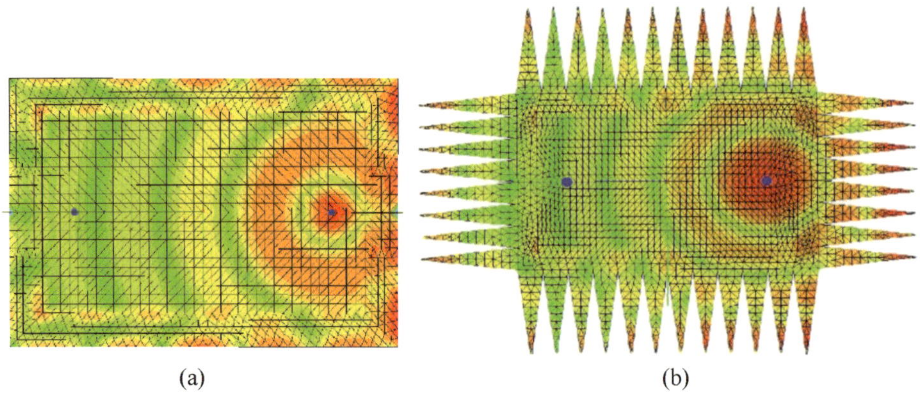

Figure 4.
The comparison of triangular wire mesh. (a) No wire mesh and (b) triangular wire mesh.

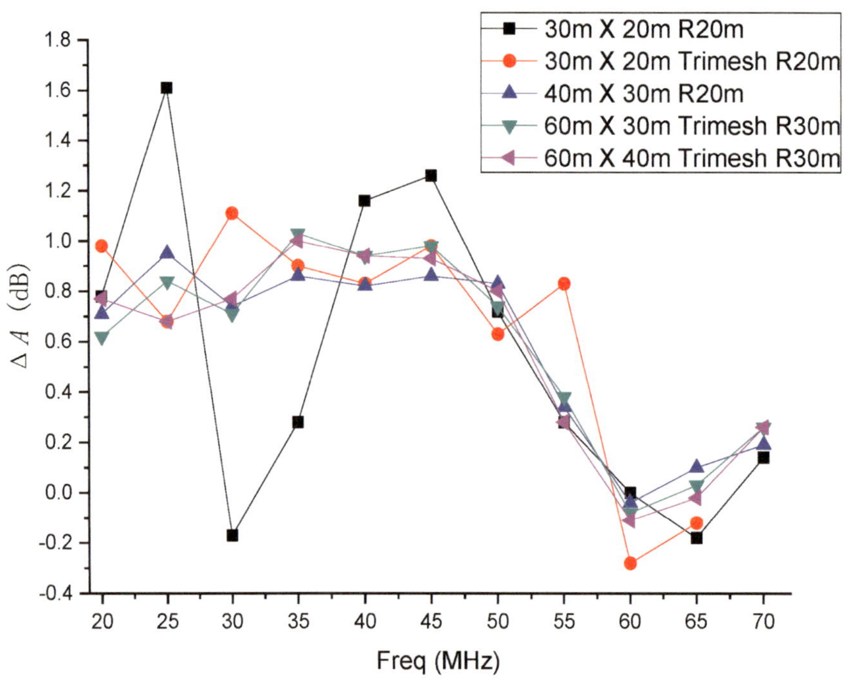

Figure 5.
The deviation of finite-size GP at VP (R = 20 m).

both the Tx and the Rx are 1.3 m above the perfect electric conductor (PEC) ground plane. The height of the isosceles triangles is 10 m, and the bottom width is 5 m. The inner area (25 m by 15 m) of the ground plane is simulated with physical optics (PO), while others are simulated with Method of Moment (MoM). The simulation model is set up in free space; in other words, no consideration is taken into account on the effect of the grounding of wire mesh into the soil. **Figure 4** shows that the surface current fluctuated rapidly near the edge of the ground plane or the tips of the mesh [7].

The deviation ΔA is shown in **Figure 5**. The means for the legends are shown below:

1. "30 m × 20 R20 m": no wire mesh, as shown in **Figure 4(a)**, GP is 30 m by 20 m, both dipole antennas are separated by 20 m;

2. "30 m × 20 Trimesh R20 m": triangular wire mesh as shown in **Figure 4(b)** is introduced based on ①; The meanings for other legends are similar.

As shown in [7], a GP of 30 m by 20 m is not qualified for measurements at R = 20 m, whose deviation is larger than 1.5 dB. However, it can be reduced to 1.2 dB at VP by introducing some triangular wire mesh shown in **Figure 4(b)**. If the GP is increased to 40 m by 30 m, the deviation is less than 1 dB. Less deviation at 30 m separation can be achieved by increasing the GP to 60 m by 40 m.

Further simulation shows that the triangular wire mesh is also effective for HP at R = 20 m, as shown in **Figure 6**.

The performance of a practical OATS should be as close as possible to the half-space model shown in **Figure 1**. The IEC standard CISPR 16–1-5:2014 provides a detailed method to validate this, which is briefly summarized as in Eq. (1).

$$\Delta A = |A_{\rm m} - A_{\rm c}| < T_{\rm SIL} - \Delta A_{\rm m}. \tag{1}$$

where ΔA is the absolute deviation of the OATS from an ideal OATS, in dB, at the specified frequency; $A_{\rm m}$ is the measured site insertion loss (SIL), in dB, between a pair of calculable dipole antennas; $\Delta A_{\rm m}$ is the SIL measurement uncertainty (k = 2), in dB, $A_{\rm c}$ the calculated SIL, in dB; $T_{\rm SIL}$ is the allowed tolerance in SIL, in dB. For a calibration test site (CALTS), $T_{\rm SIL}$ = 1.0 dB at horizontal polarization (HP).

For a reference test site (REFTS), $T_{\rm SIL}$ = 1.0 dB at HP and $T_{\rm SIL}$ = 1.5 dB at vertical polarization (VP). **Table 1** lists some famous OATS in the world. Where CR stands for the control room (CR), some data are not available (N/A) due to the limited knowledge of the author.

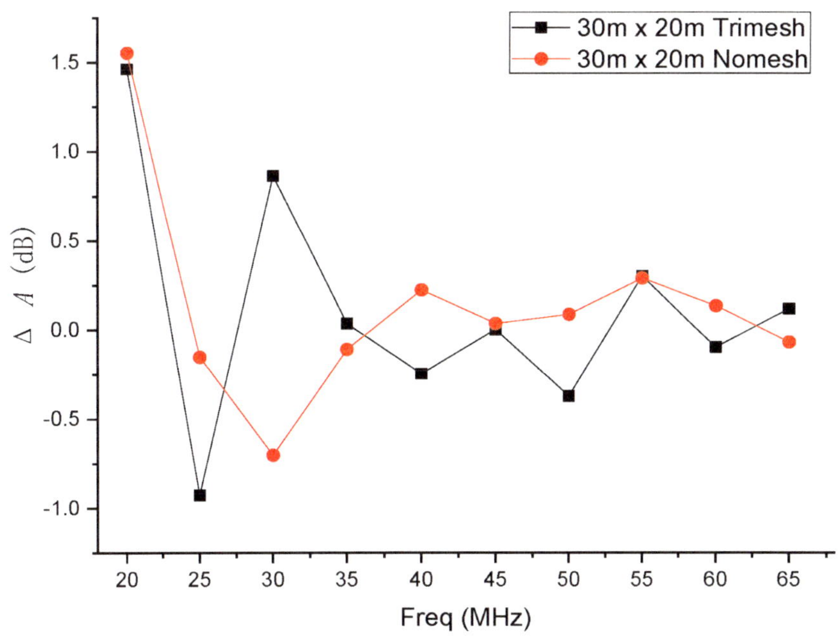

Figure 6.
The deviation of finite-size GP at HP (R = 20 m).

Lab.	Dimensions		Structure	CR Locations	ΔA^*
	Metal GP	Stainless wire mesh			
NIM (China)	60 m x 40 m x 10 mm	10 m Triangular	Floating type	Under	0.26 dB (HP) 0.34 dB (VP)
NPL (UK)	60 m x 30 m x 8 mm	Inverted L-type	Solid type	Above GP, Wooden, 25 m far away	0.3 dB (HP)
NIST (USA)	60 m x 30 m x 8 mm	Triangular	Solid type	N/A	N/A
Liberty Lab. (USA)	60 m x 30 m x 8 mm	Rectangular10 m width	Floating type	Half under	0.5 dB (HP)
ETS-LINDGREN (USA)	60 m x 30 m x 8 mm	Strip	Solid type	Under	N/A
NIMJ (Japan)	45 m x 30 m x 16 mm	Rectangular	Floating type	Above GP	N/A
KRISS (Korea)	30 m x 30 m x 2 mm	N/A	Solid type	N/A	N/A
CENAM (Mexico)	60 m x 30 m	N/A	N/A	N/A	0.5 (HP)
ARC (Austria)	20 m x 17 m	Rectangular10 m width	Solid type	Under GP, with 50 cm roof above GP	N/A

Due to the limited knowledge of the author, the data could be wrong. N/A, not available.

Table 1.
Some high-performance OATS.

3. A case study for NIM OATS

Figure 7 shows a photo of NIM OATS taken in 2014. It is located in NIM's Changpin Campus. The white part is the reflective GP, and usually, there are a motorized mast and a manual mast for holding the antennas to be measured. The masts are made of low permittivity dielectric materials. The motor of the motorized mast is located under the metal GP to avoid unwanted reflections. Obviously, these setups will imitate the theoretical model shown in **Figure 1**.

Figure 7.
Photo of NIM OATS.

BTW, in this photo, there are two more masts on the upper right corner of the GP, which is for backup, and it would better be moved away. There is also a beautiful lake in the north of the OATS. Some sand cypress trees are planted around to keep the soil wet and as fences to prevent the entrance of unintentional persons and trucks.

The structure of the OATS is illustrated in **Figure 8**.

The whole system is designed carefully. The metal GP will suffer more than 80 degrees temperature variations seriously during four seasons in Beijing. In order to cope with this critical problem, the large metal GP is placed "freely" on the top 2501 pieces (41 by 61) of the ceramics plate. The metal GP can move freely in the horizontal plane but is kept fixed in the vertical direction. The ceramics plates are mounted on adjustable bolts, which are mounted on the steel frames. These steel frames are mounted on the box-layered concrete structure. The whole concrete structure is located on 138 pile foundations, which are over 20 m deep into the rock layer, to maintain a very stable foundation.

As shown in **Figure 8**, the equipment room (ER) is located under the bottom of antenna masts, which can make sure the shortest cable routing from antenna masts

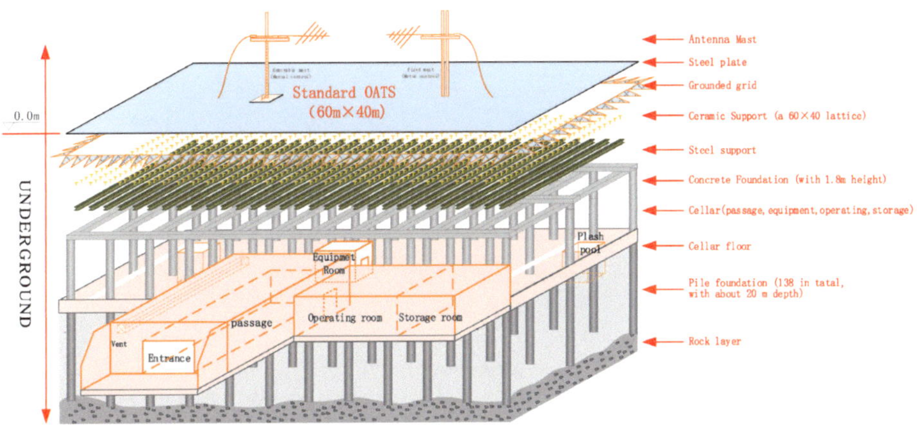

Figure 8.
The structure illustration of NIM OATS.

(a) (b)

Figure 9.
The grounding of the triangular wire mesh. (a) Triangular wire mesh and (b) tip of the triangular wire mesh.

to the ER. The control room is located under the GP, but near the entrance, to make sure a short way for operators to make measurements.

The dimesion of the GP and stainless triangular wire mesh are shown in **Figure 9**.

As shown in **Figures 4–6**, the triangular is very useful for reducing the reflections from the edge of the GP. **Figure 9(a)** is the photo for the stainless triangular wire mesh, and **Figure 9(b)** shows the tips of the triangular wire mesh connected each other and being grounded to the soil with resistance less than 1 Ω.

Lots of precise "measurements" are carried out for this OATS. The deviation ΔA shown in Eq. (1) measured with a pair of calculable dipole antennas (CDAs) at 24 resonant frequencies is shown in **Figure 10** for HP and **Figure 11** for VP [7].

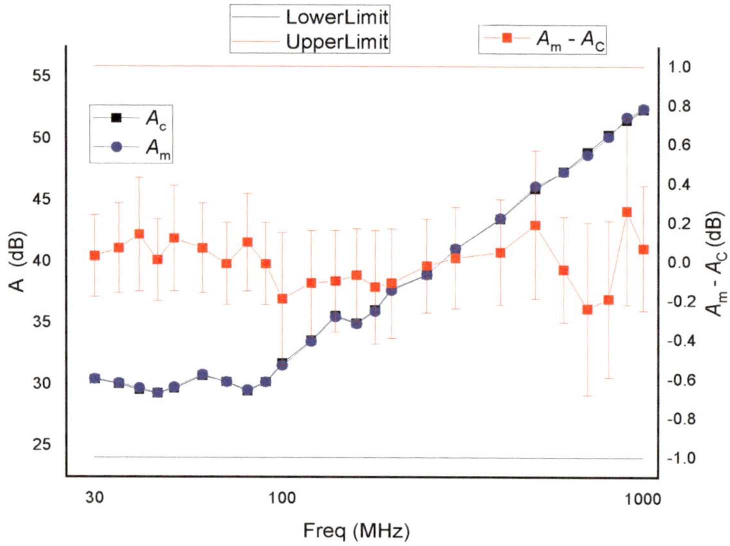

Figure 10.
The deviation of NIM's OATS at HP.

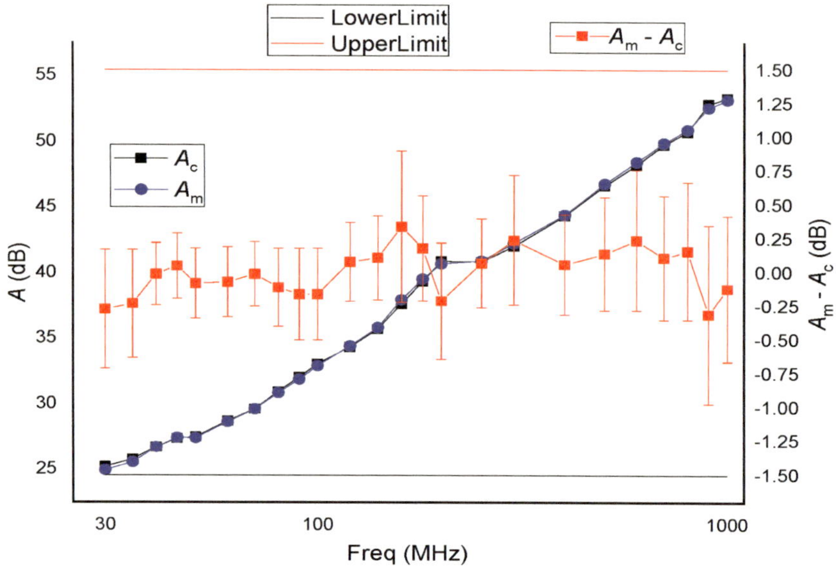

Figure 11.
The deviation of NIM's OATS at VP.

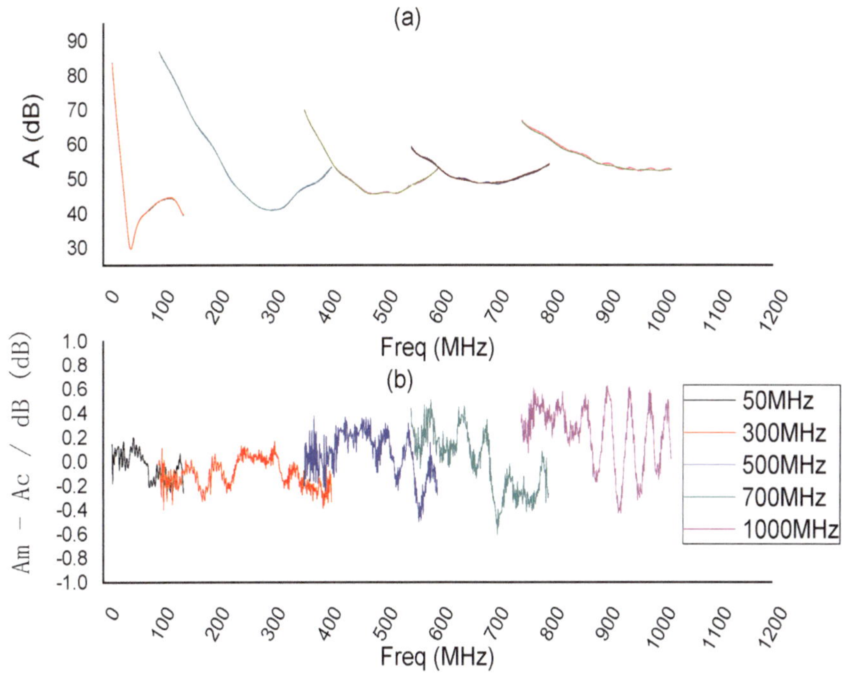

Figure 12.
Deviation of NIM's OATS over sweeping frequencies, HP, R = 10 *m. (a) the comparison of measured and calculated site insertion loss (A). (b) the difference of site insertion loss between measurements and calculations.*

The deviation is less than 0.26 dB for HP and 0.34 dB for VP. Obviously, it meets the requirements for both CALTS and REFT, though the uncertainty ΔA_m is over-estimated. Thus, we can conclude that the performance of this OATS is wonderful.

In order to investigate the results at other frequencies, sweeping measurements between a pair of broadband calculable dipole antennas separated by 10 m and HP are shown in **Figure 12**. The difference between A_m and A_c are quite less at resonant frequencies, e.g., 50 MHz, 300 MHz, 500 MHz, 700 MHz, and 1000 MHz; however, it is larger at non-resonant frequencies.

4. Uncertainty evaluation of antenna factor F_a for biconical antennas with standard site method

The setup of the standard site method (SSM) is shown in **Figure 13**. The Rx is swept from 1 m to 4 m in height to get the minimum SIL (and this is defined as site attenuation-SA). The horizontal separation R between the center (reference locations) of both antennas is 10 m.

After measuring three SAs for three pairs of antennas (#1 vs. #2, #1 vs. #3, and #2 vs. #3), respectively, the free-space antenna factor F_{a_1} of antenna #1 can be calculated with Eq. (2) [5].

$$F_{a_1} = 10\lg f_M - 24.46 + \frac{1}{2}\left[E_D^{\max} + A_{12} + A_{13} - A_{23}\right] + \Delta F_{DevFree-space} \tag{2}$$

Where f_M is the frequency in MHz; $A_{i,j}$ is the site attenuation between antenna $\# i$ and j, $(i = 1, 2, 3; j = 1, 2, 3; i \neq j)$. $E_D^{\max}$ can be calculated according to Eq. (3),

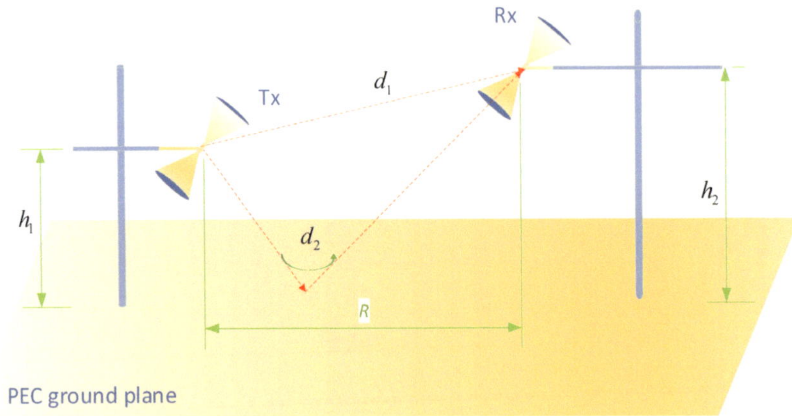

Figure 13.
The setup of the SSM.

$$E_{\mathrm{D}}^{\max} = 20\lg\left\{ \frac{\sqrt{49.2}\left[d_1^2 + d_2^2 + 2d_1d_2\cos\left[\beta(d_2 - d_1)\right]\right]^{1/2}}{d_1d_2}\Bigg|_{1\le h_2\le 4} \right\}. \tag{3}$$

Where

$$d_1 = \left[R^2 + (h_1 - h_2)^2\right]^{1/2}, \tag{4}$$

$$d_2 = \left[R^2 + (h_1 + h_2)^2\right]^{1/2}, \tag{5}$$

$$\beta = \frac{2\pi f_{\mathrm{M}}}{300}. \tag{6}$$

where $\Delta F_{DevFree-space}$ is the correction factor, as shown in Table G.1 in [5]. The meanings for other symbols are shown in **Figure 13**.

It is very hard to deduce an analysis equation for measuring antenna factor. Usually, a hybrid model (both analytical and dark box model) is adopted, as shown in Eq. (7) as an example,

$$\begin{aligned} F_{a1\mathrm{SSM}} &= 10\lg f_{\mathrm{M}} - 24.46 + \frac{1}{2}E_{\mathrm{D}}^{\max} + \left[\frac{1}{2}A_{12} + \frac{1}{2}A_{13} - \frac{1}{2}A_{23}\right] + \frac{\sqrt{3}}{2}u(\delta A_{VNA}) \\ &+ \frac{\sqrt{3}}{2}u(\delta A_{\mathrm{Impedance}}) + \frac{\sqrt{3}}{2}u(\delta A_{\mathrm{Cable}}) + \frac{\sqrt{3}}{2}u(\delta A_{Thru}) + \frac{\sqrt{3}}{2}u(\delta A_{\mathrm{AN}}) \\ &+ \frac{\sqrt{3}}{2}u(\delta A_{AntPosition}) + \frac{\sqrt{3}}{2}u(\delta A_{Site\&Mast}) + \frac{\sqrt{3}}{2}u(\delta A_{Repeat}) + \frac{\sqrt{3}}{2}u(\delta A_{Symetry}) \\ &+ \frac{\sqrt{3}}{2}u(\delta A_{X-pol}) + \Delta F_{DevFree-space} \end{aligned} \tag{7}$$

The meaning of the other symbols is shown in **Table 2**. The evaluated standard uncertainty is shown in **Table 2**, too. The combined standard uncertainty can be calculated with Eq. (8), assuming the above uncertainty sources are independent.

DOI: http://dx.doi.org/10.5772/intechopen.99727

$$
\begin{aligned}
u_c^2(F_{a1}) = {} & \left[\frac{\sqrt{3}}{2}u(\delta A_{VNA})\right]^2 + \left[\frac{\sqrt{3}}{2}u\left(\delta A_{\text{Impedance}}\right)\right]^2 + \left[\frac{\sqrt{3}}{2}u(\delta A_{\text{Cable}})\right]^2 \\
& + \left[\frac{\sqrt{3}}{2}u(\delta A_{Thru})\right]^2 + \left[\frac{\sqrt{3}}{2}u(\delta A_{\text{AN}})\right]^2 + \left[\frac{\sqrt{3}}{2}u(\delta A_{AntPosition})\right]^2 \\
& + \left[\frac{\sqrt{3}}{2}u(\delta A_{Site\&Mast})\right]^2 + \left[\frac{\sqrt{3}}{2}u\left(\delta A_{Repeat}\right)\right]^2 + \left[\frac{\sqrt{3}}{2}u\left(\delta A_{Symetry}\right)\right]^2 \\
& + \left[\frac{\sqrt{3}}{2}u\left(\delta A_{X-pol}\right)\right]^2 + u^2\left(\Delta F_{DevFree-space}\right)
\end{aligned} \tag{8}
$$

Relacing the value in **Table 2**, there will be,

$$u_c = 0.68 \text{ dB} \tag{9}$$

The expanded uncertainty can be calculated with Eq. (4) by assuming a normal distribution since there are many numbers, and their values are similar.

No.	Symbol	Source of uncertainty or quantity (X_i)	Value (dB)	Probability	Divisor	Sensitivity	Standard uncertainty (u_i dB)
1	δA_{VNA}	Uncertainty from measuring S21 with VNA	0.25	Normal	2	$\frac{\sqrt{3}}{2}$	0.11
2	$\delta A_{\text{Impedance}}$	Impedance mismatch between transmit and receive antennas	0.28	U-shaped	$\sqrt{2}$	$\frac{\sqrt{3}}{2}$	0.17
3	δA_{Cable}	Cable loss variation due to temperature variation	0.2	Rectangular	$\sqrt{3}$	$\frac{\sqrt{3}}{2}$	0.10
4	δA_{Thru}	Uncertainty from through calibration	0.2	Rectangular	$\sqrt{3}$	$\frac{\sqrt{3}}{2}$	0.10
5	δA_{AN}	Ambient noise	0.1	Normal	2	$\frac{\sqrt{3}}{2}$	0.04
6	$\delta A_{AntPosition}$	Antenna positing error	0.26	U-shaped	$\sqrt{2}$	$\frac{\sqrt{3}}{2}$	0.16
7	$\delta A_{Site\&Mast}$	Imperfection of the OATS and masts	0.6	Rectangular	$\sqrt{3}$	$\frac{\sqrt{3}}{2}$	0.30
8	δA_{Repeat}	Repeatability of the measurement system	0.3	Normal	2	$\frac{\sqrt{3}}{2}$	0.13
9	$\delta A_{Symetry}$	Symmetry of the antenna under measurement	0.8	Rectangular	$\sqrt{3}$	$\frac{\sqrt{3}}{2}$	0.40
10	δA_{X-pol}	Cross-polarization	0	Rectangular	$\sqrt{3}$	$\frac{\sqrt{3}}{2}$	0.00
11	$\Delta F_{DevFree-space}$	Deviation from the free-space antenna factor	0.4	Rectangular	$\sqrt{3}$	1	0.23
12	Combined standard uncertainty, u_c						0.68
13	Expanded uncertainty, U (k = 2)						1.4

Table 2.
Measurement uncertainty budget for F_a *of a biconical antenna with standard site method.*

$$U = ku_c \tag{10}$$

Taking $k \approx 2$, there will be

$$U = 1.4\ \text{dB}\ (k = 2) \tag{11}$$

This is the expanded uncertainty in measuring the free-space antenna factor of a biconical antenna with the standard site method, as shown in **Table 2**.

5. Conclusion

Key technologies on the design, construction and validation of a high-performance OATS have been provided, based on the author's many years of experience. Some famous OATS in the world regarding their structure, the dimensions of the ground plane (GP), the location of the control room, and performance have been summarized. A detailed case study is provided on NIM's high-performance OATS. A measurement uncertainty example has been provided in measuring the free-space antenna factor of biconical antennas.

Acknowledgements

Dr. Alexander KRIZ and Dr. Katsumi Fujii provide many details on their OATS. There are many persons involved in the design, construction, and validation of NIM's OATS. Special acknowledgments are due to TDK, Mr. Gao X. X, Dr. Cui X. H., Dr. Song Z. F., Mr. Hong L., and Ban H.

Conflict of interest

The author declares no conflict of interest.

Abbreviations

	Full name	Symbol
OATS	open-area test site	/
CALTS	calibration test site	
REFTS	reference test site	
GP	ground plane	
PEC	perfect electric conductor	
CR	control room	
ER	equipment room	
CDA	calculable dipole antenna	/
MoM	Method of Moments	/
NIM	National Institute of Metrology	/
SA	site attenuation	/
NSA	normalized site attenuation	/
HP	horizontal polarization	/
VP	vertical polarization	/
SIL	site insertion loss	/
/	calculated site insertion loss	A_c
/	measured site insertion loss	A_m
/	free-space antenna factor	F_a

Author details

Donglin Meng
National Institute of Metrology, Beijing, China

*Address all correspondence to: mengdl@nim.ac.cn

References

[1] CISPR 16-1-6, Edition 1.1 2017-01, Specification for radio disturbance and immunity measuring apparatus and methods – Part 1-6: Radio disturbance and immunity measuring apparatus – EMC antenna calibration.

[2] CISPR 16-1-5:2014+AMD1:2016 CSV, Specification for radio disturbance and immunity measuring apparatus and methods – Part 1-5: Radio disturbance and immunity measuring apparatus – Antenna calibration sites and reference test sites for 5 MHz to 18 GHz.

[3] CISPR 16-1-4, Edition 4.0 2019-01, Specification for radio disturbance and immunity measuring apparatus and methods – Part 1-4: Radio disturbance and immunity measuring apparatus – Antennas and test sites for radiated disturbance measurements.

[4] Meng D. L. "A fast way to accurately calibrate tunable dipole antennas at VHF," IEEE Transactions on Electromagnetic Compatibility 60(1): 10. 2018.

[5] ANSI C63.4a-2017 (Amendment to ANSI C63.4-2014), American National Standard for Methods of Measurement of Radio-Noise Emissions from Low-Voltage Electrical and Electronic Equipment in the Range of 9 kHz to 40 GHz.

[6] Martin Alexander, Martin Salter, Benjamin Loader, and David Knight, "Broadband calculable dipole reference antennas," IEEE Transactions on Electromagnetic Compatibility, 44(1): 45-58. PII: S 0018-9375(02)01461-8, 2002.

[7] D. L. Meng., X. Liu, and D. B. Li, "Research on unwanted reflections in an OATS for precise omni antenna measurement," Proceedings of the IEEE 6th International Symposium on Microwave, Antenna, Propagation, and EMC Technologies (MAPE), DOI: 10.1109/MAPE.2015.7510308, 2015.

Section 2

Measurements and Modelling

Chapter 5

EMC Measurement Setup Based on Near-Field Multiprobe System

Rubén Tena Sánchez, Lars Jacob Foged
and Manuel Sierra Castañer

Abstract

Multiprobe spherical near-field measurement is a potent tool for fast and accurate characterization of electrical properties of antennas. The use of fast switching in one axis, an azimuth positioner, and a near- to far-field transformation allows a substantial time reduction in antenna measurements while maintaining high-quality results. On the other hand, conventional emissions EMC measurement systems are typically based on detecting the radiated spurious emissions by a device at different frequencies. The systems usually work in far-field (or quasi-far-field conditions), performing the measurements either at 3 or 10 meters. Measurements under these conditions take space and time. Moreover, the systems are not cost-effective for pre-compliance purposes where pre-testing of the device should provide valuable information and confidence about the DUT before performing a compliance test. This chapter analyzes the possibility of cost and space reduction for EMC systems based on multiprobe near-field measurement systems in combination with OTA (over the air measurements), reference-less systems, spherical near-field transformation, phase reconstruction, modal filtering, source reconstruction, and software-defined radio receivers.

Keywords: electromagnetic compatibility, measurements, near-field, multiprobe, over-the-air

1. Introduction

Electromagnetic Compatibility systems are well established and have been widely studied during the last decades. There are a lot of references in this area, like [1]. On the other hand, multiprobe spherical near-field measurements systems have been extensively used for antenna characterization, becoming a potent tool for fast and accurate characterization of electrical properties of antennas. The use of fast switching in one axis, an azimuth positioner, and a near- to far-field transformation allows a substantial time reduction in antenna measurements while maintaining high-quality results. During the last years, combining these kinds of tools with post-processing techniques to increase the accuracy of the measurements and reduce some spurious effects like noise, leakage or echoes, has shown promising results [2]. As explained in the same book, the combination of measurements with simulations has shown the possibility of considering some effects of the measurement scenario.

On the other hand, conventional EMC systems are typically based on detecting the maximum power radiated by a device at different frequencies. The systems usually work in far-field (or quasi-far-field conditions), performing the measurements either at 3 or 10 meters, but sometimes the device's position under test (DUT) in the setup affects the measurement. Even if there are very well-established standards for EMC measurements to get these peak values and check good performance (low radiation) of the electronic devices, the measurement uncertainty is higher than the one found for antenna measurements.

Some practical radiation emission measurement solutions try to overcome the high cost of pre-compliance chambers that typically can go up to hundreds of thousands of euros. These solutions are based on small systems implemented by robotic arms to scan the volume around the DUT [3–7]. The solution is well suited for diagnostics since the field can be measured very close to the DUT. Moreover, the low-cost systems typically are not shielded, and the phase recovery capabilities are strongly setup dependent. Nevertheless, just a few of them are able to reconstruct the phase and thus compute the field at 3 or 10 meters from the near-field information.

This chapter analyzes the possibility of cost and space reduction for radiation emission EMC measurement systems based on the use of techniques already used in near-field antenna measurements, including near to far-field transformation algorithms, OTA (over the air measurements), reference-less systems, multiprobe arrays, phase reconstruction, modal filtering, source reconstruction, and software-defined radio receivers. During the last years, the authors have been working on all those topics for the complete EMC system, as can be observed in different papers in journals and conferences [8–13], and this chapter summarizes all the work included in those previous research works. The chapter describes the advantages of low-cost near-field measurement systems that could be used for EMC pre-compliance measurements, showing some practical results.

The following sections will focus on several of the essential aspects of this kind of system. Section 2 will explain the possible near-field EMC system architectures, explaining each subsystem (hardware or software). Section 3 will explain the configuration of multiprobe array systems for over-the-air (OTA) systems. Section 4 will explain the amplitude and phase calculation using cheap and integrated Software Defined Radio (SDR) receivers. Section 5 will focus on the effects of near to far-field spherical transformation algorithms, and Section 6 will introduce some of the post-processing techniques that can be included for EMC systems. As the reader can observe, all these techniques have been widely used in antenna measurements, although there is still an open research line to redefine their limitations for EMC measurements, where the objective is to detect the radiated power peak values instead of the 3D radiation pattern.

2. EMC system architecture

The near-field EMC measurement system proposed is based on the architecture of a multiprobe near-field antenna measurement system. In this case, we are using some of the conventional Microwave Vision Group measurement multiprobe setups [14]. These systems are based on wideband dual-polarized probes located on arch systems. The receiver is switching between probe and probe in order to acquire the amplitude and phase of the electromagnetic field generated by the probe. Instead of using a conventional vector network analyzer, in this case, we have replaced the receiver with an SDR platform, whose performance will be shown in the following section. For this application, the DUT is self-transmitting, and, therefore, it is not necessary to include a specific transmitter, or in that case, the transmitter is not synchronized with the receiver, as in the conventional antenna measurement systems. This is called over-the-air (OTA) systems, although the procedure we use for this specific application is called reference-less system [11, 12] since the reference is not extracted from the transmitter but from a fixed probe.

Two different architectures that can be used for phase reconstruction in a multiprobe system are presented. The first one is based on an external fixed probe whose relative position is kept fixed with respect to the DUT (**Figure 1**). In this case, the reference antenna is sensitive to amplitude and phase changes (proximity to mast, motor, cable, and multiple reflections). Nevertheless, it will be seen that excellent performance can be achieved with such a simple setup while keeping the advantages of a multiprobe system. Moreover, the solution is scalable and could be implemented in multiprobe systems of different sizes to cover larger DUTs or lower frequencies.

The second solution is based on a reference channel for phase reconstruction that uses one of the probes of the multiprobe arch. This solution does not need any additional hardware, but the complexity is transferred to the post-processing techniques. The solution is described in **Figure 2**. In this case, the setup is simplified, and the coupling between probes reduced since the reference channel belongs to the multiprobe arch.

Some extra post-processing steps are needed in order to retrieve the phase information [11]. The results obtained with this setup can be better than using an external reference antenna; nevertheless, the complexity arises for situations in which the elevation arch needs to be rotated. In this case, the top-probe is not on-axis, and further mathematical derivations are needed in order to reconstruct the near-field phase. This can be done by a non-convex iterative optimization algorithm, as explained in [15].

The next step is the near to near-to-far-field transformation. The theory included in [16], proposed by Prof. Hansen in 1973, and widely used in antenna

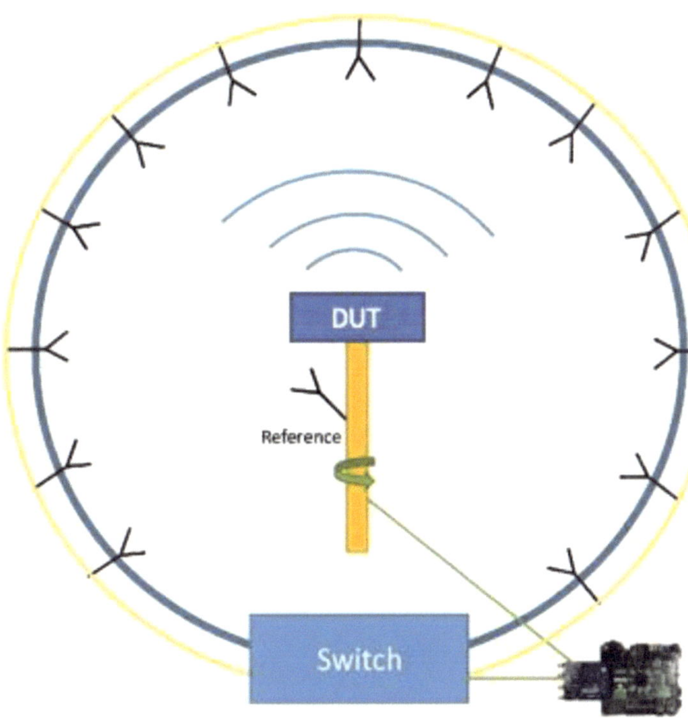

Figure 1.
Measurement setup based on external reference antenna.

near-field measurement is used. This theory is based on the decomposition of the electromagnetic field in spherical mode coefficients, using as input information the amplitude and phase of both orthogonal tangential electric field components (theta and phi) in a sphere enclosing the DUT. Once the spherical mode coefficients are calculated, the field can be computed at any desired distance, in particular 3 or 10 meters.

Some post-processing techniques appeared during the last year that use the information available of the antenna under test and the measurement system, could also be used to improve the results of these EMC measurements. These techniques are based on the filtering of the electromagnetic field in other domains: time domain, antenna electromagnetic sources, spherical modes, or cylindrical modes. A summary of all these techniques can be found in [2], and some of them have been used for extracting the results presented in this work.

Finally, the results are compared with the conventional EMC standards [17, 18] for the different cases to ensure that the DUT passes the final compliance testing in terms of radiation. In comparison with other compact and low-cost solutions, the use of a shielded and anechoic environment that is typically used for antenna characterization [14] provides accurate results when calculating the radiated field at a finite distance. For source reconstruction, the solution proposed here is based on the commercial software INSIGHT [19]. The measured tangential components of the near-field can be exported, and the equivalent currents of the DUT

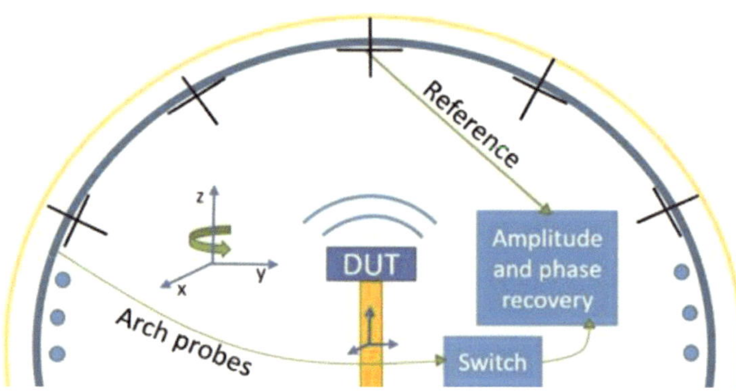

Figure 2.
Measurement setup based on the top probe as a reference antenna.

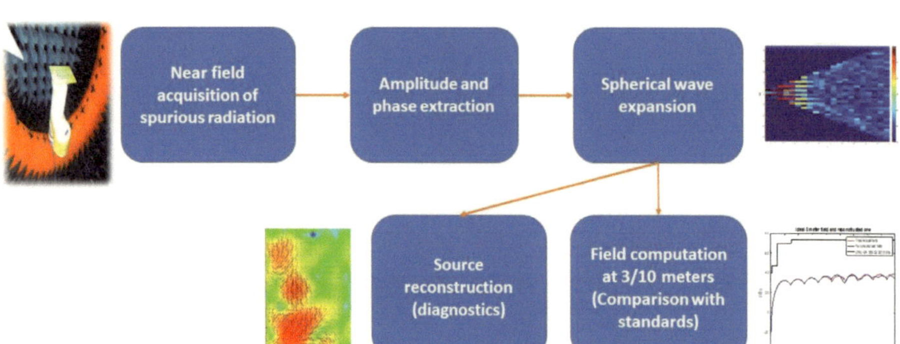

Figure 3.
EMC spurious emissions characterization with near-field multiprobe solutions.

reconstructed in order to determine which areas of the DUT might be responsible for the non-desired radiated emissions.

Figure 3 summarizes the process, and in the following sections, the different parts will be explained.

3. The configuration for multiprobe array systems as OTA system

The hardware of the EMC near-field measurement solution is based on multiprobe systems in combination with the acquisition through a low-cost receiver. The system is based on an arch where the different wideband dual-polarized probes are located. For this specific work, two systems are used: StarLab [20] and MiniLab [21]. These two systems are compact and allow a faster design process and shorter time to market times while keeping a reasonable cost. The use of the low-cost receiver reduces the complexity drastically and cost in comparison with traditional spectrum or vector network analyzers. Moreover, the systems are scalable; thus, the size or low-frequency limitation can be overcome by increasing the measurement solution using other multiprobe configurations.

Figure 4 shows both systems: MiniLab is the smallest system and works from 650 MHz up to 6 GHz. The radius of the arch is 30 cm with 12 dual linear polarized probes separated by fixed angular steps of 15 degrees. The maximum size of the device under test is 40 cm. StarLab can work from 650 MHz up to 18 GHz for a maximum size of DUT of 45 cm, with up to 15 probes (depending on the model and frequency band). In this case, the system allows a rotation of the arch to increase the number of samples in elevation. In both cases, the use of special receivers allows the working in an over-the-air mode, useful for EMC, 5G, IoT, MIMO, and other special measurement systems. A system calibration process is necessary to assure a reasonable calculation of amplitude, polarization, and phase in the multiprobe system.

In all these systems, the receiver is switching among the different probes, and the amplitude and phase of both polarizations are acquired. One antenna is used as a reference to extract the relative phase among the different probes. As it was already stated before, there are two ways to extract the phase: either by using the on-axis probe of the multiprobe system or by displacing another reference antenna. In the second case, the coupling with the system can be more significant, but still, acceptable results can be achieved [9, 10].

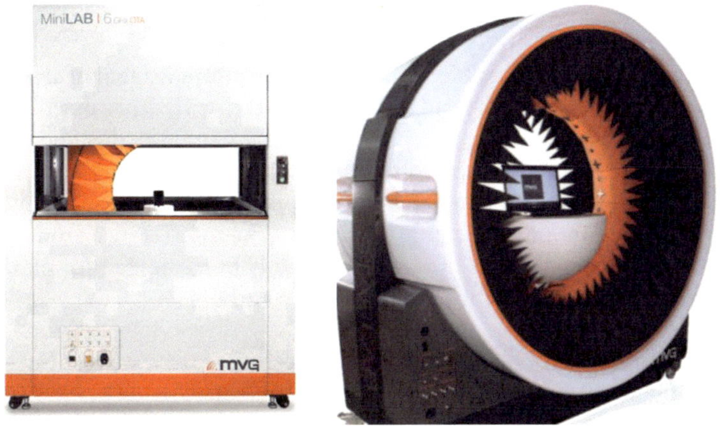

Figure 4.
Microwave vision group minilab (left) and StarLab (right).

4. Amplitude and phase extraction using SDR receivers

The low-cost reference-less receiver is implemented using the Zynq Evaluation and Development Board (Zed- Board) with the transceiver AD-FMCOMMS3- EBZ. This RF transceiver includes a configurable digital interface to an FPGA to communicate the ZedBoard and the RF module. The 2x2 transceiver module consists of a 12-bit ADC with a receiver band from 70 MHz to 6 GHz and a tunable channel bandwidth up to 56 MHz. This receiver is homodyne, and the signal is directly down-converted to baseband for digitization purposes. The I-Q samples are generated by the transceiver module. The receiver chain can be externally controlled, and it consists of two programmable low-pass filters, decimating filters and gains control for each channel.

The receiver is based on time-domain measurements. Then, the amplitude extraction is done by means of frequency-domain techniques. The power calculation is based on Parseval's theorem on the discrete-time form, see Eq. (1). In the equation, X(k) refers to the Fourier basis functions, while x(n) represents the sampled time-domain signal received by the probe. The number of samples N plays an important role if noise averaging is implemented. Post-processing steps like windowing and filtering are applied in order to get accurate results, see [12].

$$\frac{1}{N}\sum_{n=0}^{N-1}|x(n)|^2 = \frac{1}{N^2}\sum_{k=0}^{N-1}|X(k)|^2 \tag{1}$$

Phase reconstruction is not that straightforward, and the method depends on the system architecture: reference antenna independent from measurement arch or on-axis reference antenna. For this application, the transmitters are not necessary.

4.1 Phase extraction for reference antenna independent from measurement arch

If the reference antenna is displaced around the AUT in a fixed relative position, the reference channel emulates the conventional sample taken from a vector network analyzer. In that case, the relative phase between measurement points of the DUT can be extracted as described by Eq. (2). The value of k_1 determines the intermediate frequency of the computation.

$$\phi_i = arg\left\{\frac{\sum_{n=0}^{N-1}E_{multiprobe}(n)e^{-j\frac{2\pi}{N}k_1}}{\sum_{n=0}^{N-1}E_{reference}(n)e^{-j\frac{2\pi}{N}k_1}}\right\} \tag{2}$$

This method considers possible drifts in the transmitted signal of the DUT since the reference sample is taken by radiation. In contrast with other existing solutions for phase retrieval that could be used for EMC measurements, this one exploits the intrinsic advantages of multiprobe solutions that allow for better isolation of the reference antenna while providing an anechoic and shielded measurement environment. Besides, the reference is accurate as long as the interference from the reference antenna is kept small. This method has already been applied for planar as well as for spherical multiprobe measurements by the authors [12].

4.2 Phase extraction for on-axis reference antenna from multiprobe measurement arch

When the reference antenna belongs to the measurement arch, the solution is simplified. In comparison with the external reference antenna solution, there are some substantial changes:

- The multiprobe system is composed of dual-polarized probes at each sample point. The phase reference will be obtained by taking as reference only one polarization, which means that all the measured phases are relative phases with respect to the reference probe for that polarization.
- When the system rotates to measure the whole sphere, there is a change in the reference on-axis probe signal. Therefore, the relative phases for each cut are referenced to one of the top probe polarization when it is rotated.

The change in the phase reference is translated into phase unknowns for every azimuthal cut, as described by Eq. (3). These unknowns are solved by appealing to Ludwig's III definition of polarization that allows retrieving the phase unknowns of the measured signals in a direct way, as explained in [15].

$$\begin{aligned} \vec{E}_{meas}(\phi_1) &= \vec{E}_{ref} \\ \vec{E}_{meas}(\phi_2) &= \vec{E}(\phi_2)e^{j\phi_2} \\ &\vdots \\ \vec{E}_{meas}(\phi_N) &= \vec{E}(\phi_N)e^{j\phi_N} \end{aligned} \tag{3}$$

5. Near to far-field spherical transformation algorithm for EMC

Every receiving probe measures the field radiated by the DUT. The well-known transmission formula [16] gives the relation between the signal measured by the probe and the spherical modes coefficients:

$$w(r,\chi,\theta,\varphi) = \sum\nolimits_{smn\mu} Q_{smn} e^{jm\varphi} d^n_{\mu m}(\theta) e^{j\mu\chi} P_{s\mu n}(r) \tag{4}$$

being (r,θ,φ) spherical coordinates,χ the probe orientation, Q_{smn} the AUT spherical wave coefficients,$d^n_{\mu m}(\theta)$ a rotation operator and $P_{s\mu n}(r)$ the probe response constants. The summation in Eq. (4) spans for $s \in [1,2]$,$n \in [1,N]$, $m \in [-n,n]$ and $\mu \in [-V,V]$.N and V are the expansion truncation numbers for the AUT and antenna probe, respectively. In principle, this number is infinite. However, depending on the antenna, the number of SWC with significant power is finite for both AUT and probe, and there exists the following practical rule for maintaining good accuracy:

$$N = \lceil kr_0 \rceil + 10 \tag{5}$$

where k is the wavenumber, r_0 the radius of the smallest sphere circumscribing the AUT (or probe in the case of V), and the brackets indicate the largest integer smaller than or equal to the number inside them. However, this rule of Eq. (5) can be replaced by more complicated expressions depending on the signal to noise [22]. For a mode power of 40 dB below the maximum, a more accurate expression is shown in Eq. (6).

$$\mathrm{N} = \lceil kr_0 \rceil + 1.6\sqrt[3]{kr_0} \tag{6}$$

Once the spherical modes are calculated, the field generated by the probe can be calculated for each specific point at a finite or infinite distance.

A representative example of the spherical wave expansion of an EMC device can be simulated by taking as DUT a 15 cm dipole excited with 1 μA current. The field

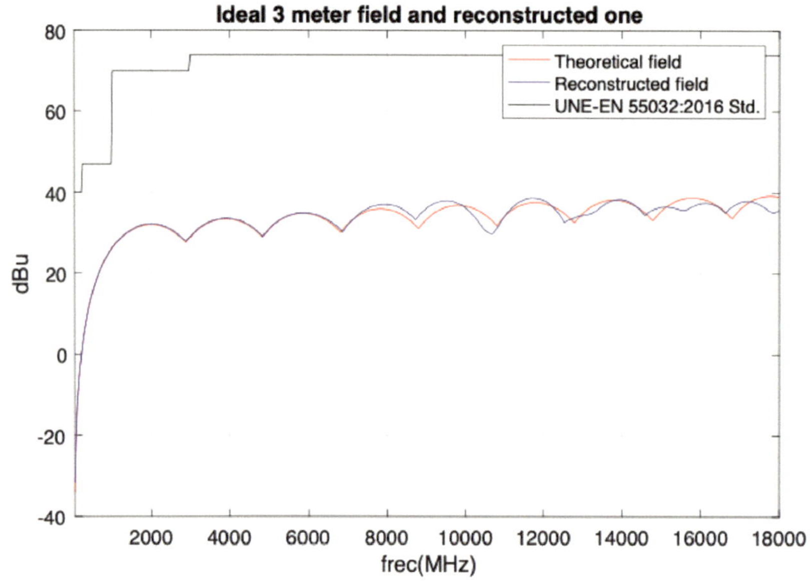

Figure 5.
Comparison of analytical electromagnetic field and field after the spherical transformation process to 3 m.

generated at 3 meters is analytically calculated and compared with the one obtained by following the near-field acquisition by an EMC multiprobe system. In this case, the probes are modeled with antenna factors of 0.1 (to represent a poor performance), the receiver includes a Gaussian noise equal to −120 dBm (the one of the SDR platform), and 12 dual linear polarized probes separated 15 degrees at a radial distance of 30 cm are considered, to emulate the MiniLab multiprobe system.

Figure 5 shows the results for the peak values, comparing them with the UNE-EN 55032:2016 Std. [18]. The standard level is included to show that the process has the required low power to ensure that the noise effect is negligible. For low frequencies, a spherical mode truncation is applied to run the transformation algorithm without numerical problems. In higher frequencies, the differences are due to undersampling effects: this is due to the limited number of samples, 15 degrees separation between probes, compared with the dimension of the device under test and higher frequencies. However, up to 6 GHz, the system works very well. For frequencies up to 18 GHz, the solution is to implement the arch rotation and use a non-convex optimization for phase retrieval [15].

6. Practical implementation of radiated spurious near-field multiprobe measurement system

To validate the system for radiated EMC measurements, two experiments have been performed. In those experiments, different post-processing techniques are used to improve the results. The same reference DUT sample is used in both cases: this consists of a PCB of 150 mm by 225 mm with a substrate thickness of 2 mm. One of the traces of the PCB is excited through an external transmitter, while the other ones are excited through coupling. The DUT represents a bad radiator, so it is a good sample of a typical EMC device.

The first experiment was done in MVG Italy by using StarLab in one of the possible EMC configurations: a reference antenna independent from the arch and SDR receiver. The DUT was also measured in a certified laboratory. In particular, the DUT was measured in CATECHOM [23] (University of Alcalá, Spain). According to the technical specifications marked by the European Standards (EN), the laboratory is certified to perform EMC measurements, providing the certification process required for the CE mark of a product. A comparison of both configurations can be seen in **Figure 6**.

The characteristics of each measurement solution are described in **Table 1**. The main difference is the measurement time, which can be drastically reduced with the multiprobe EMC solution while keeping a good accuracy, as will be seen. The processing time for the NF transformation to 3/10 meters distance is negligible.

In order to measure the multiprobe EMC system, an electric sleeve dipole was used as a reference antenna. Three different signals are compared: as a reference, the measurement in the conventional multiprobe configuration of Starlab (45 cm distance), using the vector network analyzer to feed the DUT and thus, having access to the reference signal to measure the amplitude and phase. Then, the multiprobe EMC configuration measurement is performed, where the reference antenna does the phase retrieval, and the SDR is used to reconstruct the amplitude and phase near-field pattern. Finally, the conventional EMC laboratory (CATECHOM) measurement at a 3 meters distance is done. The flowchart of the comparison procedure and post-processing steps are shown in **Figure 7**.

The results herein presented correspond to 2 GHz. In the flowchart, CST [24] was used before performing the final comparison. This was necessary to include the effect of the wooden table in the device's radiation when measured in the certified EMC laboratory. This is also an added value of the multiprobe EMC solution since the radiated emission can be computed, including different scatterers around the DUT. Moreover, source reconstruction was done by using Insight [19]. This source reconstruction is able to calculate the currents on the PCB structure, filtering out all the contributions out of the PCB itself. As shown in [2], this technique can improve the results of the measurements. This allows comparing how accurate the EMC multiprobe solution could be compared to the ideal source reconstruction that can be achieved with the vector network analyzer measurements.

Figure 6.
EMC test-case: Compliance laboratory (CATECHOM), and multiprobe EMC system.

Measurement system	Measurement distance	Geometry	Measurement time
CATECHOM	3 m	Height variation and azimuth sweep	≈3 hours
EMC Multiprobe	45 cm (NF to NF/FF transformation)	Azimuth rotation, multiprobe in elevation	≈10 min

Table 1.
Measurement characteristics: Conventional EMC measurement and multiprobe EMC solution.

The radiation pattern results at 3 m are shown in **Figure 8**. In this case, from the near-field, the sources are reconstructed, and once these currents are calculated, the electromagnetic field at 3 meters is calculated using the commercial software CST for both horizontal and vertical polarizations. It was done in this way to consider the table used to support the PCB. The results show that the difference between using a conventional VNA (most accurate system) and the SDR platform is minimal. This opens the possibility of low-cost receivers for this kind of system. Second, the differences with respect to the measurements with a conventional EMC setup are within the uncertainty of these systems. The main advantages of using near-field systems are the lower uncertainty in the measurement process, due to the easier control of the environment, and the possibility of including some external setups, using commercial electromagnetic software such as CST.

Nevertheless, the peak error for the maximum of the radiation is below 2 dBV/m. The comparison of the pattern is suggesting that the angular variation of the radiated emission is appropriately reconstructed. This would be translated into a good correlation of the currents' distribution. The source reconstruction comparison between the measurements done with the vector network analyzer and the multiprobe EMC setup can be seen in **Figure 9**.

Another experiment was performed in the MiniLab system of Microwave Vision Group in Pomezia (Italy). This system was used in a different architecture for EMC multiprobe solutions (**Figure 2**), with the on-axis top-probe used as a reference channel. In order to measure with this architecture, hardware modifications are needed since one of the connected signals to the conventional switching matrix of a multiprobe system is connected to one input of the SDR receiver. Thereby, a

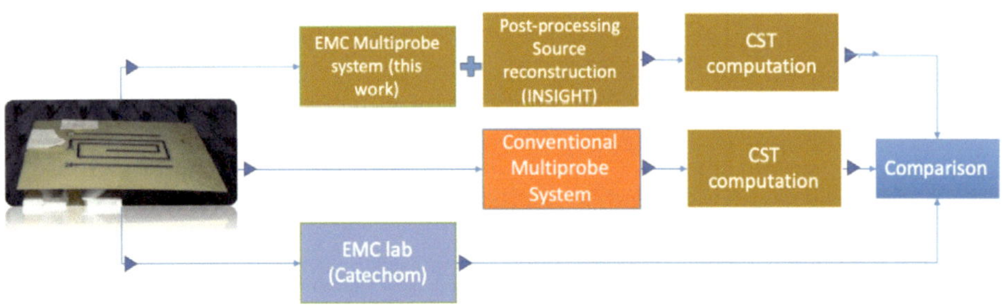

Figure 7.
Process for comparison of the measurements of the PCB structure.

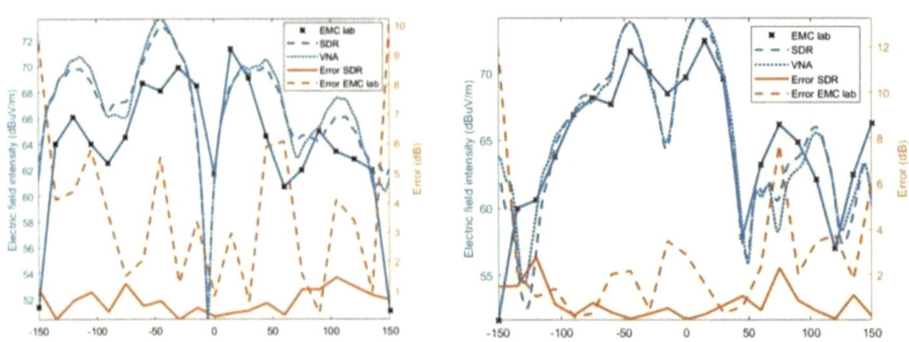

Figure 8.
Comparison between conventional measurement and measurement using near-field procedure: With conventional VNA and SDR platform (H component left and V component right).

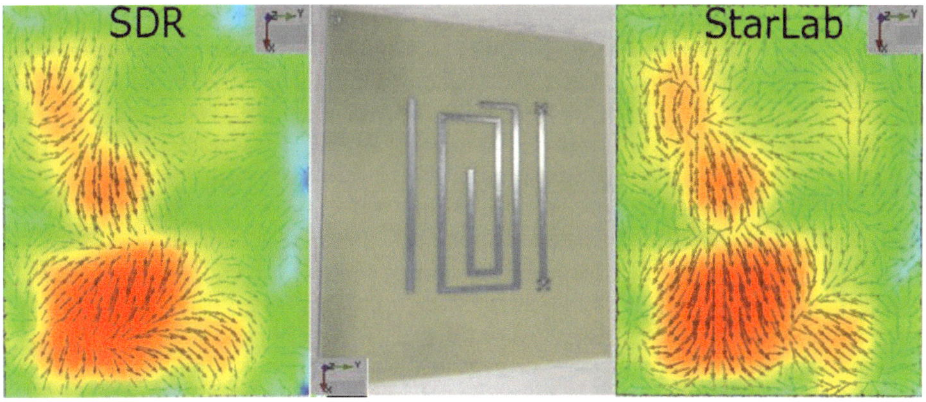

Figure 9.
Source reconstruction on top of DUT.

calibration is needed in order to account for differences between the switching matrix RF path and the direct connection of the on-axis probe.

The goal of the experiments was to verify the performance of the multiprobe EMC solution based on the on-axis probe as a reference when it comes to modulated EMC signals. Some other experiments were performed with continuous-wave signals to validate the system. As was the case for the antenna independent from the measurement arch, good results were obtained, and some results can be found in [11].

It is well known that some DUTs could work with modulated signals. The extrapolation of radiation parameters for modulated signals and how to measure them is not that clear [25]. Some experiments have been done in order to validate EMC multiprobe measurements of modulated signals. In that sense, let us assume there is an IoT device transmitting a modulated signal. Let us also assume that the signal is LTE FDD type. Emulation of this scenario was done by exciting a known antenna with a modulated LTE FDD signal of a predefined bandwidth (see **Figure 10**).

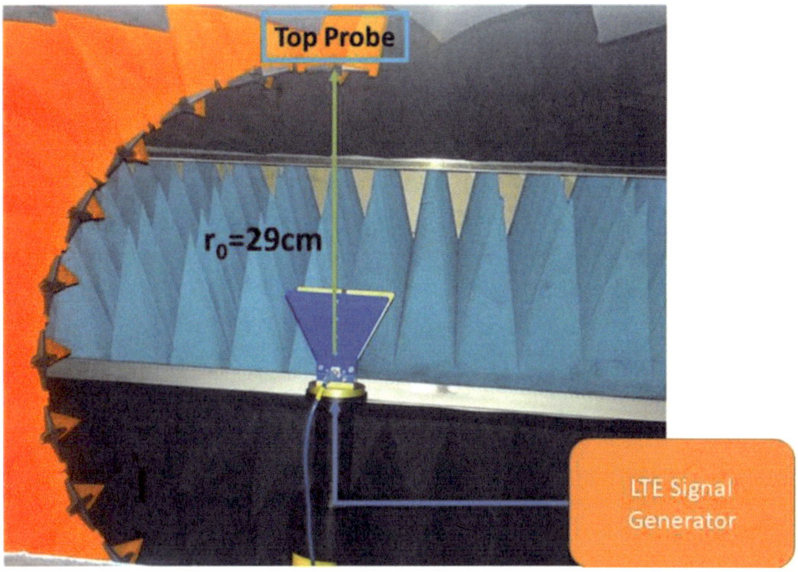

Figure 10.
On-Axis architecture of multiprobe EMC system for LTE measurements.

The SDR was optimized in order to extract the time domain near-field amplitude and phase of the transmitted signal. In the particular case of the results herein presented, an LTE FDD signal of 5 MHz bandwidth is analyzed. The parameters extraction is done in a similar way it is done for a continuous wave signal. For the power, the spectrum under interest is integrated. For the phase, linearity is assumed in such a way that the average (intermediate) phase over the whole bandwidth represents, ideally, the radiation pattern of the DUT at the central frequency. This statement is mathematically described by Eq. (7). In the equation, the measured phase at equispaced frequencies from the central one (f_c) cancels out, giving the measured phase at the central frequency as a result. This is true under the assumption that the radiation pattern of the DUT is not changing over the given bandwidth, which is true for most practical cases.

$$\frac{1}{N+1}\sum_{i=-N/2}^{N/2}\phi_{meas}(f_c+i\Delta_f)=\phi_{meas}\left(f_c-\frac{N}{2}\Delta_f\right)+\ldots+\phi_{meas}(f_c)+\ldots+\phi_{meas}\left(f_c+\frac{N}{2}\Delta_f\right) \tag{7}$$

The experiment was conducted at 1 GHz. First, the reference continuous wave signal radiated by the antenna was measured. Then the LTE signal of different bandwidths was measured using the EMC multiprobe architecture and the parameters at the central frequency extracted. Some comparisons for different bandwidths can be seen in **Figure 11**. The continuous wave (CW) is the reference curve. A deeper analysis of the error pattern can be seen in **Figure 12**. The near-field mean error is very low for both components, below −45 dB. This shows the low error introduced for the field reconstruction.

The correlation between the different signals is very good, demonstrating that the technique used could be suitable to characterize the radiation of EMC devices when modulation is applied. The optimized measurement corresponds to the results

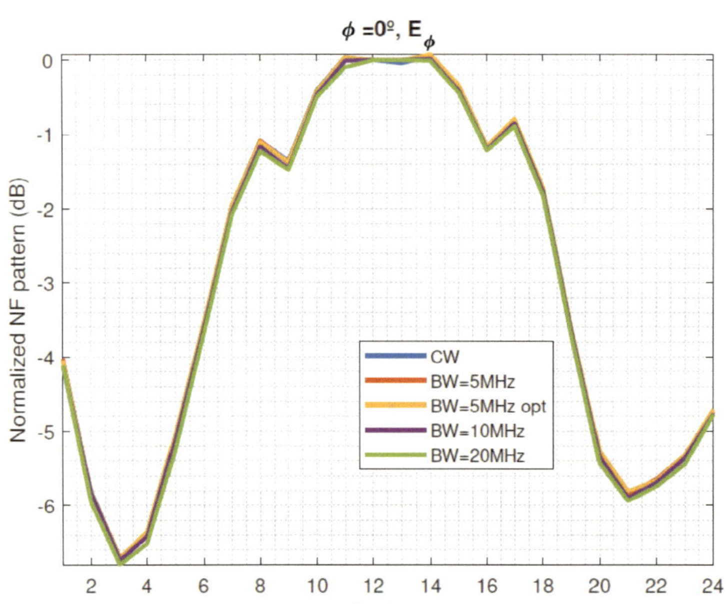

Figure 11.
Main cut near-field reconstruction of LTE signal for different bandwidths.

DOI: http://dx.doi.org/10.5772/intechopen.99604

obtained when optimizing the post-processing steps for retrieving the amplitude and phase information of the modulated signal. The same procedure would be applied afterward: near-field to near/far-field transformation at 3 or 10 meters to compare with EMC standards and diagnostics. In this case, the EMC measurement process is explained in **Figure 3**. This test was done on the same PCB used in the previous experiment and was used to validate the performance with a low signal-to-noise ratio, including different signal attenuations, and lower frequencies (400 MHz). Spherical modes are calculated from the acquired near-field, and some spherical mode filtering is applied. This spherical mode filtering consists of canceling those modes that cannot correspond to the signal itself.

The results are shown in **Figure 13**, where the power of the different spherical modes (under m and n index) are shown for the reference case, 30 dB and 65 dB of extra attenuation. It is observed that for 30 dB attenuation, the results are very good, and even for 65 dB attenuation, the results are acceptable. This is also reflected in the right part of **Figure 13**, where the field for different theta angles is shown. Again, the differences are within the typical uncertainty values for EMC setups, even with very low power values, and a frequency lower than the specifications of the multiprobe system for antenna measurements.

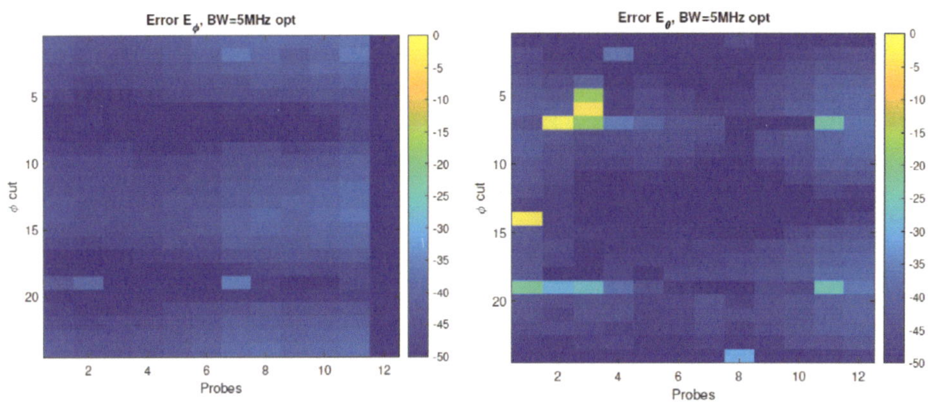

Figure 12.
Near-field error between CW signal and EMC multiprobe measurement of 5 MHz LTE FDD signal left (Eϕ), right (Eθ).

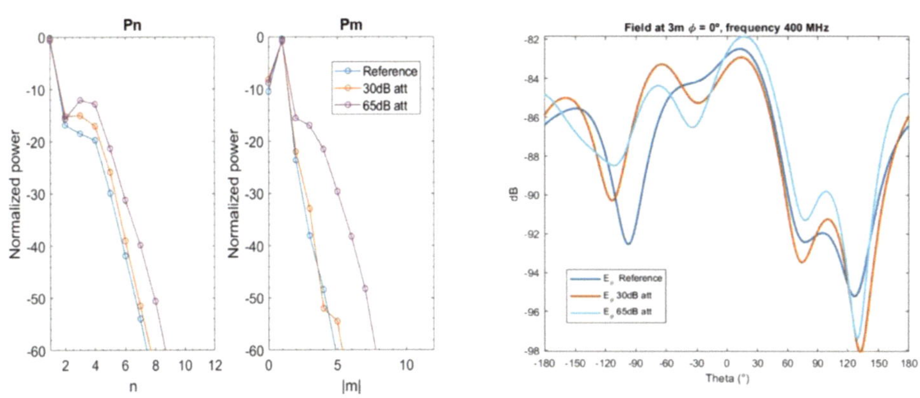

Figure 13.
Measurements at 400 MHz, using spherical near-field transformation and mode filtering for different attenuations of the received signal.

These results for EMC multiprobe solution for modulated signals show the potential of this setup, not only for traditional EMC measurements where the spurious emissions are characterized at particular frequencies, but also represents a low cost, accurate and fast solution for addressing pre-compliance of new self-transmitting devices using modulated signals.

7. Conclusions

This chapter has shown the use of antenna measurement techniques to measure the radiation of devices under tests in EMC. The proposed system has been based on the following technologies: multiprobe measurement systems, software-defined radio receivers, phase recovery, spherical near- to far-field transformation software, post-processing techniques as source reconstructions or calculation of equivalent currents and spherical mode filtering, and combination of measurements and simulations.

This process has been validated with measurements in a Microwave Vision Group StarLab system at 400 MHz and 2 GHz, comparing the results with measurements in a conventional EMC setup. The results have shown good agreement, and this is the first step to advance in the use of new technologies of other disciplines, as antenna measurement, to improve the results obtained with the conventional EMC setups.

This chapter has only introduced a summary of the different works done by the authors during the last few years, but most information can be found in some of the references included at the end of the chapter.

Acknowledgements

This work is included in a collaboration between Microwave Vision Group and Universidad Politécnica de Madrid. The Spanish Government also supports it, Ministry of Economy, National Program of Research, Development, and Innovation through the project FUTURE-RADIO “Radio systems and technologies for high capacity terrestrial and satellite communications in a hyperconnected world” (project number TEC2017-85529-C3-1-R).

Conflict of interest

The authors declare no conflict of interest.

Author details

Rubén Tena Sánchez[1,2], Lars Jacob Foged[1] and Manuel Sierra Castañer[2*]

1 Microwave Vision Group, Pomezia, Rome, Italy

2 Universidad Politécnica de Madrid, ETSI Telecomunicación, Information Processing and Telecommunications Centre, Madrid, Spain

*Address all correspondence to: manuel.sierra@upm.es

References

[1] Henry W. Ott. Electromagnetic Compatibility Engineering. Wiley, 2009. ISBN: 978-0-470-18930-6

[2] M. Sierra Castañer, Lars J. Foged. "Post-processing techniques in Antenna Measurements." Scitech Publishing (IET). 2019. ISBN: 978-1-78561-537-5.

[3] Amber Precision Instruments. http://www.amberpi.com

[4] EMSCAN, EMxpert ERX+. "Wideband Very-Near-Field Array of Probes for Efficient EMI Measurements." Presented at COST Action. www.emscan.com

[5] LUXONDES. http://www.luxondes.com

[6] Batscanner (Nexio Group). https://nexiogroup.com

[7] ICEy. The future of Interference & Compatibility Evaluations. Speag. www.speag.com

[8] Rubén Tena Sánchez, Manuel Sierra Castañer. "Evaluation of Software Defined Radio Receiver for Phaseless Near-Field Measurements" Proceedings of the 40th Annual Meeting and Symposium of the Antenna Measurement Techniques Association. Williamsburg, Virginia (EEUU), November 4-9, 2018.

[9] Rubén Tena Sánchez, Manuel Sierra Castañer, L.J. Foged. "Use of Software Defined Radio Receivers for Antenna Measurements." 2019 PhotonIcs & Electromagnetics Research Symposium — Spring (PIERS — SPRING), Rome, Italy, 17–20 June 2019, pp. 1862-1869.

[10] Rubén Tena Sánchez, Manuel Sierra Castañer, L.J. Foged, D. Gray "EMC Measurement System Based on Software Defined Radio and Diagnostic Techniques" Proceedings of the 41st Annual Meeting and Symposium of the Antenna Measurement Techniques Association. San Diego, California (EEUU), October 6-11, 2019.

[11] Rubén Tena Sánchez, Manuel Sierra Castañer, L.J. Foged. "Relative Phase Reconstruction Based on Multiprobe Solutions and Post-Processing Techniques." Proceedings of the 14th European Conference on Antennas and Propagation (EuCAP), Copenhagen (Denmark), Virtual Edition May-June 2020.

[12] Rubén Tena Sánchez, Manuel Sierra Castañer, and L. J. Foged. "A Referenceless Antenna Measurement System Based on Software-Defined Radio" IEEE Antennas and Propagation Magazine, Volume: 62, Issue: 5, Oct. 2020. Page(s): 108 – 118. DOI: 10.1109/MAP.2020.3012897

[13] Rubén Tena Sánchez, Manuel Sierra-Castañer , Ricardo Albarracín-Sánchez , Fernando Rodríguez Varela, Alessandro Scannavini , Lars J. Foged. "Feasibility Study of Near-field Multiprobe System as EMC Measurement Setup". Proceedings of the 15th European Conference on Antennas and Propagation (EuCAP), Düsseldorf (Germany), Virtual Edition 22-26 March, 2021.

[14] Microwave Vision Group Webpage. https://www.mvg-world.com/en/products?category=Antenna%20Measurement

[15] Rubén Tena Sánchez, Fernando Rodríguez Varela, Lars J. Foged, Manuel Sierra Castañer. "Reconstruction of Relative Phase of Self-Transmitting Devices by Using Multiprobe Solutions and Non-Convex Optimization" Sensors 2021, 21, 2459. https://doi.org/10.3390/s21072459.

[16] J. E. Hansen, "Spherical Near-Field Antenna Measurements," London, U.K.: Peter Peregrinus, 1988.

[17] EN 55016-2-3: 2010 Standard. Specification for radio disturbance and immunity measuring apparatus and methods – Part 2-3: Methods of measurement of disturbances and immunity. Radiated disturbance measurements.

[18] UNE-EN 55032:2016 Standard. Electromagnetic compatibility of multimedia equipment - Emission Requirements.

[19] Insight. https://www.mvg-world.com/es/products/antenna-measurement/software/insight

[20] StarLab system. https://www.mvg-world.com/es/products/antenna-measurement/multi-probe-systems/starlab

[21] MiniLab system. https://www.mvg-world.com/es/products/antenna-measurement/multi-probe-systems/minilab-6-ghz-ota

[22] F. Jensen, A. Frandsen, "On the number of Modes in Spherical Wave Expansions." 26th meeting and symposium of the AMTA, Stone Mountain Park, GA, USA, Oct 2004, pp. 489–494.

[23] Catechom. http://www3.uah.es/catechom/index.php?lang=en

[24] CST Studio Suite: https://www.3ds.com/products-services/simulia/products/cst-studio-suite/

[25] A. P. Mynster and P. T. Jensen, "EMC for the IoT," 2016 International Symposium on Electromagnetic Compatibility - EMC EUROPE, 2016, pp. 144-149

Chapter 6

Modeling Grounding Systems for Electromagnetic Compatibility Analysis

Antonio Carlos S. Lima, Pedro H.N. Vieira, Marco Aurélio O. Schroeder and Rodolfo Antônio R. Moura

Abstract

In recent years, the development of smart grids for power distribution and the increasing usage of 5G communication networks have played a large impact on the resilience and reliability of grounding systems. Unexpected electromagnetic coupling between a communication tower and the one used for the electric power networks may pose a threat to the suitable performance of either system as one must assure that electromagnetic compatibility together with unexpected transient issues is within reasonable parameters. This requires wideband modeling of a grounding system, typically carried out using numerical approaches based on the Method of Moments. This modeling is implied in numerous segments to represent the conductors involved and the numerical solution of a double integral for each one of these segments. The modified nodal formulation used to obtain system voltages and branch currents is first solved in the frequency domain, leading to a heavy computational burden and a time-consuming simulation. This chapter briefly reviews the procedure used to model grounding grids and presents some results to illustrate the typical behavior. Afterward, a more complex system comprising a case of electromagnetic coupling is then analyzed to illustrate the impact of nearby grounding grids.

Keywords: grounding systems, method of moments, frequency domain analysis, transient response, computational methods for electromagnetism

1. Introduction

This chapter focuses on the transient analysis of grounding systems and the impact that the associated responses might have with respect to the electromagnetic compatibility in nearby power apparatuses, installations, and people. It is assumed that the reader is familiar with the electromagnetic field theory in both frequency and time domains and some mathematical tools such as Method of Moments [1] and Numerical Laplace Transform [2–6].

The development of smart grids for electric power distribution and the widespread use of 5G communication networks will demand an accurate, efficient, and resilient grounding system to avoid electromagnetic compatibility issues such as interference or noise in apparatuses due to poor power quality, that is, harmonic distortion in voltages and currents. Furthermore, an injected current due to

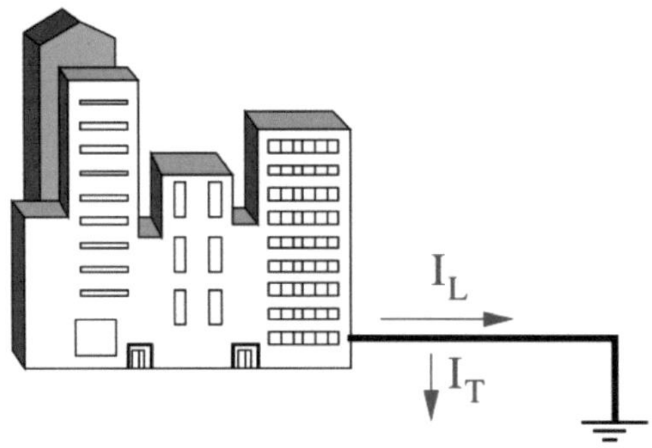

Figure 1.
Schematic of an industrial facility with a simple bare conductor acting as the grounding system.

lightning-related phenomena may lead to a ground potential rise (GPR) that could damage several devices and play an important aspect in the reduction of personal safety in the area surrounding the apparatuses. The most common approach to overcome these possible challenges is to design a feasible, reliable, and efficient grounding system. Therefore, it is of utmost importance to properly understand the transient behavior of a given grounding system in an electrical power or communication installation and its interaction with its surroundings, be it a sensitive electronic device, another grounding system, people, or even the induced voltage at ground level. In general terms, the main function of a grounding system is to provide a pathway with the lowest possible impedance for faulty currents, leading to the least possible voltage increase at the injection point and in its surrounding area. By faulty current, one may consider the one associated with phenomena such as lightning, switching, misoperation, and more recently, the combined harmonic currents related to the presence of power electronic converters in the electric power network.

To illustrate the general idea, consider the schematic presented in **Figure 1**, where a small facility has a rather simple grounding system. It consists of a bare horizontal conductor buried near the ground surface, for example, 0.5 m. The conductor should be long enough that its longitudinal current I_L decays as smoothly as possible, preferably monotonically decreasing, and reaches to zero before reaching the end of the conductor. The bare conductor is open-ended at the far end. The shunt current I_T represents the injected current in the soil, and it is the main contributor to the voltage increase in the surrounding area. Both currents are distributed in nature, and to account for their behavior, one must consider the frequency dependency of conductor and ground. Even in this simple scenario, one needs to divide the conductor into small segments so the Method of Moments can be applied, thus leading to a large order for all the matrices involved. The segmentation of any given conductor to very small segments is also needed for representing the electromagnetic propagation throughout the conductor and surrounding media.

In actual installations, the scenario is even direr as more complex configurations need to be considered. Transmission towers will demand a counterpoise configuration involving an arrangement of conductors that are not parallel for some extension of their lengths. In some scenarios, there are horizontal and vertical conductors to be considered. **Figure 2(a)** depicts the basic structure of a counterpoise for grounding a power transmission tower. Electric power substations demand a grounding grid, which is complex and large in dimensions, representing a challenge for an accurate simulation of transient behavior. **Figure 2(b)** shows a nonuniform grounding grid typically found in electric power substations.

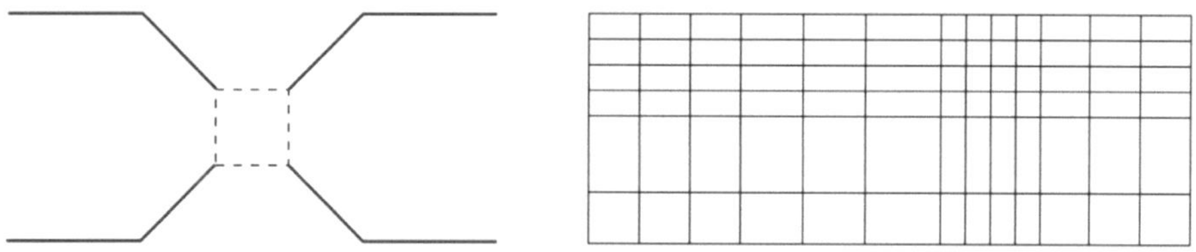

(a) Counterpoise configuration for overhead lines transmission towers

(b) Nonuniform Grounding Grid

Figure 2.
Two grounding grids configuration. (a) Counterpoise configuration for overhead line transmission towers and (b) nonuniform grounding grid.

Although the grounding grid is an element to ensure equipment and people safety, it may have some drawbacks. For instance, an unexpected electromagnetic coupling between a communication tower and the one used for the power network may pose a threat for the suitable performance of electronic equipment as one must assure that electromagnetic compatibility, together with unexpected transient issues, is within reasonable parameters.

In the following sections, we present a mathematical approach based on the Method of Moments to accurately represent a given grounding system. It is a formulation suitable for the analysis related to electromagnetic compatibility issues, as well as the associated current and voltage transient analysis. It is based on the frequency domain formulation of a modified nodal admittance matrix with time responses being obtained using the Numerical Laplace Transform.

2. Mathematical modeling of the equivalent circuit

As commented before, for an accurate assessment of the transient response of the grounding system, a wideband representation is commonly required. The frequency range of lightning-related phenomena goes from a few hertz up to tenths of MHz. Thereby, in the frequency domain, a detailed representation is warranted by using approaches such as the Finite Element Method (FEM) [7] or the Method of Moments (MoM) [8–10]. It is also worth noticing that there are methods that properly solve this numerical issue by solving Maxwell equations directly in the time domain, such as the so-called Finite Differences Time Domain (FDTD) [11, 12]. However, these approaches usually take even more computational time and do not consider the frequency dependence of the soil. In this scenario, considering full-wave frequency domain techniques, there are essentially two approaches to do so. The first one relies on solving Maxwell equations (or the associated vector and scalar potentials) numerically, while the second one applies circuit theory approximation whenever possible. For the latter, there are two approaches in the literature that, although developed independently, share a "common kernel." The first one is called partial element equivalent circuit (PEEC) [13, 14] and has focused mainly on electromagnetic compatibility issues. The second one is the so-called hybrid electromagnetic model (HEM) [10] and was initially developed to analyze underground bare conductors such as the ones found in electrical grounding. It is important to highlight that the so-called HEM is a particular case of the PEEC in the frequency domain, considering only cylindrical conductors.

Given that both approaches lead to an equivalent circuit derived from the behavior of the electromagnetic fields, these methods can be understood as hybrid models. Regardless of the adopted approach, one needs to divide the conductors

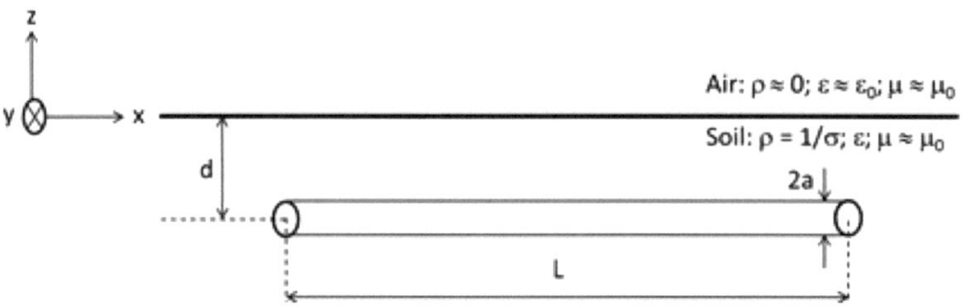

Figure 3.
Horizontal grounding electrode: Real physical system.

involved in several shorter segments to assume a uniform electric current through it and another uniform current, leaving the segment into the surrounding media. Thus, all methods present a heavy computational burden, demanding an improvement of their numerical performance.

In general, hybrid models consist of segmenting the grounding systems into small segments and estimating the electric current that circulates along each segment (here called longitudinal current or I_L) and the current that flows from the electrode to the soil (here called transversal current or I_T). Since these currents are not known previously, it is necessary to apply the MoM to estimate these parameters. To clarify this procedure, consider, as an example, the horizontal grounding electrode shown in **Figure 3**. In this simple case, there are only two semi-infinite media: air and soil. The electrode, with length L, radius a is buried at depth d in a linear, isotropic, and homogeneous soil, with electrical resistivity ρ (electrical conductivity $\sigma = 1/\rho$), electrical permittivity ε, and magnetic permeability μ. In general, according to data [15], the magnetic permeability of the soil is close to that of the vacuum ($\mu \approx \mu_0$). The air has $\rho_{air} \to \infty$ ($\sigma_{air} \to 0$), $\varepsilon_{air} \approx \varepsilon_0$, and $\mu_{air} \approx \mu_0$.

The first step is to divide the electrode into N segments with length $\ell = L/N$ to solve the problem. Then, the electromagnetic coupling is calculated for each pair of elements considering the contribution of both I_T (1) and I_L (2). The numerical values are obtained considering a simplification of the traditional MoM, that is, considering a piecewise pulse base function (more details about this basis function can be found in [16]). At this point, there are two approaches:

- consider the equivalent circuit obtained and solving it either in time or frequency domain
- obtaining a modified nodal analysis (MNA) [17, 18] and solve the system.

$$u_{mn} = \frac{I_{Tn}}{4\pi[\sigma + j\omega\varepsilon]L_n L_m} \int_{L_m}\int_{L_n} \frac{e^{-\gamma r}}{r} d\ell_n d\ell_m \tag{1}$$

$$\Delta U_{mn} = -j\omega\mu \frac{I_{Ln}}{4\pi} \int_{L_m}\int_{L_n} \frac{e^{-\gamma r}}{r} d\vec{\ell}_n \cdot d\vec{\ell}_m \tag{2}$$

As in PEEC, HEM considers that both I_T and I_L do not vary along the electrode, i. e., it is uniform for each segment. Since the currents do not vary along the electrode, it is possible to represent a linear system concentrating half of each I_{Tn} in each node that composes a particular segment and considering Kirchhoff Current Law (KCL) in each of these nodes, i.e., considering a pi-equivalent system, similar to the one illustrated in **Figure 4(a)**. Another possibility is to consider a T-equivalent circuit (similar to the one illustrated in **Figure 4b**) and apply the KCL.

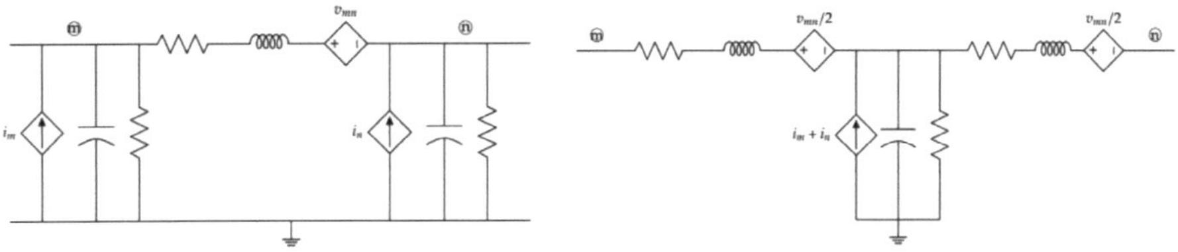

(a) PEEC pi-equivalent circuit (b) PEEC T-equivalent circuit

Figure 4.
Equivalent circuits obtained to apply the circuit theory in PEEC-type simulation. (a) PEEC pi-equivalent circuit and (b) PEEC T-equivalent circuit.

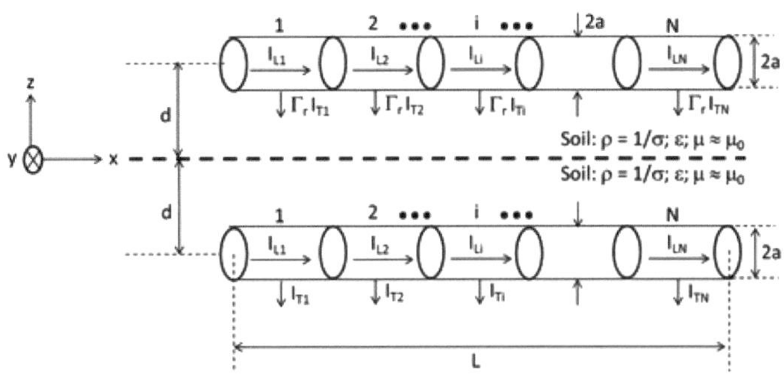

Figure 5.
Horizontal grounding electrode: corresponding model.

Furthermore, the interface must be considered. Hence, a classical solution in hybrid models is using the so-called "Image Methods" (IM) [19–21]. **Figure 5** illustrates the equivalent physical system obtained by applying the IM already considering the electrode subdivided into *N* segments and considering uniform currents in each element.

Additionally, it is important to comment that two main bottlenecks guarantee computational inefficiency:

1. one is related to segmenting the electrode, leading to large and dense matrices;

2. other is associated with a large number of numerical integration of terms in the form of $(e^{-\gamma r}/r)$, leading to a high time-consuming process.

Moreover, these two computationally intense tasks are to be performed at every frequency sample. To overcome such problems, there are some technical propositions in the literature. In [22], the possibility of increasing each segment length is discussed to reduce the matrices' size, leading to a reduction in the computational burden. In [23], it is presented an alternative to the problem by presenting an average exponential term, that is, approximating the $(e^{-\gamma r}/r)$ as an average value along the electrode, this leads to the necessity of solving the numerical integral only once for each segment. In [24, 25], it is proposed to use the first term in the MacLaurin series expansion of the integrand to obtain a closed form approximation. Since it is necessary to solve the system just once in both cases, it reduces the computational time. Note that these approximations have presented a feasible alternative in most cases (considering that the inject current has a limit frequency band below the 10 MHz). If the frequency spectrum is superior to 10 MHz, these approximations should be avoided. The Appendix presents a relationship between electromagnetic fields and more practical parameters (potential difference, voltage drop, and step voltage). If the reader is not familiar with the potential/voltage concepts, the authors recommend that the reader goes to the Appendix.

3. Case studies and results

Some case studies were selected to show the response in both frequency and time domains. For the frequency domain analysis, harmonic analysis is performed by injecting a $1A$ current in every frequency. In the time domain, a double exponential function with unit amplitude was injected to simulate a $0.1/50$ μs wave, illustrated in **Figure 6** and given by (3), in A. The investigated quantities are the harmonic impedance, GPR, and step voltage along a $1-$ m straight line in the $+x$-direction, calculated both by the potential difference Δu and by the voltage drop U_{p1-p2}, as defined in the Appendix. Note: the $-z$-direction is the one considered downward, that is, deeper in the soil.

$$i(t) = e^{-2\times10^4\ t} - e^{-10^7\ t} \tag{3}$$

3.1 Horizontal electrode

The first investigation is of a horizontal electrode, 15 m long with a 7 mm radius. The conductor starts at $x = 2.5$ m and ends at $x = 17.5$ m. A current of $1A$ is injected in the $x = 2.5$ m point of the conductor. Two cases are considered: a soil which low-frequency conductivity σ_0 is 1 mS and other which it is 10 mS. In both cases, the Alipio-Visacro soil model is used with mean values [26].

This case is very similar to the one presented by Alipio et al. [27] but has an important difference: The soil models used therein were both with constant parameters and one presented by Portela et al. [28]. Therefore, a difference in their results is expected because the Portela soil model overestimates the soil conductivity in higher frequencies compared to the Alipio-Visacro soil model used here.

The harmonic impedance is shown in **Figure 7**. According to these results, in the low-frequency spectrum, it is mainly resistive, thus being approximately modeled by a resistor. However, up to a few megahertz, its capacitive and inductive

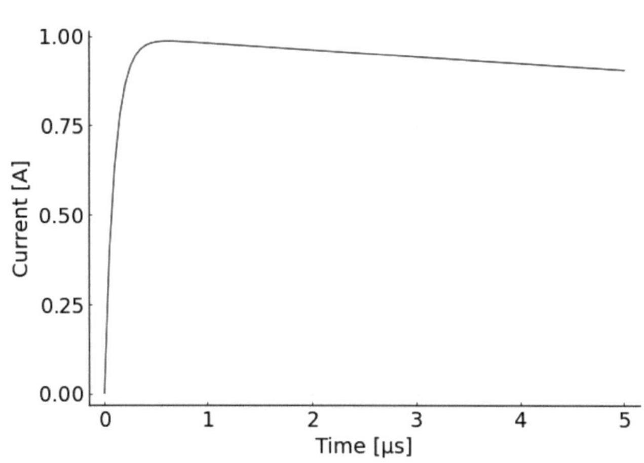

Figure 6.
Injected current for time-domain analysis.

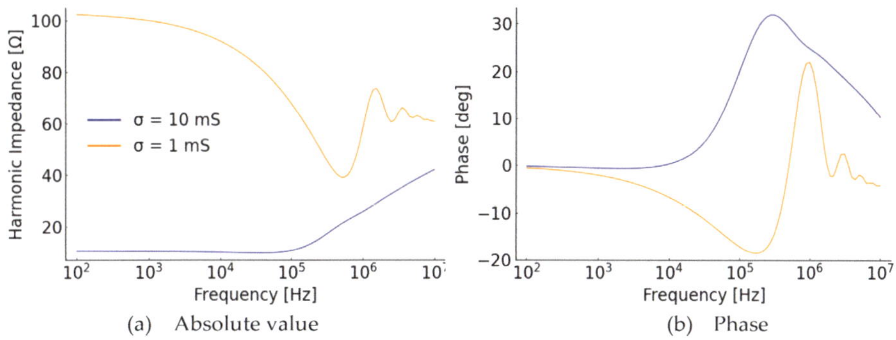

Figure 7.
Harmonic impedance of a horizontal electrode. (a) Absolute value and (b) phase.

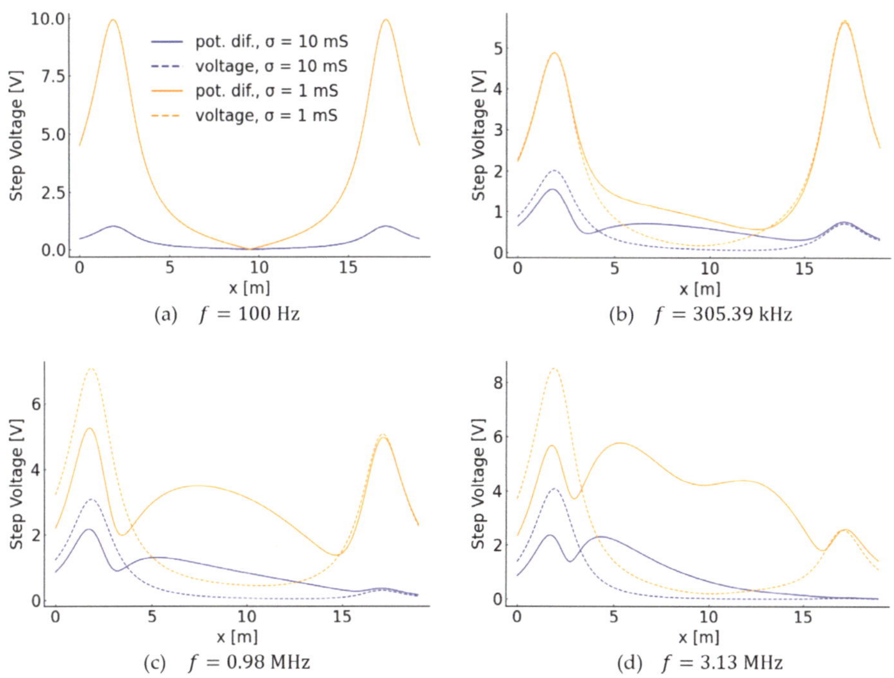

Figure 8.
Harmonic step voltage above a horizontal electrode comparing the potential difference $\mathbf{\Delta u}$ *and voltage drop* $\mathbf{U_{p1-p2}}$*. (a) f = 100 Hz, (b) f = 305:39 kHz, (c) f = 0:98 MHz, and (d) f = 3:13 MHz.*

natures start to play an important role. These pieces of information can be seen in both module and phase values. This impact is even more pronounced in low conductivity soil.

The harmonic step voltage absolute value is shown in **Figure 8**. In low frequencies, the potential difference Δu coincides with the voltage drop U_{p1-p2}. Then, after a few kHz, they begin to differ significantly. This difference is accentuated in higher frequencies. Hence, it is important to consider the nonconservative component of the electric field for high-frequency phenomena.

The GPR is greater for low conductivity soils for the time-domain simulation, as shown in **Figure 9**.

The transient step voltage, illustrated in **Figure 10**, shows a difference between the potential difference Δu and the voltage drop U_{p1-p2}, which is greater in the first

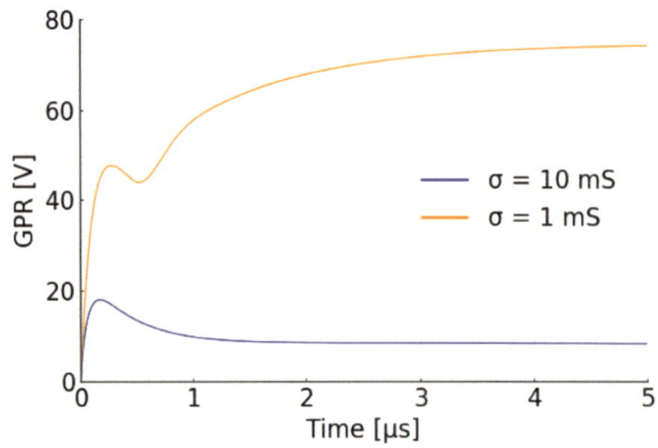

Figure 9.
GPR of a horizontal electrode.

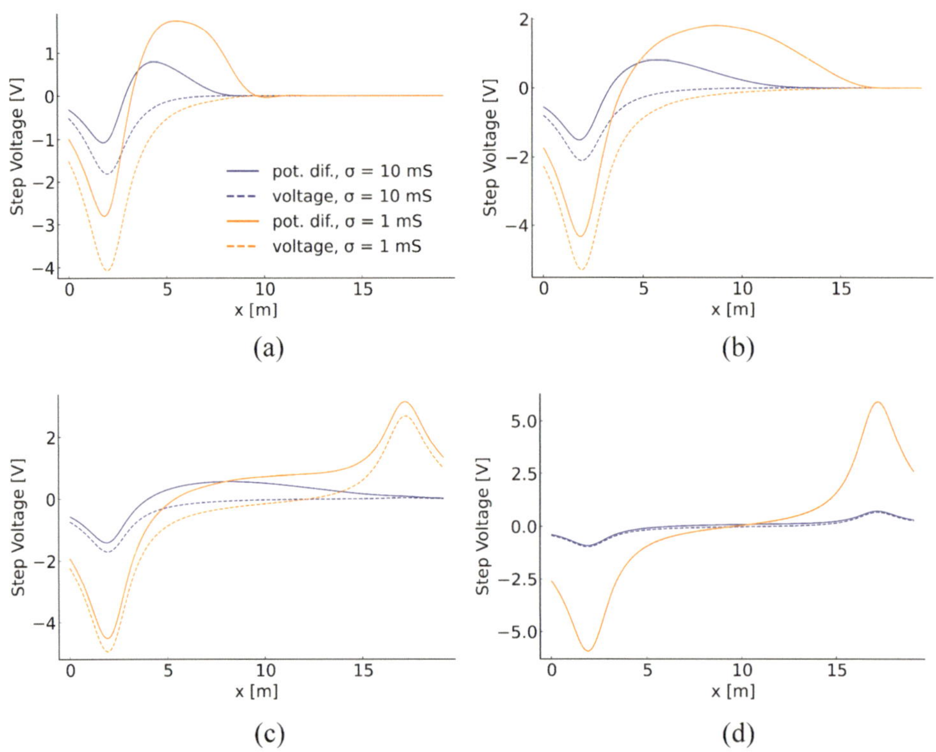

Figure 10.
Transient step voltage above a horizontal electrode comparing the potential difference Δu *and voltage drop* U_{p1-p2}. *(a) t = 0.1 us, (b) t = 0.2 us, (c) t = 0.4 us and (d) t = 1.2 us.*

time steps. This difference diminishes when the high-frequency components of the injected current go to zero in later time steps.

3.2 Vertical rod

Another simple case that was investigated is of a vertical rod, but its harmonic impedance and GPR are very similar to that of a horizontal electrode. Moreover, because the longitudinal current is downward in the $-z$-direction, the nonconservative electric field only has a component in the z-direction. Therefore, the potential difference Δu is equal to the voltage drop U_{p1-p2}, independently of the frequency. Hence, this case is omitted.

3.3 Rectangular grid

A rectangular grounding grid was investigated, illustrated in **Figure 11**. It is a $20 \times 16\ \mathrm{m}^2$ grid divided into $4 \times 4\ \mathrm{m}^2$ squares buried 0.5 m deep in the soil. It is the same geometric configuration as the one presented in [29], but the soil parameters used here are different. Similarly, to the previous section, the Alipio-Visacro soil model is used with mean values [26] considering, in one case, a low-frequency conductivity σ_0 of 1 mS and 10 mS for the other one.

For this case, the step voltage was calculated in the $+x$-direction along the line $y = 8$, that is, along the middle of the grid. The current was injected in the grid's corner at $x = y = 0$.

The harmonic impedance of that grid is shown in **Figure 12**. Because the total conductor area is bigger and goes further from the injection point, the harmonic impedance of the grid is smaller than the horizontal electrode. This effect is stronger

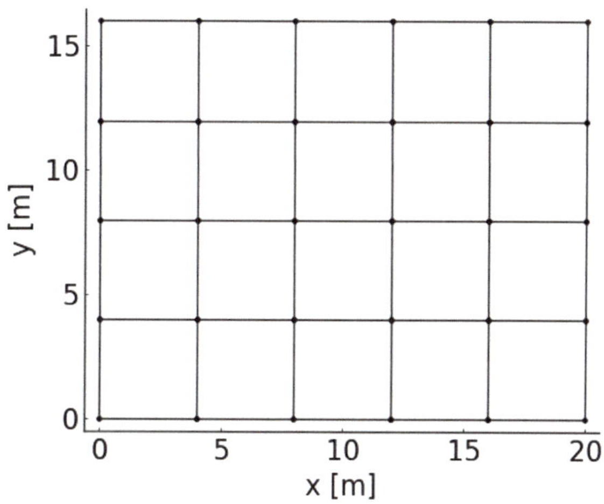

Figure 11.
Simulated grounding grid.

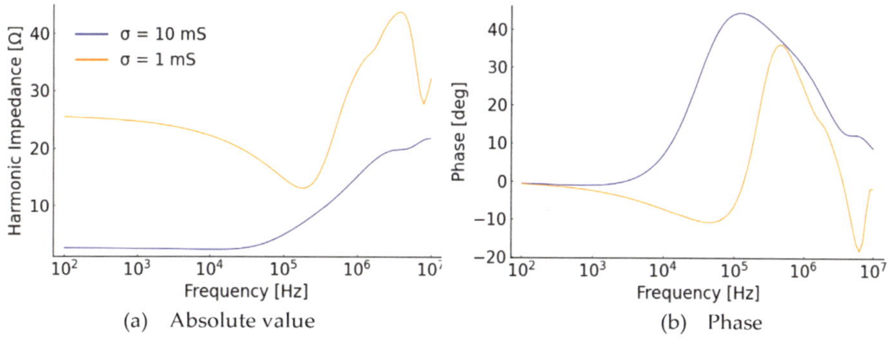

Figure 12.
Harmonic impedance of a grounding grid. (a) Absolute value and (b) phase.

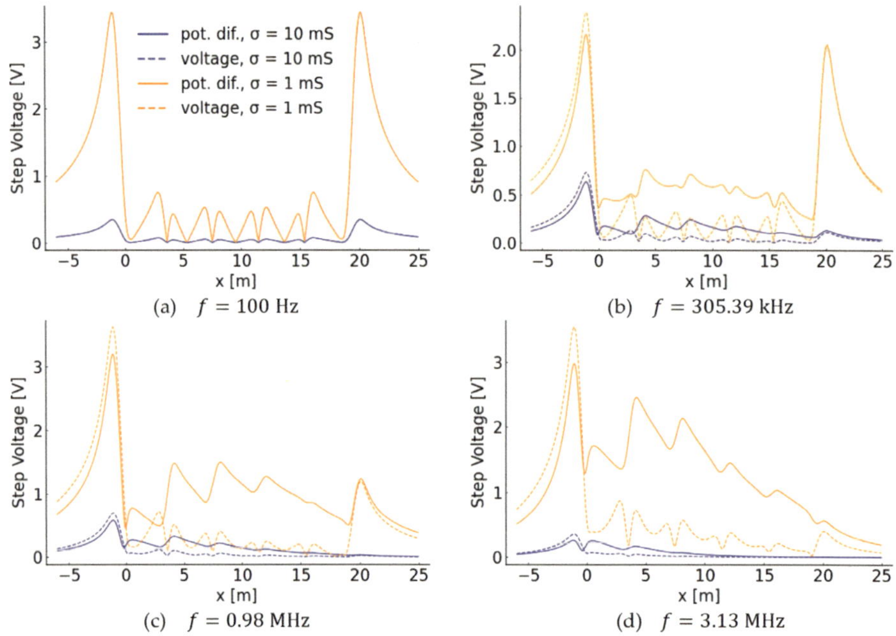

Figure 13.
Harmonic step voltage above a grounding grid comparing the potential difference $\boldsymbol{\Delta u}$ *and voltage drop* $\boldsymbol{U_{p1-p2}}$. *(a)* $f = 100$ *Hz, (b)* $f = 305.39$ *kHz, (c)* $f = 0.98$ *MHz, and (d)* $f = 3.13$ *MHz.*

for less conductive soil. The phase behavior of the impedance, however, is very similar.

The harmonic step voltage is shown in **Figure 13**. For low frequencies, the maxima are near the grid edges and peaks when crossing conductors (namely at $x = 0, 4, 8, 12, 16, 20$). Nearer the injection point, the voltage drop U_{p1-p2} is greater than the potential difference Δu. The voltage drop quickly becomes smaller than the potential difference further away from the injection point. The step voltage in a higher conductivity soil is much smaller than that for a low conductivity soil.

The GPR, illustrated in **Figure 14**, is smaller than the one observed for the horizontal conductor due to the smaller harmonic impedance.

The transient step voltage is illustrated in **Figure 15**. There is little coincidence between the potential difference Δu and voltage drop U_{p1-p2}. However, the difference between these quantities takes longer to fade for the grid. The voltage drop has

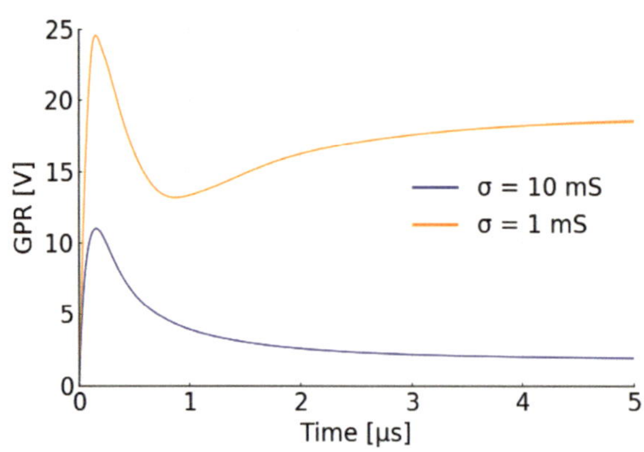

Figure 14.
Ground potential rise (GPR) of a grounding grid.

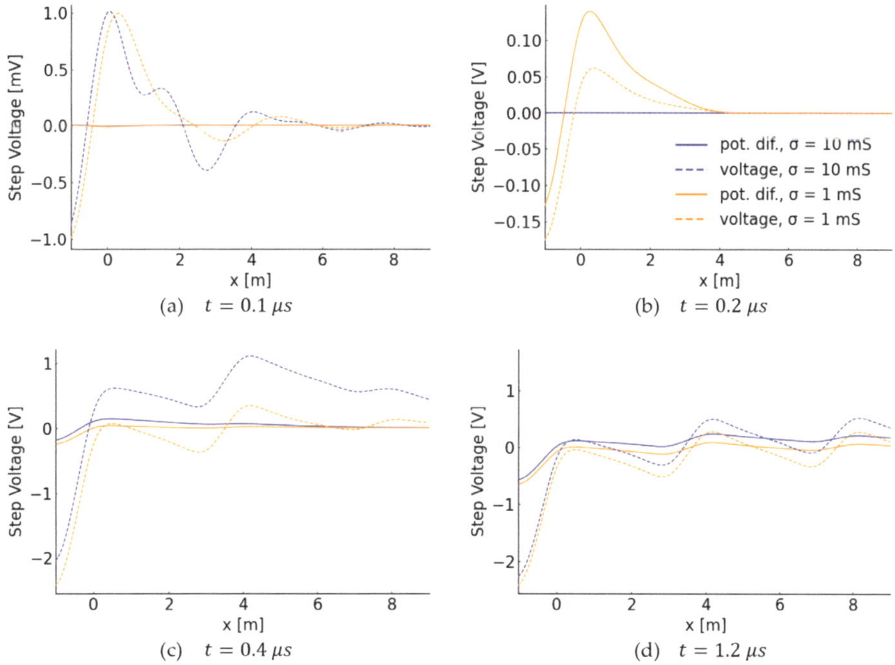

Figure 15.
Transient step voltage above a grounding grid comparing the potential difference Δu *and voltage drop* U_{p1-p2}. *(a)* $t = 0.1\ \mu s$, *(b)* $t = 0.2\ \mu s$, *(c)* $t = 0.4\ \mu s$, *and (d)* $t = 1.2\ \mu s$.

a negative offset due to the nonconservative electric field's direction associated with the conductors' longitudinal currents.

3.4 Three communication towers

This case is about three communication towers in close proximity to each other. All of them are grounded by $12 \times 12\ m^2$ square grids buried 0.5 m deep. Each grounding grid is 6 m apart from the next one, as illustrated in **Figure 16**. Four cases are considered: having all the grids connected or isolated from each other while the soil is represented by the Alipio-Visacro model [29] with mean values and low-frequency conductivity σ_0 of 1 mS and 10 mS. The current is always injected at the first grid's lower-left corner (the grids are enumerated 1 to 3, from left to right). These connections are a common practice since it reduces the global grounding impedance for low-frequency phenomena.

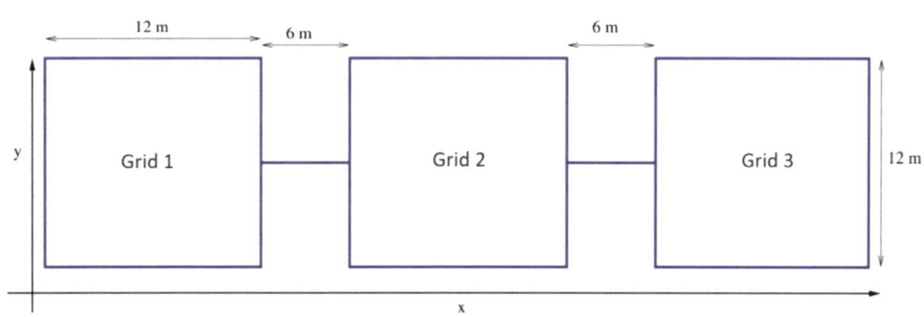

Figure 16.
Grounding grids from the communication towers.

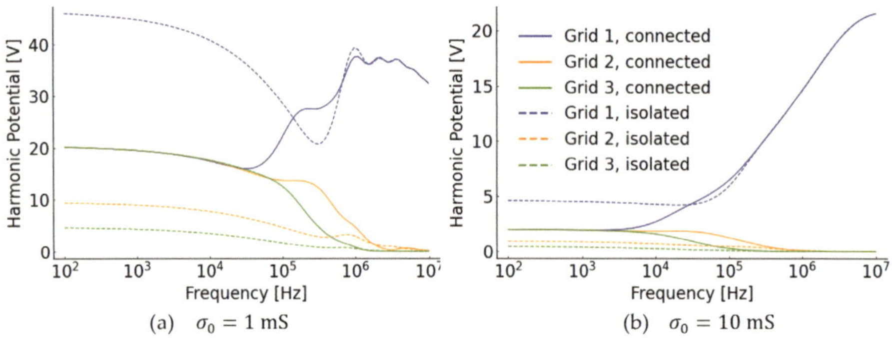

Figure 17.
Harmonic potential at the lower-left corner of each tower's grounding grid. (a) $\sigma_0 = 1$ mS and (b) $\sigma_0 = 10$ mS.

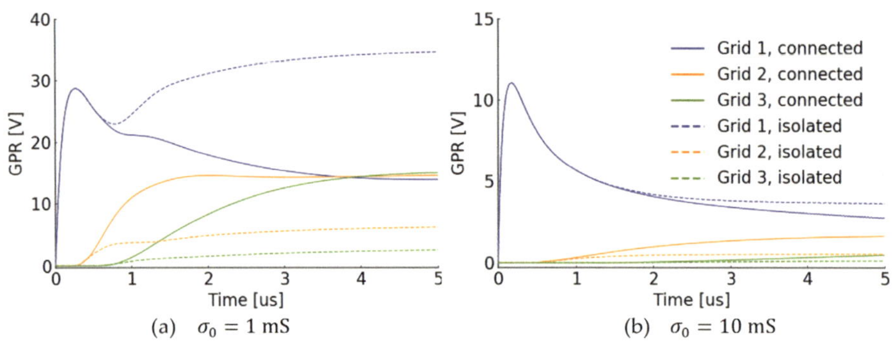

Figure 18.
GPR at the lower-left corner of each tower's grounding grids. (a) $\sigma_0 = 1$ mS and (b) $\sigma_0 = 10$ mS.

The absolute values of the harmonic potentials that appear at the corner of each grid are shown in **Figure 17**. Connecting the grounding grids of the towers has the effect of equalizing their potentials in low frequencies, reducing the potential that arises in the tower that injects the current in the ground, but raising the potential in the other ones. However, after a few MHz, connecting has practically no effect in reducing harmonic potential. This can be explained based on electromagnetic wave theory. With the increase in both frequency and conductivity, the propagation constant has a greater attenuation constant (real part of the propagation constant). Hence, the nearby grounding grids do not guarantee an actual impact on the potential.

These results show that there is an effective length[1] of the conductors, indicating that for some current excitation, the interconnection between grids does not reduce GPR. Thus, adding more or longer conductors to reduce potential has no practical impact in high frequencies and high conductivity soils, although it is very beneficial

[1] The current dispersed to the soil along the grounding electrode shows nonuniform distribution. This nonuniformity is more pronounced at high frequencies. In this case, the attenuation effects are intense. Associated with this attenuation there is a critical electrode length, such that if a longer electrode is considered there will be no additional current dispersion. At this critical length is given the name of effective length, which depends on the soil conductivity and the frequency spectrum of the injected current.

(and extensively used in grounding design) in steady-state operations. In time-domain analysis, it is possible to notice this phenomenon in the first-time steps. Take **Figure 18(a)**, for instance. Considering Grid 1, until around 0.5 μs, the curves associated with the connected system and the isolated one are overlapped. However, after that point, the values start to drift away from each other. This is less pronounced in higher conductivity media; see **Figure 18(b)**, for instance.

As mentioned before, connecting the grids has the downside of raising the GPR in the other grids as all of them become equipotential. Thus, it is of utmost importance to properly design the grounding grid, taking into account the equipment that will be connected in these grids as well as the possibility of protecting such elements from unexpected transferred potential. For example, consider **Figure 18(a)**. If a fast current strikes Grid 1 and Grid 2 has sensible electronics equipment connected to it, the potential rise on this equipment will be around three times higher if the grids are connected, leading to a faulted equipment.

As expected from the analysis of previous cases, the harmonic step voltage in the $+x$-direction calculated by scalar potential difference Δu and calculated by voltage drop U_{p1-p2} has the same numerical values for low frequencies but differ in higher frequencies, as illustrated in **Figures 19** and **20**. That difference is greater when the grids are connected because the current can then travel farther from its injection point and, therefore, cause a stronger electric field above the other grids.

The transient step voltage in the $+x$-direction is shown in **Figures 21** and **22**. In the first moments, while the injection current is rising, the step voltage has high-frequency components, and the step voltage is zero, far from the first grid.

The step voltage calculated by the voltage drop U_{p1-p2} has a negative offset compared to the potential difference Δu because of the current direction in the

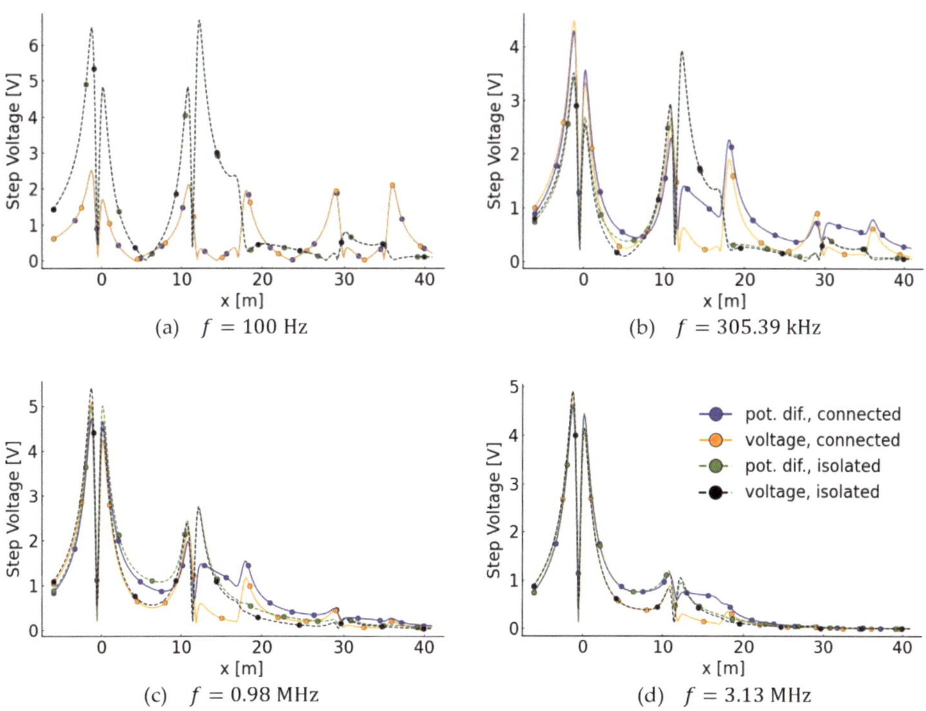

Figure 19.
Harmonic step voltage on the ground surface along the line $\mathbf{y} = \mathbf{6}$ **m** *comparing the potential difference* $\boldsymbol{\Delta u}$ *and voltage drop* $\mathbf{U_{p1-p2}}$ *above the tower's grounding grids for soil with* $\boldsymbol{\sigma_0} = \mathbf{1}$ *mS. (a)* $f = 100$ *Hz, (b)* $f = 305.39$ *kHz, (c)* $f = 0.98$ *MHz, and (d)* $f = 3.13$ *MHz.*

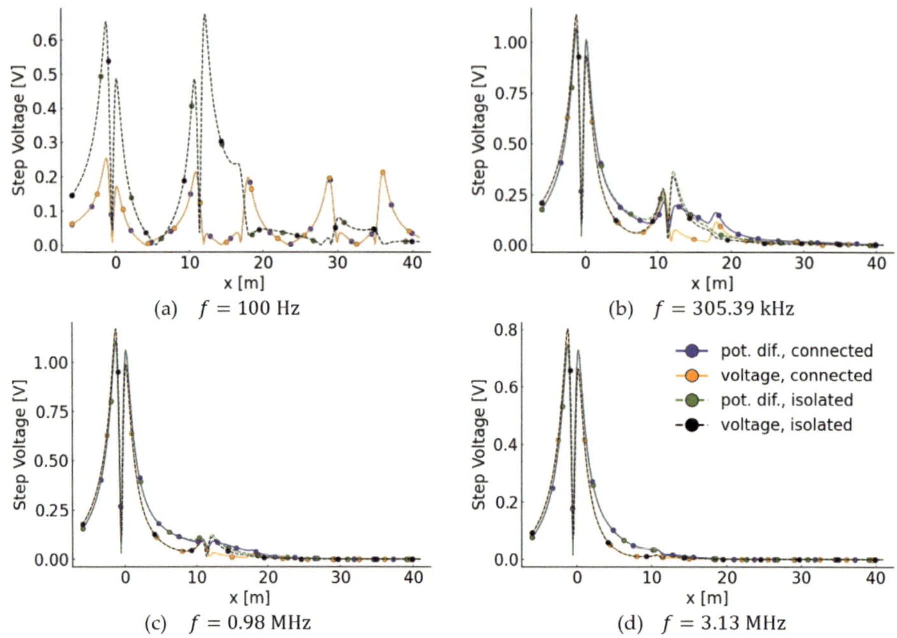

Figure 20.
Harmonic step voltage on the ground surface along the line $\mathbf{y} = \mathbf{6}$ m comparing the potential difference Δu and voltage drop $\mathbf{U_{p1-p2}}$ above the tower's grounding grids for soil with $\sigma_0 = 10$ mS. (a) $f = 100$ Hz, (b) $f = 305.39$ kHz, (c) $f = 0.98$ MHz, and (d) $f = 3.13$ MHz.

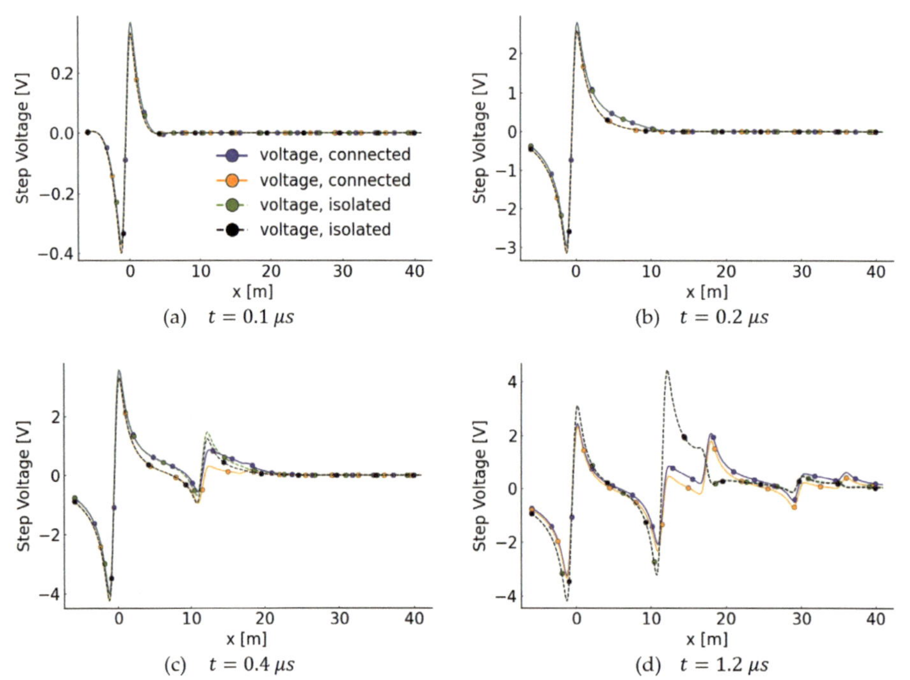

Figure 21.
Transient step voltage on the ground surface along the line $\mathbf{y} = \mathbf{6}$ m comparing the potential difference Δu and voltage drop $\mathbf{U_{p1-p2}}$ above the tower's grounding grids for soil with $\sigma_0 = 1$ mS. (a) $t = 0.1$ µs, (b) $t = 0.2$ µs, (c) $t = 0.4$ µs, and (d) $t = 1.2$ µs.

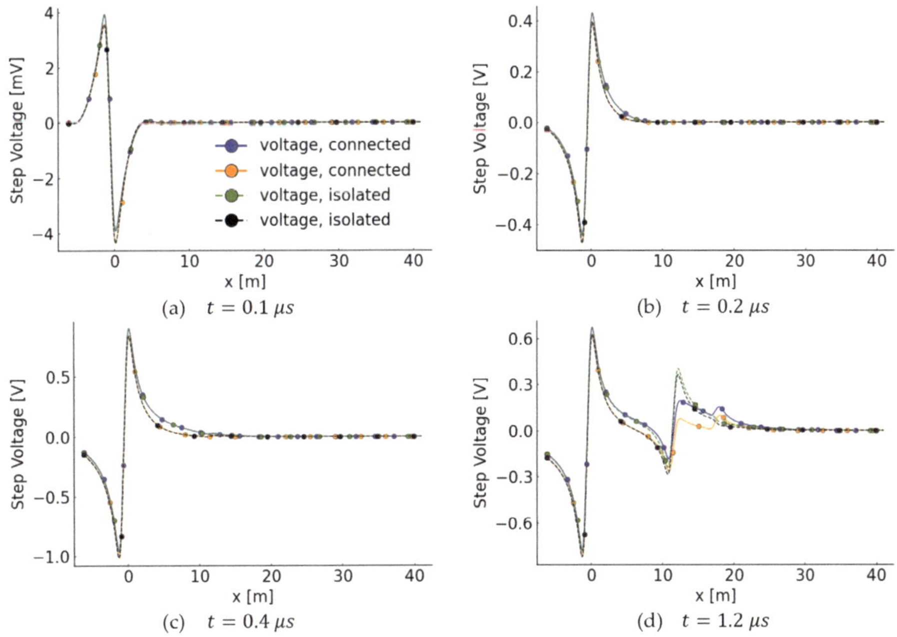

Figure 22.
Transient step voltage on the ground surface along the line $y = \mathbf{6\ m}$ comparing the potential difference $\boldsymbol{\Delta u}$ and voltage drop $\mathbf{U_{p1-p2}}$ above the tower's grounding grids for soil with $\sigma_0 = 10$ mS. (a) $t = 0.1\ \mu s$, (b) $t = 0.2\ \mu s$, (c) $t = 0.4\ \mu s$, and (d) $t = 1.2\ \mu s$.

conductors, particularly in the conductors that connect the grids. The nonconservative component of the electric field from the currents in the conductors that connect the grids has a $-x$-direction.

4. Discussions and conclusions

When making an analysis or project, it is essential to be aware of the used mathematical model's simplifications (and limitations thereof). It is often desirable to reduce computational times, usually through mathematical approximations, which restrains the model's applicability. For instance, the results and modeling are presented here to consider all conductors as thin wires.

One approximation that is often made in electromagnetic compatibility and grounding projects is to consider voltage drop (integral effect of the electric field along a given path) as equal to the scalar electric potential difference. However, this will only be true for low-frequency phenomena. As it was shown as an example in the previous sections, the step voltage calculated by integrating the total electric field is different from the scalar potential difference (i.e., considering only the conservative component of the field).

Simplifying the voltage calculation along a path by computing only the scalar potential difference is desirable because that significantly reduces computational time. Unfortunately, that simplification cannot be made except for some low-frequency phenomena. The simulations presented here demonstrate that the voltage drop can be higher or lower than the electric potential difference, depending on the studied case. Hence, more careful approach should be made when calculating the voltage due to lightning, and the total electric field should be integrated.

A common practice is to connect multiple grounding structures that are close to each other. However, the results have shown that connecting multiple grounding structures may not always be beneficial, depending on the intention. In low conductivity soils, that connection has the advantage of reducing the ground potential rise (GPR) of a structure when it is subject to current discharges, but it inadvertently raises the GPR in the other structures. It also has the consequence of raising the step voltage near those other grounding structures. Therefore, the common practices and standard recommendations should be inquired if they are the best for a given project and intention.

A. Appendix: Electromagnetic modeling for determining potential difference, voltage drop, and step voltage

After determining the current distributions in the grounding electrodes, as presented in Section 2, it is possible to calculate the total electric field, the potential difference, and voltage drop. This Appendix presents a brief quantification of these quantities.

A.1 Total electric field $\vec{E}_{\text{Total}}$ and its components

Knowing I_T and I_L, it is possible to calculate the total electric field ($\vec{E}_{\text{Total}}$) at a generic point in soil, as well as its conservative ($\vec{E}_{\text{C}}$) and nonconservative ($\vec{E}_{NC}$) components. Here, it is called conservative electric field component, the part associated with the transversal current, that is, the one that contemplates the electromagnetic field's divergence nature. On the other hand, here it is called nonconservative electric field component, the part associated with the longitudinal current, that is, the one that contemplates the curling nature of the electromagnetic field.

A.1.1 Conservative electric field component ($\vec{E}_C$)

To illustrate, consider a point current source I_{T-p}, located in the soil. This current generates an electric scalar potential V_P given by:

$$V_P = \frac{I_{T-p}}{4\pi[\sigma + j\omega\varepsilon]} \frac{e^{-\gamma r}}{r} \tag{4}$$

where r is the distance between the source point and the observation point P.

The electric field associated with this current source ($\vec{E}_C$) is straightforwardly obtained (at any generic point P, into the ground) by the gradient of V_P [27]:

$$\vec{E}_C = -\vec{\nabla} V_P = \frac{I_{T-p}}{4\pi[\sigma + j\omega\varepsilon]} \frac{(1+\gamma r)\, e^{-\gamma r}}{r^2} \vec{a}_e \tag{5}$$

where $\vec{a}_e$ is the unit vector that defines the direction of $\vec{E}_C$.

The electric field generated by the total transverse current I_T from a given segment of length $\ell = L/N$ (see **Figure 5**) can be determined by the cumulative effect of point current sources along the segment, considering a continuous and uniform current distribution with linear density I_{Ti}/L_i. Then, the electric field $\vec{E}_{Ci}$ due to these currents is given by [27]:

$$\vec{E}_{Ci} = \frac{1}{4\pi[\sigma + j\omega\varepsilon]} \int_{\ell_i} \frac{I_{Ti}\,(1+\gamma r_i)\,e^{-\gamma r_i}}{L_i r_i^2} d\ell_i\,\vec{a}_e \tag{6}$$

where r_i is the distance between the infinitesimal element $d\ell_i$ and point P; $\vec{a}_{ei}$ is the unit vector that defines the direction of $\vec{E}_{Ci}$.

With similar reasoning, it is possible to determine the conservative electric field component associated with the transverse current of the image segment $\vec{E}_{Ci-\text{image}}$.

$$\vec{E}_{Ci-\text{image}} = \frac{1}{4\pi[\sigma + j\omega\varepsilon]} \int_{\ell_i} \frac{\Gamma_r I_{Ti}\left(1+\gamma r_{i-\text{image}}\right) e^{-\gamma r_{i-\text{image}}}}{L_i r_{i-\text{image}}^2} d\ell_{i-\text{image}}\,\vec{a}_{e-\text{image}} \tag{7}$$

where $r_{i-\text{image}}$ is the distance between the in the infinitesimal element $d\ell_{i-\text{image}}$ and point P; $\vec{a}_{e-\text{image}}$ is the unit vector that defines the direction of $\vec{E}_{Ci-\text{image}}$ and Γ_r is the reflection coefficient given by [19, 20] (see **Figure 5**):

$$\Gamma_r = \frac{\sigma + j\omega(\varepsilon - \varepsilon_0)}{\sigma + j\omega(\varepsilon + \varepsilon_0)} \tag{8}$$

By considering the system linear and applying the superposition theorem, one can determine the total conservative component $\vec{E}_{Ci-\text{Total}}$ associated with the segment i (see **Figure 5**):

$$\vec{E}_{Ci-\text{Total}} = \vec{E}_{Ci} + \vec{E}_{Ci-\text{image}} \tag{9}$$

Finally, the total conservative component $\vec{E}_{C-\text{Total}}$ is obtained *via* the sum of all the conservative field components associated with the transverse currents in the N segments:

$$\vec{E}_{C-\text{Total}} = \sum_{i=1}^{N} \left(\vec{E}_{Ci} + \vec{E}_{Ci-\text{image}}\right) \tag{10}$$

A.1.2 Nonconservative electric field component ($\vec{E}_{NC}$)

Consider the longitudinal current I_{Li} flowing along segment i (see **Figure 5**). This current source generates a magnetic vector potential $\vec{A}_{Pi}$ at a generic point P of the medium given by:

$$\vec{A}_{Pi} = \frac{\mu_0}{4\pi} \int_{\ell_i} I_{Li} \frac{e^{-\gamma r_i}}{r_i} d\vec{\ell}_i \tag{11}$$

where: $d\vec{\ell}_i$ is the length vector element of segment i, which defines the direction of $\vec{A}_{Pi}$.

In the frequency domain, the associated nonconservative electric field $\vec{E}_{NCi}$ is given by [27]:

$$\vec{E}_{NCi} = -j\omega\vec{A}_{Pi} = -j\omega\frac{\mu_0}{4\pi}\int_{\ell_i} I_{Li}\frac{e^{-\gamma r_i}}{r_i}d\vec{\ell}_i \tag{12}$$

Similarly, the component corresponding to the image segments $\vec{E}_{NCi-\text{image}}$ is given by [27]:

$$\vec{E}_{NCi-\text{image}} = -j\omega\vec{A}_{Pi-\text{image}} = -j\omega\frac{\mu_0}{4\pi}\int_{\ell_i} I_{Li}\frac{e^{-\gamma r_{i-\text{image}}}}{r_{i-\text{image}}}d\vec{\ell}_{i-\text{image}} \tag{13}$$

In this case, the image method is applied as shown in [21], where the reflection coefficient is equal to the unit, that is, $\Gamma_r = 1$.

Finally, for the total nonconservative component of the electric field $\vec{E}_{NC-\text{Total}}$, Eq. (14) is obtained.

$$\vec{E}_{NC-\text{Total}} = \sum_{i=1}^{N}\left(\vec{E}_{NCi} + \vec{E}_{NCi-\text{image}}\right) \tag{14}$$

A.1.3 Total electric field ($\vec{E}_{\text{Total}}$)

The total electric field $\vec{E}_{\text{Total}}$ is determined by the vector sum of Eqs. (10) and (14):

$$\vec{E}_{\text{Total}} = \sum_{i=1}^{N}\left(\vec{E}_{C-\text{Total}} + \vec{E}_{NC-\text{Total}}\right) \tag{15}$$

A.2 Potential difference, voltage drop, and step voltage

In several cases, potential difference, voltage drop, and step voltage are more practical parameters. However, these definitions depend on the electric field's nature. For instance, in studying personal protection or electromagnetic compatibility, these parameters may be more interesting and applicable. Here follows a brief definition of each one of them.

A.2.1 Potential difference (Δu) – path independent

The potential difference corresponds to the line integral of $\vec{E}_{C-\text{Total}}$, which depends only on the start (p_1) and end (p_2) points, not depending on the integration path. Thus, it is defined according to Eq. (16).

$$\Delta u = -\int_{p_1}^{p_2}\vec{E}_{C-\text{Total}}\cdot d\vec{\ell} = u_2 - u_1 \tag{16}$$

A.2.2 Voltage drop (U_{p1-p2}) – path dependent

The voltage drop corresponds to the line integral of $\vec{E}_{\text{Total}}$, which depends on the start and endpoints and the integration path. Thus, it is defined according to Eq. (17).

$$U_{p_1-p_2} = \int_{p_1}^{p_2}\vec{E}_{\text{Total}}\cdot d\vec{\ell} \tag{17}$$

Acknowledgements

This work was financed in part by the Coordenação de Aperfeiçoamento de Pessoal de Nível Superior – Brasil (CAPES), Finance Code 001. It also was partially supported by INERGE (Instituto Nacional de Energia Elétrica), CNPq (Conselho Nacional de Desenvolvimento Cientıfico e Tecnológico), FAPERJ (Fundação Carlos Chagas Filho de Amparo à Pesquisa do Estado do Rio de Janeiro), and FAPEMIG (Fundação de Amparo à Pesquisa do Estado de Minas Gerais).

Author details

Antonio Carlos S. Lima[1*], Pedro H.N. Vieira[1], Marco Aurélio O. Schroeder[2] and Rodolfo Antônio R. Moura[2]

1 Federal University of Rio de Janeiro, Rio de Janeiro, Brazil

2 Federal University of São João del-Rei, São João del-Rei, Brazil

*Address all correspondence to: acsl@coppe.ufrj.br

References

[1] Harrington R. Origin and development of the method of moments for field computation. IEEE Antennas and Propagation Magazine. 1990;**32**(3):31-35

[2] Wedepohl LM, Mohamed SET. Multiconductor transmission lines. Theory of natural modes and Fourier integral applied to transient analysis. Proceedings of the Institution of Electrical Engineers. 1969;**116**(9): 1553-1563

[3] Wilcox DJ. Numerical Laplace transformation and inversion. International Journal of Electrical Engineering. 1978;**15**:247-265

[4] Uribe FA, Naredo JL, Moreno P, Guardado L. Electromagnetic transients in underground transmission systems through the Numerical Laplace Transform. International Journal of Electrical Power & Energy Systems. 2002;**24**:215-221

[5] Moreno P, Ramirez A. Implementation of the Numerical Laplace Transform: A review task force on frequency domain methods for EMT studies. IEEE Transactions on Power Delivery. 2008;**23**(4):2599-2609

[6] Gomez P, Uribe FA. The Numerical Laplace Transform: An accurate technique for analyzing electromagnetic transients on power system devices. International Journal of Electrical Power & Energy Systems. 2009;**31**:116-123

[7] Akbari M, Sheshyekani K, Alemi MR. The effect of frequency dependence of soil electrical parameters on the lightning performance of grounding systems. IEEE Transactions on Electromagnetic Compatibility. 2013;**55**(4):739-746

[8] Grcev L, Dawalibi F. An electromagnetic model for transients in grounding systems. IEEE Transactions on Power Delivery. 1990;**5**(4):1773-1781

[9] Visacro S, Portela CM. Modelling of earthing systems for lightning protection applications, including propagation effects. In: 21st International Conference on Lightning Protection (ICLP). Berlin; 1992. pp. 129–132

[10] Visacro S, Soares A. HEM: A model for simulation of lightning-related engineering problems. IEEE Transactions on Power Delivery. 2005; **20**(2):1206-1208

[11] Tanabe K. Novel method for analyzing dynamic behavior of grounding systems based on the finite-difference time-domain method. IEEE Power Engineering Review. 2001;**21**(9): 55-57

[12] Tsumura M, Baba Y, Nagaoka N, Ametani A. FDTD simulation of a horizontal grounding electrode and modeling of its equivalent circuit. IEEE Transactions on Electromagnetic Compatibility. 2006;**48**(4):817-825

[13] Ruehli AE. Equivalent circuit models for three-dimensional multiconductor systems. IEEE Transactions on Microwave Theory and Techniques. 1974;**22**(3): 216-221

[14] Ruehli A. Partial element equivalent circuit (PEEC) method and its application in the frequency and time domain. In: Proceedings of Symposium on Electromagnetic Compatibility. IEEE; 1996. pp: 128–133

[15] Alipio R, Visacro S. Impulse efficiency of grounding electrodes: Effect of frequency-dependent soil parameters. IEEE Transactions on Power Delivery. 2014;**29**(2):716-723

[16] Sadiku MNO. Solutions Manual for Numerical Techniques in Electromagnetics. Florida, US: CRC Press; 1992

[17] Ho CW, Ruehli A, Brennan P. The modified nodal approach to network analysis. IEEE Transactions on Circuits and Systems. 1975;**22**(6):504-509

[18] Wedepohl L, Jackson L. Modified nodal analysis: An essential addition to electrical circuit theory and analysis. Engineering Science and Education Journal. 2002;**11**(3):84-92

[19] Takashima T, Nakae T, Ishibashi R. Calculation of complex fields in conducting media. IEEE Transactions on Electrical Insulation. 1980;**EI-15**(1):1-7

[20] Takashima T, Nakae T, Ishibashi R. High frequency characteristics of impedances to ground and field distributions of ground electrodes. IEEE Power Engineering Review. 1981;**1**(4):60-61

[21] Arnautovski-Toseva V, Grcev L. On the image model of a buried horizontal wire. IEEE Transactions on Electromagnetic Compatibility. 2016;**58**(1):278-286

[22] Schroeder MAO, Moura RAR, Machado VM. A discussion on practical limits for segmentation procedures of tower-footing grounding modeling for lightning responses. IEEE Transactions on Electromagnetic Compatibility. 2020;**62**(2):2520-2527

[23] Lima ACS, Moura RAR, Vieira PHN, Schroeder MAO, Correia de Barros MT. A computational improvement in grounding systems. Transient analysis. IEEE Transactions on Electromagnetic Compatibility. 2020;**62**(3):765-773

[24] Moura RAR, Schroeder MAO, Lima ACS, Vieira PHN. Closed-form approximation for horizontal grounding electrodes transient analysis. Journal of Control, Automation and Electrical Systems. 2020;**31**(4):1063-1073

[25] Moura RAR, Schroeder MAO, Lima ACS, Vieira PHN, Alipio R. Closed-form approximation for grounding grids transient analysis. Journal of Control, Automation and Electrical Systems. 2021;**32**:796-806

[26] Alipio R, Visacro S. Modeling the frequency dependence of electrical parameters of soil. IEEE Transactions on Electromagnetic Compatibility. 2014;**56**(5):1163-1171

[27] Alipio RS, Schroeder MAO, Afonso MM, Oliveira TAS, Assis SC. Electric fields of grounding electrodes with frequency dependent soil parameters. Electric Power Systems Research. 2012;**83**(1):220-226

[28] Portela CM, Tavares MC, Pissolato FJ. Accurate representation of soil behaviour for transient studies. IEE Proceedings - Generation, Transmission and Distribution. 2003;**150**(6):736-744

[29] Visacro S, Alipio R, Pereira C, Guimarães M, Schroeder MAO. Lightning response of grounding grids: Simulated and experimental results. IEEE Transactions on Electromagnetic Compatibility. 2015;**57**(1):121-127

Chapter 7

Study of Electromagnetic Radiation Sources Using Time Reversal: Application to a Power Electronic Converter

Sassia Hedia, Bessem Zitouna,
Jaleleddine Ben Hadj Slama and Lionel Pichon

Abstract

Recently, modern power electronic systems have been introduced in different applications, such as in avionics and wireless communication. The increasing technological complexity of these systems is posing serious challenges regarding electromagnetic compatibility (EMC) issues. Indeed, the radiation emitted from electronic circuits can induce harmful effects on nearby devices. Thus, several research works have been conducted using the nearfield technique to deal with electromagnetic interferences (EMI) that might occur, especially due to rapidly changing currents and voltages. In the present work, a detailed study about the characterization of the electromagnetic nearfield-radiated emissions is established using a time-domain analysis to provide an equivalent model constituted of a set of electromagnetic dipole parameters. Source reconstruction has been obtained using electromagnetic time reversal (EMTR), which has proven successful and efficient in identifying transient disturbance sources in power electronics. Experimental measurements of the magnetic nearfield have been carried out under an AC/DC flyback converter. The accuracy of the proposed method has been confirmed by visualizing measured magnetic field components, which are in good agreement with the calculated maps. The results of a reasonable computing time have shown that, particularly in transient signals with a wide frequency band, the suggested inverse method is an adequate alternative to overcome frequency domain limitations.

Keywords: electromagnetic compatibility (EMC), electromagnetic time reversal (EMTR), nearfield, power electronics, radiated emissions, time-domain analysis

1. Introduction

In any electronic circuit, designers face serious challenges to ensure product efficiency and make it fit into a large community of electronic systems where harmful interferences can occur. Indeed, electromagnetic compatibility aims to achieve a peaceful coexistence of electronic equipment, especially when subjected to severe electromagnetic disturbances. In recent high-density PCBs (printed circuit boards), components are becoming increasingly sensitive to unwanted phenomena, such as radiated and conducted emissions characterized by sudden variations in voltage and current [1, 2]. Thus, EMC researchers and engineers have conducted multiple works using various relevant approaches to predict these emissions with a reasonable tradeoff between guaranteeing a good accuracy and a reduced computing time. For instance, to deal with radiated emissions, authors in [3–5] have proposed to use matrix inversion methods that have ended up with a need for a large set of measurements. As a solution, hybrid optimization methods based on artificial intelligence techniques such as genetic algorithms (GA) and artificial neural networks (ANN) have been progressively introduced [6, 7], and this has contributed considerably to speed up the algorithm convergence at a fixed operating frequency.

The common point in the research mentioned above has been using the frequency domain (FD) to model electromagnetic fields through a set of equivalent dipoles [2–8]. To obtain a behavioral model of the device under test (DUT), particularly in power circuits of a fast switching nature [2, 5], a direct alternative way of proceeding is to perform measurements in the time domain (TD), which can also recover the frequency spectrum using a further signal processing. Indeed, time-domain measurement has numerous advantages over frequency-domain measurements. On the one hand, TD devices provide faster and simpler measurements because they use high-speed time sampling techniques and offer the possibility of capturing the signal in TD and compute its different spectral components in a shorter period (e.g., spectrum analyzers, which are more expensive than oscilloscopes, usually have a narrow resolution bandwidth). On the other hand, better precision is obtained using TD measurements because all the frequencies are simultaneously measured, mainly in the case of a one-time transient where only a single frequency is captured and in the case of a periodic signal (with repetitive events) where the frequency step could limit the identification of all the events of interest. In the literature, few TD-based works have been presented to obtain the different equivalent parameters of radiating sources [9–15]. In [11–13], authors have evaluated the performance of TD methods to deal with EM transients using a microstrip device. More studies have been performed in [11–14].

Nevertheless, these approaches select the frequency band, and complex calculations require powerful machines with sufficient memory [11]. In [10], the temporal electromagnetic inverse method has been developed and implemented using GA, and thus, it depends on initial GA options. Therefore, researchers have developed alternative TD methods based on electromagnetic time reversal (EMTR) for radiating source identification in literature. Indeed, several studies have been achieved in the field of electromagnetics using the time-reversal (TR) technique for different purposes such as in source location identification [14, 15], acoustics [16–18], and recently power electronics [19–27]. Several studies have demonstrated TR's robustness and efficiency [14, 27, 28]. However, a handful of studies investigated the use of TR for EMC applications in the time domain until now. Rachidi et al. have presented using electromagnetic time reversal in lighting location and faults detection in power networks [14].

Most of the existing TR studies dedicated to sources characterization have been applied in the far-field region [17, 19, 20]. While in the case of power electronic EM radiation measurement, a nearfield (NF) scan has proven to be more advantageous because it is less dependent on different test conditions such as application range and equipment. Recent studies, as in [3] and [10], report that different issues in the converter circuits, basically linked to transient disturbances with a quite short duration, should be investigated in the TD using an NF test bench. In [22, 23], the proposed method has been applied to identify the equivalent model of radiation of simple structure, as an interesting basic study but not representative enough to deal with an advanced power circuit. Hence, this work applies the full time-domain method to a complicated board containing several bulky components with several radiating sources based on the electromagnetic time-reversal technique. Current and voltage high-speed commutations can create serious non-intentional transient disturbances covering a wide frequency band in structures such as power electronic converters. Knowing that in electromagnetic interferences (EMI) mitigation techniques (such as shielding), it is essential to predict radiating sources parameters at the early design stage to reduce emissions related to switching activities and parasitic interconnections. In this present work, we have studied the application of the EMTR technique based on time-domain analysis to reconstruct the radiation behavior and obtain an accurate equivalent model of the device under test (DUT), which emits critical non-sinusoidal signals. This methodology aims to help designers estimate radiated EM fields at different measurement distances and study the possible couplings between components.

Section 2 gives a detailed overview of the NF measurement test bench in the time domain. Moreover, the selected studied structure, which is an AC/DC flyback converter, is described. Section 3 discusses the theoretical investigations of the electromagnetic time-reversal technique and details the suggested implementation of the method for evaluating radiated emissions and sources reconstruction issues. Section 4 provides an application of the proposed EMTR-based method, and an adequate equivalent model that emits the same radiation behavior is studied.

Furthermore, a comparison study has been carried out for experimental validation purposes. The obtained results are presented in the form of time-dependent mappings of the magnetic nearfield. The last section outlines the main conclusions.

2. Time-domain measurements test bench

A TD test bench is employed to evaluate EM disturbances in the NF for power systems in this study. A single measurement (one scan) is carried out for multiples radiation frequencies at once.

2.1 TD measurement technique

TD measurement uses high precision measurement devices with a wide bandwidth, such as oscilloscopes, to capture temporal signals using a magnetic field probe. This measuring probe is nothing else than an academic small coil probe shielded sung copper [11] or a commercial magnetic loop of a radius equal to 1.6 mm generating a voltage from the varying magnetic flux. It scans the surface above the DUT, in the nearfield region, with a fixed displacement step, which prevents capacitive couplings and noise while measuring. The probe is connected to the scope using a shielded coaxial cable through an SMA connector, and it is calibrated as explained in [11, 29].

Indeed, based on the Faraday and Lenz laws and transformation equations, the variable magnetic field is obtained through the voltage measured at the terminals of the loop using the following equation:

$$\mathrm{H}(t) = -\frac{1}{\mu_0 \times S}\int_0^t \mathrm{V}(t)dt \tag{1}$$

Where $S = \pi \times r^2$, is the loop surface and $\mu_0 = 4 \times \pi \times 10^{-7}\ \mathrm{N/A^2}$ is the permeability in the free space.

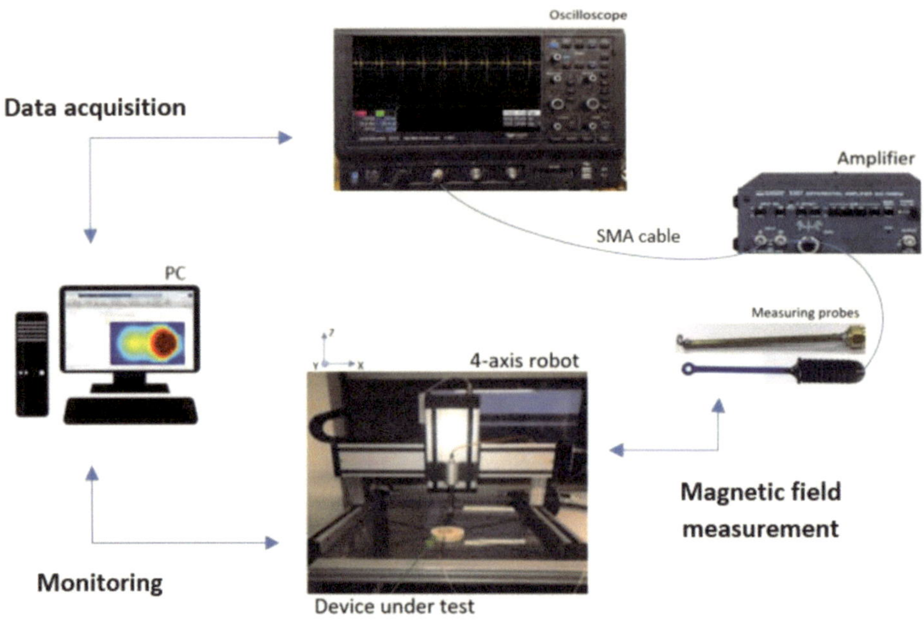

Figure 1.
Automated scanning test stands for NF measurement in the time domain.

The studied test bench provides a three-dimensional field measurement, but according to our needs, we only consider measuring the normal component of the magnetic field (H_z). Synchronous acquisitions are performed using a trigger, chosen to be a periodic and a repetitive reference signal. The following **Figure 1** presents the proposed nearfield test bench using a LeCroy WaveRunner 104XI oscilloscope.

2.2 Device under test: AC/DC flyback converter

In this work, the studied structure is a flyback AC/DC converter based on the principle of switched-mode power supplies (SMPS), as in [30]. Indeed, despite the superior efficiency that they can offer, these switching converters generate a lot of noise and EMI because of their high changing dv/dt and di/dt events. These power supplies have a complex structure regarding EM emissions. It generates different types of EMI, such as harmonics of switching frequency and high-frequency noise due to phase voltage ringing or the reverse recovery. In industry, this converter topology is typically used in both low and medium power as in automotive, battery charging and avionics, etc.

In practice, the design of the SMPS circuit is of great importance because it has to meet the configuration specifications in terms of energy storage and losses and avoid transient oscillations that generate EM interference modes covering a wide frequency band of several tens of MHz.

In our study, the DUT is a low power converter of 5 W. The top and bottom faces of the studied flyback AC/DC converter are shown in **Figure 2**. Its corresponding configuration parameters are presented in **Table 1**, and the schematic is depicted in **Figure 3**. To guarantee a synchronous acquisition, transient radiation signals at the scan surface have been measured with respect to the input of the optocoupler of the converter as a reference signal, **Figure 3**.

The NF radiation distribution of the DUT is obtained through a rigorous scan of the bottom face of the board, which allows avoiding bulky components and having a better recoding of EM emissions at a reasonable measuring distance.

DOI: http://dx.doi.org/10.5772/intechopen.100611

Figure 2.
Studied flyback AC/DC converter: top and bottom faces.

Input voltage	Output voltage	Switching frequency	Efficiency
220 V	5 V DC	60 kHz	79%

Table 1.
Parameters of the studied configuration.

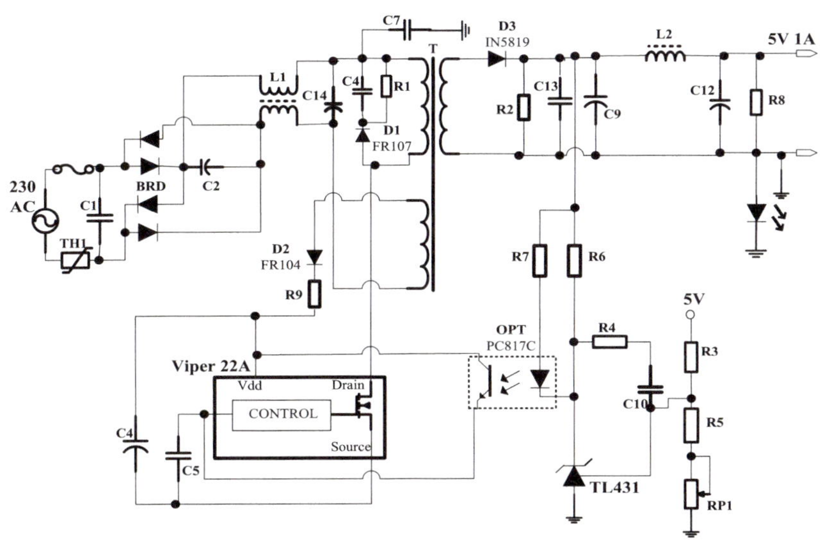

Figure 3.
Schematic of the studied flyback AC/DC converter.

Scanning area	**(38 mm × 54 mm)**
Height of measurement	1.2 cm
Scanning resolution	2 mm
Scanning points	560 positions
Signals duration	20 μs

Table 2.
Measurement setup.

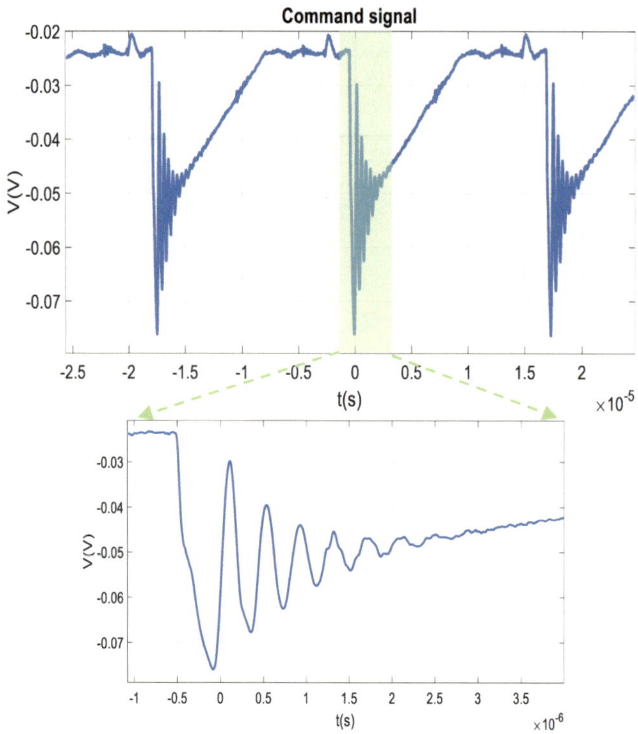

Figure 4.
Control signal.

Table 2 provides the measurement setup adopted in the present work, and **Figure 4** shows the control signal that ensures the smooth running of the converter.

A measured radiated signal is depicted in **Figure 5**. As we can notice in **Figure 6**, sources are not radiating simultaneously to obtain an intense radiation area at one time step but not in another. The radiation behavior of EM signals changes over time, and it is not evident to capture these events using frequency analysis only.

3. Basis of electromagnetic time-reversal technique

3.1 Theoretical principles

The time-reversal technique has been known for a long with great success in acoustics and ultrasound applications [17]. This has encouraged researchers to investigate the use of this method in the vast field of electrical engineering, such as in power line communications, fault location, and EMC issues [14, 15, 18–27]. Time and space focusing of waves is the specific feature of time-reversal theory [28]. Indeed, when a random source is in a pulse mode, generated signals are propagated indefinitely in the environment until they vanish (forward propagation). To measure these emissions, a set of transceivers is placed in a defined configuration called a time-reversal mirror (TRM). Then, recorded waves are time-reversed and injected into the same context (time reverse + back-propagation). After that, back-propagated waveforms will try to merge to form a maximum peak at a certain time step and position (focusing). In fact, the obtained location and parameters correspond to the actual source due to the reciprocity theorem and the reversibility in time of the wave equation. Hence, in an electromagnetic context, the performance of sources excited by high-frequency generators is easily evaluated or, more precisely, for an EMC

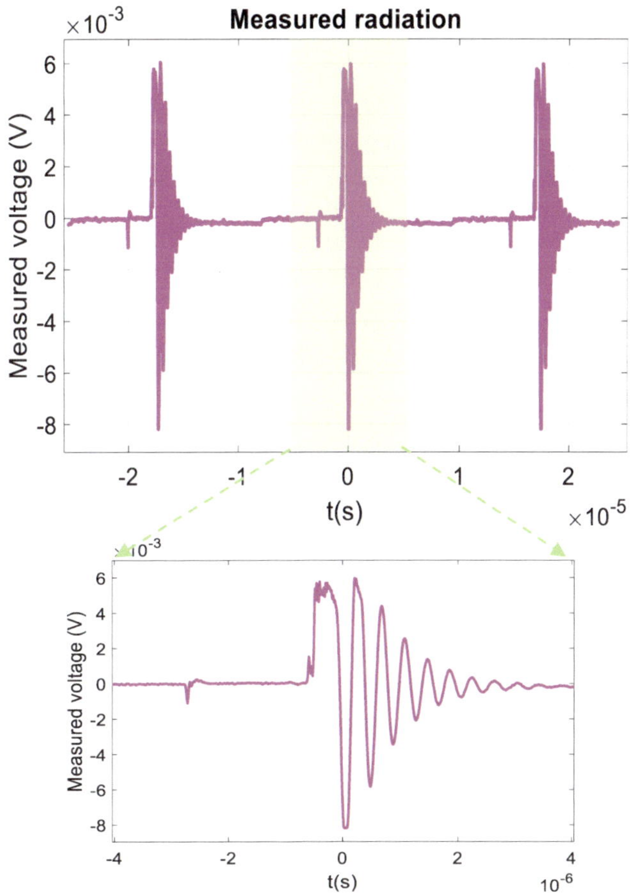

Figure 5.
Captured waveforms: (a) whole signal, (b) window.

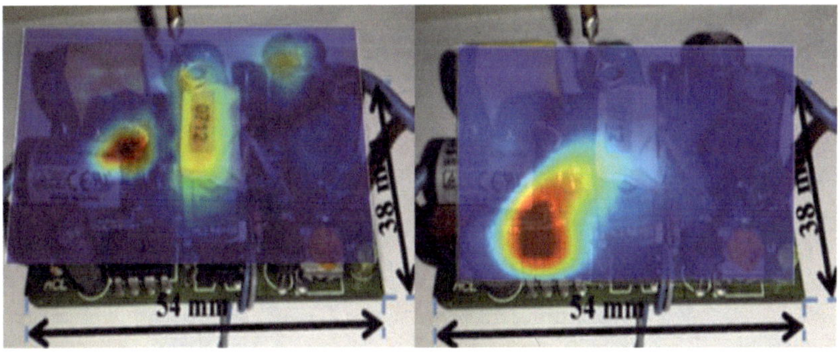

Figure 6.
Measured radiation field maps above the DUT at two different time steps: $t_1 = 10.25\ \mu s$ *and* $t_2 = 10.5\ \mu s$.

application. The EM emission behavior of a radiating source that exhibits high-level transients can be reconstructed in both space and time [14–27].

Based on the basic electromagnetic wave equations in a homogeneous medium, which is written as follows:

$$\frac{1}{c^2}\frac{\partial^2\Phi}{\partial t^2} = \nabla^2\Phi \tag{2}$$

Where Φ stands for electric E or magnetic H field, and c is the speed of light in free space.

If $f(t)$ is a solution of Eq. (1), then $g(t) = f(-t)$ is also a solution. In other words, theoretically, EM waves may propagate backward from the time step $t = \mathrm{T}$ to $t = 0$ s. In practice, the measured magnetic (or electric) field $H(r, t)$ at probe position r and for each time step t, $t \in [0; \mathrm{T}]$ where T is the time period and the back-propagated magnetic (or electric) field $H(r, \mathrm{T} - t)$ are both solutions. However, it has been demonstrated that under the action of time inversion, the magnetic field H is of an odd parity [15, 31]. Accordingly, assuming that TR represents the time inversion operator, we have:

$$\mathrm{TR}\,\{\mathrm{H}(r,t)\} = -\mathrm{H}(r,-t) \tag{3}$$

It is interesting to note that when a current $I(t)$ flows through a source, a magnetic field $H(t)$ is created. In the case of a conductive loop, the following relation gives the measured field:

$$\mathrm{H}_i(t) = h(r_0 \rightarrow r_i, t) \otimes \mathrm{I}(t) \tag{4}$$

Where ⊗ denotes a convolution product, $1 \le i \le \mathrm{N}$ number of receiving antennas, and $h(r_0 \rightarrow r_i, t)$ is the impulse response of the system at a position r_i and for a pulse generated by r_0. Indeed, $h(t)$ represents the transfer matrix that rules the transmission and the reception procedures between the TRM transducers (area discretized into N positions) and a defined virtual set of point-like sources placed on the DUT surface and driven by a Dirac delta function. As a result, the focused signal is obtained as follows:

$$\mathrm{H_{TR}}(r_0, t) = \sum_{\mathrm{i}=1}^{\mathrm{N}} h(r_i \rightarrow r_0, t) \otimes \mathrm{H}_{\mathrm{x,y,z_i}}(-t) \tag{5}$$

Where, $\mathrm{H_{TR}}\ (r_0, t)$ represents the focused signal at the position $\mathrm{r_0}$ and time step t. In the literature [14, 15], the excitation signal has commonly been identified as the reversed version of the extracted focusing signal at the source position r_0 using the following equation:

$$\mathrm{Max}(r_0) = \ \max_{t \in \mathrm{T}}(|\mathrm{H_{TR}}(r_0, t)|) \tag{6}$$

3.2 Equivalent model determination

In power systems with high current levels, the magnetic field has a strong predominance effect. In the literature, particularly during the switching activity, disturbances resulting from high dI/dt and leading to a high-frequency current can occur. Thus, to model these transients, the inverse problem resolution has been proposed and applied to a nearfield scanning experiment. Indeed, an equivalent model in the nearfield region is defined as a set of equivalent dipoles that reproduce the same EM radiation behavior as the DUT. For an equivalent magnetic dipole, **Figure 1**, the main characteristics are the center position (X_d, Y_d, Z_d), orientation angles (θ, φ), diameter (the loop surface $S = \pi \times r^2$, where r is the radius), and the current I flowings in the loop [32]. The magnetic dipole moment M_d is the following:

$$\overrightarrow{\mathbf{M}_\mathrm{d}} = I \times \vec{\mathbf{S}} \tag{7}$$

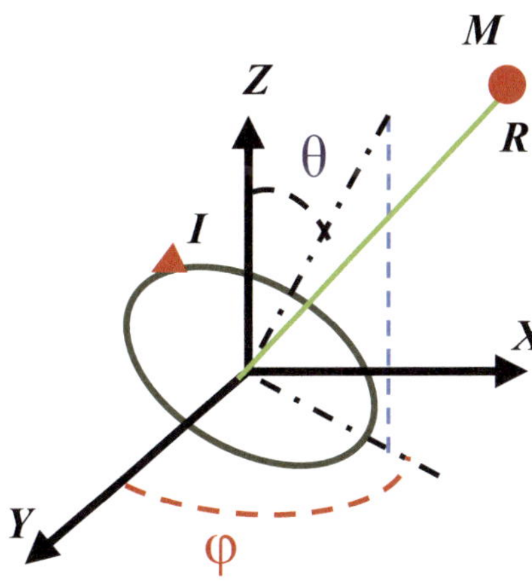

Figure 7.
Definition of an equivalent magnetic dipole.

From the formula for the equivalent field radiated by a magnetic loop in the nearfield, expressed by (8), we can obtain the three-dimensional distribution map of the radiated field (H_x, H_y and H_z) in a defined height of measurement using a time-domain representation (**Figure** 7).

$$\mathrm{H}_{x,y,z} = A\left[\left(B_1 C_{x,y,z}\right) - \left(B_3 D_{x,y,z} E\right)\right] \tag{8}$$

Where:

$$A = \frac{1}{4\pi R} \tag{9}$$

$$B_{m=1,3} = \left(\left(\frac{1}{c^2}\frac{\partial^2 M_d(t')}{\partial t^2}\right) + \left(\frac{m}{cR}\frac{\partial M_d(t')}{\partial t}\right) + \left(\frac{1}{R^2}M_d(t')\right)\right) \tag{10}$$

$$C_{x,y,z} = (\sin(\theta)\cos(\varphi), \sin(\theta)\sin(\varphi), \cos\theta) \tag{11}$$

$$D_{x,y,z} = \frac{1}{R^2}((X_d - X_0), (Y_d - Y_0), (Z_d - Z_0)) \tag{12}$$

$$E = \begin{pmatrix} \cos(\theta)(Z_d - Z_0) + \\ \sin(\theta)\sin(\varphi)(Y_d - Y_0) + \\ \sin(\theta)\cos(\varphi)(X_d - X_0) \end{pmatrix} \tag{13}$$

$$\mathrm{R} = \sqrt{(X_d - X_0)^2 + (Y_d - Y_0)^2 + (Z_d - Z_0)^2} \tag{14}$$

$$t' = t - \frac{R}{\mathrm{c}} \tag{15}$$

3.3 Applying EMTR to radiating source identification

This work focuses on the characterization of EM emissions in the nearfield using an equivalent radiation model. An algorithm based on electromagnetic time-reversal technique is applied to identify radiating sources parameters using the time-domain analysis. Indeed, in practice, active sources are not radiating

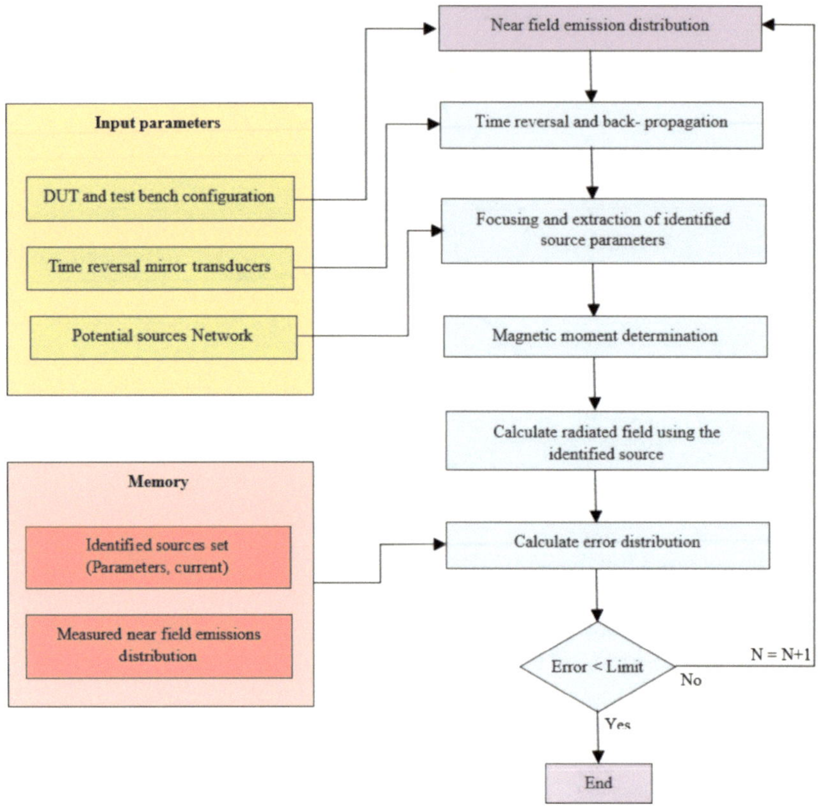

Figure 8.
Flowchart of the proposed method based on the EMTR technique.

simultaneously and the emission distribution at each time step corresponds only to the contributions of the different sources at this specific time. In the proposed method, an elimination process is carried out over time, starting from the most radiating sources (a hot spot in the map where the radiation level is significantly greater than the neighboring regions in the scanning area), using (5) and (6). A reconstructed field distribution is obtained using the identified source parameters, Eq. (8). Then, the difference between the measured and the estimated scans is evaluated using an error distribution at each iteration. To guarantee the convergence of the method and obtain a physical meaning model, an error limit is defined as a threshold value. Ideally, the estimated scan corresponds to the measured one, and then, the error limit equals zero. Apart from that, as in an NF experiment, the threshold corresponds to the measurement errors magnitudes (can be estimated when making measurement tests without a load). The effect of the different configuration parameters involved in this method has been studied in [33]. The flowchart of the whole procedure is shown in **Figure 8**.

4. Application of the proposed method

4.1 Modeling of the flyback converter

An adequate equivalent radiation model has been identified by applying the proposed EMTR method to the measured emissions of the AC-DC converter

N	Max moment (A × m^2) ×e − 7	Coordinates (mm)	Orientations (rad)	
# 1	0.1545	10.9; 11.6; − 0.67	0	1, 5708
# 2	0.2477	5 e − 3; 2 e − 2; 0.0	0	0
# 3	0.0708	−2.7; − 11.6; − 0.7	0	−1, 5708
# 4	0.0163	5.4; − 15.4; − 1.2	1, 5708	1, 5708
# 5	0.0900	13.6; − 3.8; − 0.37	1.5708	0
# 6	0.0179	−5.4; 3.9; 0.0	0	0
# 7	0.3468	13.6; − 7.7; − 0.61	0	1, 5708

Table 3.
The obtained equivalent model parameters.

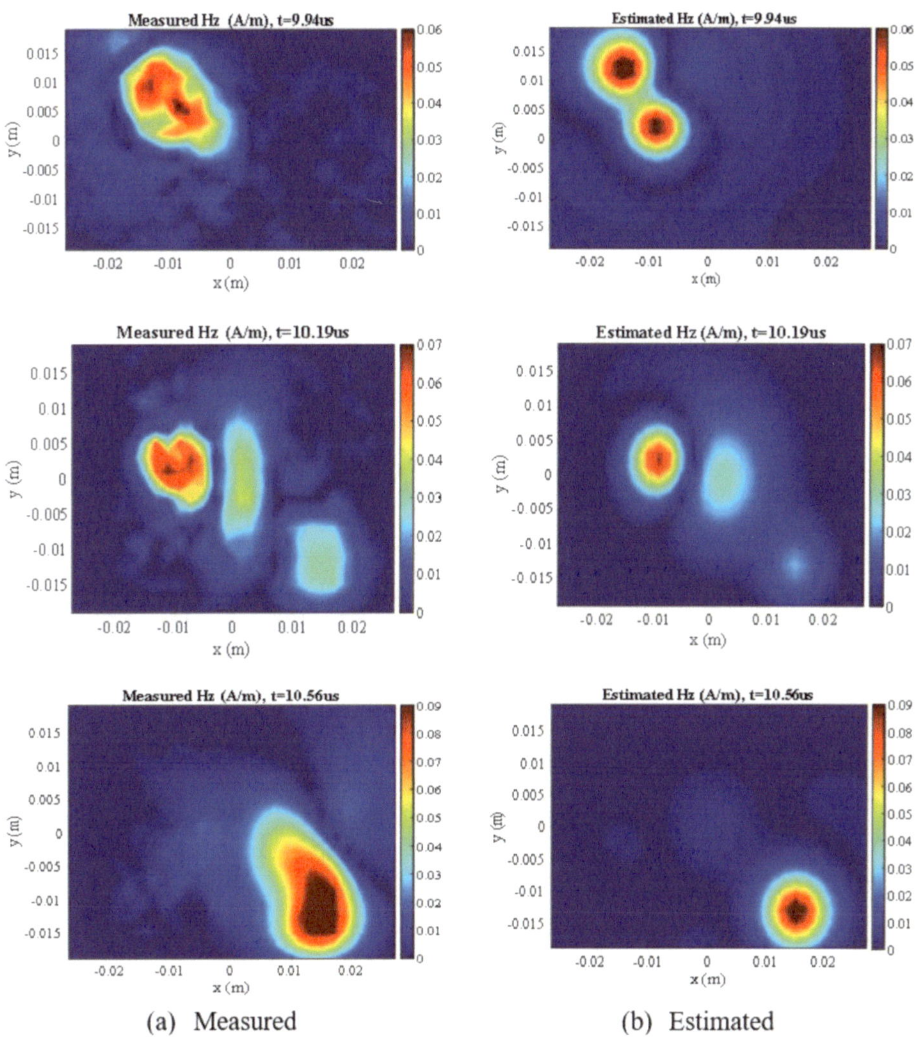

Figure 9.
Magnetic nearfield maps of z-component in TD: (a) measured. (b) Estimated.

obtained, as previously explained. Indeed, by comparing measured and estimated magnetic nearfield distribution H_z, using identified sources parameters given in **Table 3**, at a constant height of measurement and scan area, we can notice a good

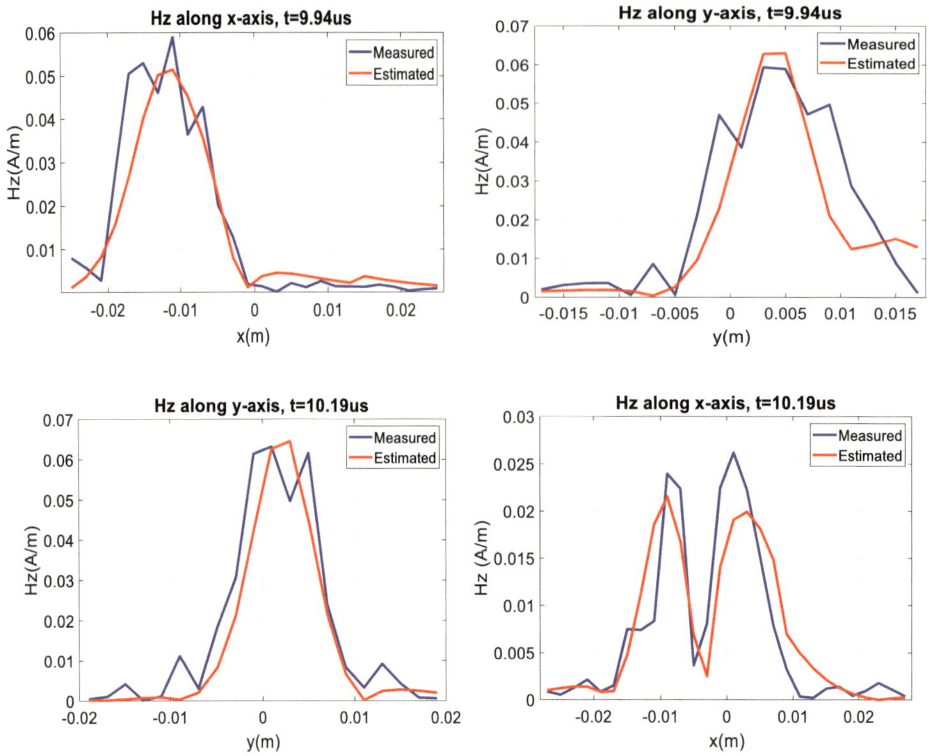

Figure 10.
Cuts of measured and simulated H_z *at* $Z = 0$.

agreement between the two sets of results along time, as shown in **Figure 9**. Particularly, the obtained equivalent sources perform valid waveforms with less noise as if they have been captured when scanning the circuit without measurement errors (which is inevitable in practice).

In **Figure 10**, estimated radiated field cuts at $Z = 0$ are compared to the reference signals along the x- and y-axis. Moreover, besides H_z reconstruction, the prediction of H_x and H_y is easy by including the obtained sources parameters in the analytical expressions of the magnetic field, as in (8).

It is worth noting that in order to the enhance the obtained results, it would be great to refine the mesh of the scanning area further and reduce even more the error limit (corresponding to the measurement errors including the test stand inaccuracies, spatial probe motion, and the coupling effects between the probe and the DUT, etc.). Nevertheless, both a large memory and a considerable computing time are needed.

4.2 Validation of the obtained model

In this study, we propose to confirm the efficiency of the obtained results using the EMTR-based method in TD with a standard inverse method performed in the frequency domain. Thus, we do not make measurements using an FD test bench, but we instead proceed by applying a fast Fourier transform to the measured magnetic field signals in TD. Indeed, in **Figure 11**, we notice the existence of several harmonics covering tens of megahertz in the frequency spectrum of the obtained FD magnetic field at different probe positions. This is among the major

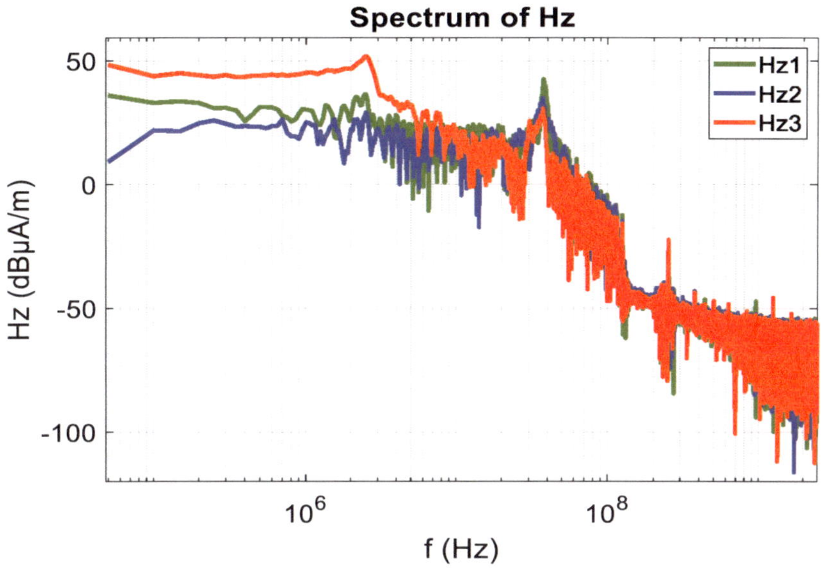

Figure 11.
The spectrum of the measured magnetic field above DUT at different positions.

disadvantages of FD measurements, especially in multisource structures with different characteristics and radiation patterns. We have implemented an optimization method based on genetic algorithms to apply the frequency inverse method and obtain an equivalent model for each studied frequency [8, 9]. In fact, the iterative process needed in this procedure can be reduced when choosing only significant frequencies; otherwise, it is time and memory-consuming. Moreover, to guarantee and accelerate the convergence of the method, a fine configuration has to be employed, particularly having a proper choice of the variable boundaries and population size. In **Figure 12**, mappings of the measured and estimated radiated field are depicted for different frequencies.

A good agreement is achieved between FD maps of measured fields and estimated using an optimization algorithm at different frequencies. Similar source positions have been obtained, which confirms the previous results. Looking at **Figures 9** and **12**, we notice that the proposed method based on EMTR matches the measurements better. Indeed, TD investigations provide rapid results, making modeling easier to use in critical test cases. It is worth noting that FD maps are only depicted at a single frequency for each measurement, and then, we cannot observe the behavior of all existing radiating sources without performing several measurement tests. However, a single temporal measurement is required for a TD method, as we have explained.

For further validation purposes, we propose to reconstruct the x-component of the EM field (H_x) using the obtained equivalent model and then compare the generated maps to the measured radiation distributions acquired with the magnetic probe dedicated for tangential components capturing. **Figure 13** shows an example of mappings at a defined time step. A good agreement is noticed between measured and estimated H_x. Hence, we conclude that the obtained equivalent magnetic dipoles are in good positions and that the proposed method is efficient in providing an equivalent method reproducing a three-dimensional reconstruction of the radiated field describing the overall behavior of the studied converter.

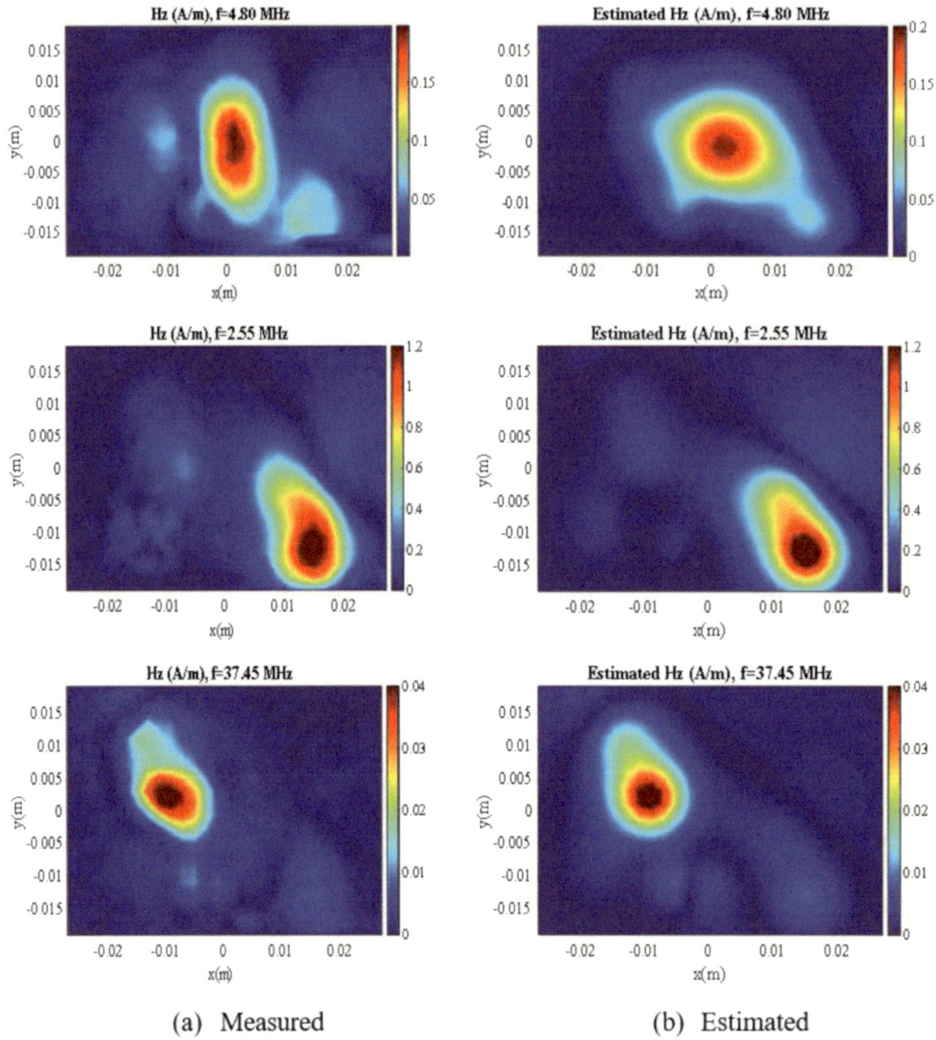

Figure 12.
Magnetic NF maps of z-component in FD: (a) measured. (b) Estimated.

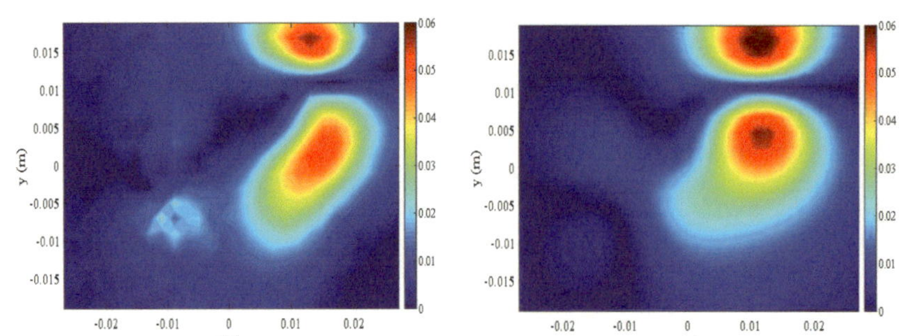

Figure 13.
Measured and estimated magnetic NF maps of x-component at $t = 10.50\ \mu s$.

5. Conclusion

The motivation of the presented work was the characterization of electromagnetic radiating sources commonly found in power converter circuits. Time-domain measurement of the radiated field has been carried out in the nearfield using a specific test bench, including high-precision tools. An electromagnetic inverse method based on the EMTR technique has been studied to obtain an efficient equivalent model of the device under test. In case of transient disturbances that occur, for instance, during the turn-on of the transistor of a flyback AC/DC converter, a satisfactory agreement has been achieved between measured and estimated radiation maps, along with a reasonable computation time and memory usage. By comparing these results to a standard frequency domain inverse method based on the genetic algorithm, the proposed TD method has been shown to be advantageous over the FD method for EMC studies and to deal with related EMI issues in advanced electronic systems such as in power electronics and those developed for avionics and wireless communication. Furthermore, we conclude that the EMTR-based method is a good alternative for studying electromagnetic radiation behavior with a three-dimensional reconstruction in the case of multisource structures.

Author details

Sassia Hedia[1,2,3*], Bessem Zitouna[1], Jaleleddine Ben Hadj Slama[1] and Lionel Pichon[2,3]

1 LATIS, Laboratory of Advanced Technology and Intelligent Systems, ENISo, National Engineering School of Sousse, University of Sousse, Tunisia

2 Group of Electrical Engineering-Paris, CNRS, CentraleSupélec, Université Paris-Saclay, Gif-sur-Yvette, France

3 Group of Electrical Engineering-Paris, CNRS, Sorbonne Université, CNRS, Paris, France

*Address all correspondence to: hdia.sassia@gmail.com

References

[1] Clayton RP. *Introduction to Electromagnetic Compatibility*. Vol. 184. Hoboken, NJ, USA: Wiley, 2006

[2] Beghou L, Costa F, Pichon L. Detection of electromagnetic radiations sources at the switching time scale using an inverse problem-based resolution method—application to power electronic circuits. IEEE Transactions on Electromagnetic Compatibility. 2015; **57**(1):52-60

[3] Vives-Gilabert Y, Arcambal C, Louis A, de Daran F, Eudeline P, Mazari B. Modeling magnetic radiations of electronic circuits using near-field scanning method. IEEE Transactions on Electromagnetic Compatibility. 2007; **49**(2):391-400

[4] Essakhi B et al. Characterization of radiated emissions from power electronic devices: synthesis of an equivalent model from nearfield measurement. The European Physical Journal-Applied Physics. 2007;**38**(3): 275-281

[5] Shall H, Riah Z, Kadi M. A 3-D near-field modeling approach for electromagnetic interference prediction. IEEE Transactions on Electromagnetic Compatibility. 2014; **56**(1):102-112

[6] Benyoubi F, Pichon L, Bensetti M, Le Bihan Y, Feliachi M. An efficient method for modeling the magnetic field emissions of power electronic equipment from magnetic near field measurements. IEEE Transactions on Electromagnetic Compatibility. 2017; **59**(2):609-617

[7] Saidi S, Ben Hadj Slama J. A near-field technique based on PZMI, GA, and ANN: application to power electronics systems. IEEE Transactions on Electromagnetic Compatibility. 2014; **56**(4):784-791

[8] Sivaraman N. *Design of Magnetic Probes for Near Field Measurements and the Development of Algorithms for the Prediction of EMC*. Diss: Université Grenoble Alpes; 2017

[9] Ravelo B, Yang L. Computation of transient nearfield radiated by electronic devices from frequency data. In: Fourier Transform Applications. Rijeka: InTech; 2012

[10] Zitouna B, Ben Hadj Slama J. Enhancement of time-domain electromagnetic inverse method for modeling circuits radiations. IEEE Transactions on Electromagnetic Compatibility. 2016;**58**(2):534-542

[11] Liu Y, Ravelo B. Fully time-domain scanning of EM nearfield radiated by RF circuits. Progress In Electromagnetics Research. 2014;**57**:21-46

[12] Liu Y, Ravelo B, Jastrzebski AK. Time-domain magnetic dipole model of PCB nearfield emission. IEEE Transactions on Electromagnetic Compatibility. 2016;**58**(5):1561-1569

[13] Ravelo B, Liu Y, Jastrzebski AK. PCB near-field transient emission time-domain model. IEEE Transactions on Electromagnetic Compatibility. 2015; **57**(6):1320-1328

[14] Rachidi F. Rubinstein M, Paolone M. Electromagnetic Time Reversal: Application to Electromagnetic Compatibility and Power Systems. Hoboken, NJ, USA: Wiley; 2017

[15] El Baba I, Lalléchère S, Bonnet P. Electromagnetic time-reversal for reverberation chamber applications using FDTD, Proceedings of ACTEA 2009, International Conference on Advances in Computational Tools for Engineering Applications, Lebanon; 2009. pp. 157-167, ISBN: 978-1-4244-3833-4

[16] Fink M. Time reversal of ultrasonic fields. I. Basic principles. IEEE Transactions on Ultrasonics, Ferroelectrics, and Frequency Control. Sept. 1992;**39**(5):555-566

[17] Minonzio J, Davy M, de Rosny J, Prada C, Fink M. Theory of the time-reversal operator for a dielectric cylinder using separate transmit and receive arrays. IEEE Transactions on Antennas and Propagation. 2009;**57**(8):2331-2340

[18] Conti SG, Roux P, Kuperman WA. Nearfield time-reversal amplification. The Journal of the Acoustical Society of America. 2007;**121**(6):3602-3606

[19] Wang Z et al. A full-scale experimental validation of electromagnetic time reversal applied to locate disturbances in overhead power distribution lines. IEEE Transactions on Electromagnetic Compatibility. 2018; **60**(5):1562-1570

[20] Spirlet M, Broun V, Camus P, Geuzaine C, Beauvois V, Molenberg I. Modelling time reversal applications in a reverberation chamber using the current image method. In: 2013 International Symposium on Electromagnetic Compatibility, Brugge. 2013. pp. 359-364

[21] Kosmas P, Rappaport CM. FDTD-based time reversal for microwave breast cancer detection-localization in three dimensions. IEEE Transactions on Microwave Theory and Techniques. 2006;**54**(4):1921-1927

[22] Hedia S, Zitouna B, Ben Hadj Slama J, Pichon L. Electromagnetic time reversal for radiating source identification in time domain. In: *2018 15th International Multi-Conference on Systems, Signals & Devices (SSD),* Yassmine Hammamet, Tunisia. 2018. pp. 531-536

[23] Hedia S, Zitouna B, Ben Hadj Slama J, Pichon L. A full time domain methodology based on near field time reversal for equivalent source identification. In: 2018 IEEE International Symposium on Electromagnetic Compatibility and 2018 IEEE Asia-Pacific Symposium on Electromagnetic Compatibility (EMC/APEMC), Singapore. 2018. pp. 141-146

[24] El Baba I, Lalléchère S, Bonnet P. Time reversal for electromagnetism: applications in electromagnetic compatibility. Trends in Electromagnetism: From Fundamentals to Applications. 2012;**177**:177-206

[25] Ungureanu A, Vuong TP, Ndagijimana F. Electromagnetic point source reconstruction by reversed-TLM method. Applied Computational Electromagnetics Society Journal. 2011; **26**(9):754-759

[26] Chabalko MJ, Sample AP. Electromagnetic time reversal focusing of near field waves in metamaterials. Applied Physics Letters. 2016;**109**(26): 263901

[27] De Rosny J, Lerosey G, Fink M. Theory of electromagnetic time-reversal mirrors. IEEE Transactions on Antennas and Propagation. Oct. 2010;**58**(10): 3139-3149

[28] De Rosny J, Fink M. Focusing properties of nearfield time reversal. Physical Review A. 2007;**76**(6):065801

[29] Benyoubi F, Feliachi M, Le Bihan Y, Bensetti M, Pichon L. Implementation of tools for electromagnetic compatibility studies in the near field. In: *International Conference on Electrical Sciences and Technologies in Maghreb (CISTEM)*, Marrakech. 2016. pp. 1-6

[30] Zitouna B, Ben Hadj Slama J. Time domain inverse method based on the near field technique to solve electromagnetic interference problems: Application to an AC/DC flyback

converter. IET Power Electronics. 2018; **11**(13):2133-2139

[31] Jackson JD. Classical Electrodynamics. Hoboken, New Jersey, USA: John Wiley & Sons; 1999. pp. 841-842

[32] Balanis CA. Antenna Theory: Analysis & Design. Hoboken, NJ, USA: Wiley, 1997

[33] Hedia S, Zitouna B, Ben Hadj Slama J, Pichon L. Characterization of radiating sources in the NEAR field using EMTR technique: A parametric study. In: 2020 6th IEEE International Energy Conference (ENERGYCon). 2020. pp. 104-109

Section 3

Absorbers

Chapter 8

Analysis and Design of Absorbers for Electromagnetic Compatibility Applications

Shiva Hayati Raad

Abstract

Absorbers are one of the key components in the realm of electromagnetic compatibility. Depending on the frequency range of interest, different types of absorbers can be utilized for this purpose. This chapter introduces the analysis and modeling of ferrite-based absorbers for low-frequency applications (below 1 GHz) and discusses the issues encountered in their installation, resulting in air gaps. Later, different kinds of pyramidal absorbers, commonly used in the broadband microwave frequency range (above 1 GHz), are presented, and analytical and numerical approaches for predicting their performance are reviewed. The combination of the ferrite tile and pyramidal dielectric absorbers is also provided. Then, some practical aspects of designing hybrid absorbers, including the influence of carbon loading and matching layer on their performance, are mentioned. Finally, the absorber operating frequency extension to the millimeter-wave spectrum using metamaterial structures or graphene material is presented.

Keywords: ferrite absorber, dielectric absorber, hybrid absorber, matching layer, air gaps, absorber modeling, millimeter-wave

1. Introduction

The electromagnetic compatibility of the electronic devices is mainly considered in two ways, measuring radiated emission (RE) or radiated immunity (RI), where the test procedures are specified in different standards [1, 2]. The frequency spectrum of RE/RI tests starts from tens of kHz. Since the absorber technology cannot cover such low frequencies, the concentration of the absorber design goes around 70 MHz test frequencies [3]. Anechoic chambers (AC) in the form of fully or partially covered rooms with electromagnetic absorbers that simulate the open area test site (OATS) are the most common indoor facilities for the electromagnetic compatibility (EMC) tests, where the quality of the installed absorbers influences the precision of the tests [4]. Moreover, there are other types of shielded enclosures such as compact renege rooms, transverse electromagnetic (TEM) cells, and gigahertz transverse electromagnetic (GTEM) cells. Properly lining these test facilities with absorbing materials is a key factor for their expected operation [5–8].

When an electromagnetic wave illuminates an environment, it is reflected, transmitted, or absorbed. An efficient absorber can be realized by minimizing the contributions of the former two components [9]. For covering the test frequencies in different applications, broadband absorbers are necessary. Various approaches are proposed for bandwidth enhancement of the electromagnetic absorbers, multi-resonance and multilayered structures being two important groups, realized by merging multiple closely spaced resonances [10, 11]. Broadband absorbers in the EMC field are commonly attained by tapered geometries, such as wedge or pyramidal configurations, or by parameter gradient flat structures to provide gradual impedance matching [12]. In all cases, the metallic backside prevents wave transmission [5]. Unit cell analysis, transmission-line model, homogenization method, finite difference time domain (FDTD) technique, and mode-matching technique are some of the approaches that have been used for the performance prediction of the absorbers [13–18]. In this chapter, the analysis and design of some important types of EMC absorbers are reviewed. The chapter also includes millimeter-wave absorbers for future EMC applications.

2. Analysis and design of absorbers for the electromagnetic compatibility applications

Different types of absorbing materials and geometries are used for EMC applications depending on the operating frequency range. Below 1 GHz, ferrite-based materials having dispersive lossy permeability are used in different planar configurations. Moreover, above 1 GHz, electric losses or magnetic losses provided by lossy dielectric or lossy magnetic materials, respectively, can be used. To cover both of the above ranges, hybrid absorbers constructed by combining the ferrite tiles and pyramidal absorbers are a solution. In the higher microwave/THz frequencies, metamaterial absorbers can be used. In the following subsections, a detailed discussion of each category is presented. Note that the required absorber's reflectivity is dedicated by the frequency range and its application. Specifically, the military standard requires −6 dB normal incidence reflectivity for 50–250 MHz frequencies and −10 dB normal incidence reflectivity above 250 MHz. Moreover, for immunity tests, −18 dB normal incidence reflectivity for 80–1000 MHz spectrum is essential. For emission tests in 3-meter chambers, −18 dB normal incidence reflectivity for 30–1000 MHz region and −12 dB at 45° for 30–1000 MHz spectrum fulfill the requirements, while in 10-meter chambers, −20 dB normal incidence reflectivity for 30–1000 MHz spectrum and −15 dB at 45° for 30–1000 MHz range are required [19]. Typical reflectivity of absorbers in different frequency bands can be found in the data sheets provided by the manufacturers, from which an appropriate absorber model can be specified. **Table 1** is given as an example [20]. Finally, depending on the operating frequency range, NRI arch, waveguide, coaxial line, time-domain method, and free-space approach can be used for the performance evaluation of the absorbers [17, 21]. The test procedures are specified in IEEE Standard 1128 [22]. The difference between measured reflection coefficient (S_{11}) with and without the absorber is the reflectivity in the former method. In the latter approach, reflectivity is attained by comparing the received power with that using the perfect electric conductor (PEC) plate [20, 23].

2.1 Ferrite absorbers

The simplest type of ferrite absorbers is ferrite tiles, where the whole desired surface can be covered by installing multiple tiles next to each other. This type of absorber is shown in **Figure 1(a)**, backed by a PEC layer. Ferrite tile ceramics are realized by heating powders under pressure at 1000–1500°C. Thus, brittle and thin tiles between 4 and 7 mm are commonly used because of the heavyweight of iron,

Model number	80 MHz	120 MHz	200 MHz	300 MHz	500 MHz	L-BAND 1–2 GHz	S-BAND 2–4 GHz	C-BAND 4–8 GHz	X-BAND 8–12 GHz	KU-BAND 12–18 GHz	K-BAND 18–40 GHz
EHP-3PCL								−30 dB	−40 dB	−45 dB	−45 dB
EHP-5PCL							−30 dB	−40 dB	−45 dB	−50 dB	−50 dB
EHP-8PCL						−30 dB	−40 dB	−45 dB	−50 dB	−50 dB	−50 dB
EHP-12PCL						−35 dB	−40 dB	−45 dB	−50 dB	−50 dB	−50 dB
EHP-18PCL					−30 dB	−40 dB	−45 dB	−50 dB	−50 dB	−50 dB	−50 dB
EHP-24PCL			−20 dB	−30 dB	−35 dB	−40 dB	−50 dB	−50 dB	−50 dB	−50 dB	−50 dB
EHP-36PCL	−11 dB	−13 dB	−25 dB	−30 dB	−40 dB	−45 dB	−50 dB	−50 dB	−50 dB	−50 dB	−50 dB
EHP-48PCL	−15 dB	−20 dB	−30 dB	−35 dB	−40 dB	−45 dB	−50 dB	−50 dB	−50 dB	−50 dB	−50 dB
EHP-72PCL	−20 dB	−30 dB	−40 dB	−40 dB	−45 dB	−50 dB	−50 dB	−50 dB	−50 dB	−50 dB	−50 dB
EMC-24PCL	−6 dB	−6 dB	−7 dB	−30 dB	−35 dB	−45 dB	−50 dB	−50 dB	−50 dB	−50 dB	−45 dB

Table used with permission of ETS-Lindgren.

Table 1.
Typical reflectivity of a specific absorber series in different frequency bands for normal incidence [20].

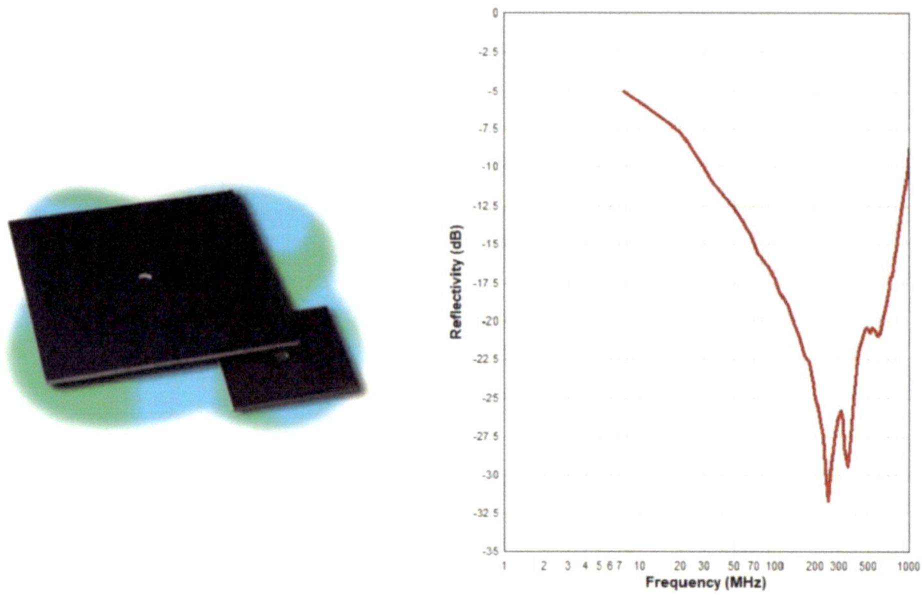

(a) (b)

Figure 1.
(a) Ferrite tile absorber [24] and (b) its typical reflectivity [25] (The photos used with permission of ETS-Lindgren).

manganese, and other metal oxides [1]. The typical surface area of the tiles is 100 mm^2, and they are usually installed onto plywood panels leading to an extremely fire-retardant installation [5]. Due to the low operation frequency, the homogenization of the entire structure and later using the transmission line method gives reasonable results for the associated absorption rate. Thus, the surface impedance of the shortened ferrite slab is [14]:

$$\eta_t = \eta_0 \sqrt{\frac{\mu_r}{\varepsilon_r}} \tanh\left(\frac{2\pi t}{\lambda_0}\sqrt{\mu_r \varepsilon_r}\right) \tag{1}$$

where t is the thickness of the tile, η_0 is the free-space intrinsic wave impedance, $\varepsilon_r = 14$ is the relative permittivity of the tile, and μ_r is the relative permeability calculated by the Drude model, which reads as [26]:

$$\mu_r = 1 + \frac{1101.4}{1 + j\frac{f}{7.066 \times 10^6}} \tag{2}$$

where f is the operating frequency. Using the aforesaid surface impedance, the reflectivity of the ferrite tile absorbers can be calculated readily. **Figure 1(b)** shows the typical reflectivity of a ferrite tile.

An inherent problem of installing the ferrite tiles is the small-sized air gaps between the adjacent cells. The air gaps result in performance degradation due to the significant difference between the permeability of the air gaps and ferrite material. The impact of air gaps on the performance of the ferrite tile absorbers can be approximated *via* their similarity to the air gaps in the core of power transformers to obtain the equivalent permeability as [14]:

$$\mu_r^e = \frac{\mu_r}{1 + \frac{\Delta}{l}(\mu_r - \mu_{air})} \tag{3}$$

Maxwell Garnett formula	$\mu_{eff} = \mu_b \left[1 + \frac{2g(\mu_i - \mu_b)}{(1-g)\mu_i + (1+g)\mu_b}\right]$
Bruggeman formula	$(1-g)\frac{\mu_b - \mu_{eff}}{\mu_b + \mu_{eff}} + g\frac{\mu_i - \mu_{eff}}{\mu_i + \mu_{eff}} = 0$
Liu formula	$\mu_{eff} = \frac{\mu_i}{1 + (1-\sqrt{g})\left(\frac{\mu_i}{\mu_b} - 1\right)}$
Hashin–Shtrikman bounds	$\mu_b \frac{(1-g)\mu_b + (1+g)\mu_i}{(1+g)\mu_b + (1-g)\mu_i} \le \mu_{eff} \le \frac{(2-g)\mu_b + g\mu_i}{g\mu_b + (2-g)\mu_i}\mu_i$

Table 2.
Different homogenization approaches for modeling ferrite tiles with air gaps [29].

where Δ is the size of the air gap perpendicular to the magnetic field and l is the width of the ferrite tile. Some gap-filling materials, such as ferrite rubber or developed plastic material called REC 65, are proposed to obviate this issue [27]. Moreover, ferrite absorbers are fabricated in other geometries such as grids, where they have an additional degree of freedom to adjust the resonant frequency through the filling factor. Therefore, synthetized capacitance and synthetized inductance models can be used to determine the effective permeability of the ferrite grids and ferrite grids with air gaps [28]. Furthermore, due to the lower amount of utilized ferrite material in the fabrication, the performance sensitivity to the air gaps is lower in the ferrite grids with respect to the ferrite tiles. Different homogenization approaches for modeling ferrite tiles with air gaps are summarized in **Table 2**. Note that these approximate mixing rules do not always work well, especially for hollow and wedge shapes (or other complicated shapes). A full numerical model using a unit cell/periodic boundary (Floquet ports) and full-wave modeling is fast and easy. The full-wave model can account for air gaps and complicated shapes [29]. As a final comment, due to ferrite tile's heavyweight and expensive cost, its minimum usage, which is predicted by the optimization tools, is preferred [30–32].

2.2 Pyramidal absorbers

Different types of dielectric absorbers having lossy permittivity (electric loss) or permeability (magnetic loss) are commonly used for the wideband electromagnetic wave absorption above 1 GHz [5]. Organic materials (such as rice straw, oil palm empty fruit bunch, sugar cane bagasse, and coconut shell [33]), magnetic materials (such as NiZn spinel ferrite, Co_2Z hexaferrite, and RuCoM hexaferrite [34]), and dielectric materials (such as foam (carbon-based), polyurethane, polystyrene, polyethylene, and thin film [30]) can be used for this purpose. The absorption capability of each material can be evaluated using its constitutive parameters. To this end, the attenuation constant of the incoming wave is calculated to guarantee that the wave can penetrate the device. Therefore [2],

$$\alpha = \omega\sqrt{\varepsilon_0\mu_0}\left(a^2 + b^2\right)^{1/4} \sin\left(\frac{1}{2}\tan^{-1}\left(\frac{a}{b}\right)\right) \quad (4)$$

where prime and double prime, respectively, denote the real and imaginary parts of the permittivity (ε) and permeability (μ). Also, the parameters a and b in Eq. (4) are defined as follows:

$$a = \left(\varepsilon_r'\mu_r' - \varepsilon_r''\mu_r''\right) \quad (5)$$

$$b = \left(\varepsilon_r'\mu_r'' + \varepsilon_r''\mu_r'\right) \quad (6)$$

Note that the real and imaginary parts of the constitutive parameters cannot change arbitrarily, and the Kramers–Kronig relationship should be satisfied [2]. Above 1 GHz, absorbers are fabricated in the form of pyramidal (standard, twisted, or hollow), wedge, convolved, or multilayered geometries, the pyramidal shape being the most common one [5]. Pyramidal absorbers are fabricated in different lengths, depending on the lower limit of the operating band. Specifically, with the optimized carbon loading, the one and eight wavelength absorber's reflective is around −33 dB and −51 dB [5]. Also, for mechanical reasons, the base-to-height ratio is about 2.5 [1]. The product is commonly black after fabrication and is painted with blue latex to improve light reflectance. The tips are retained unpainted to prevent absorption degradation in the millimeter-wave band. The measured data do not seem to support such a common practice, and the whole absorbers should be unpainted to achieve improvements at the millimeter band [35]. Moreover, pressure-sensitive adhesives can be used for the installation, and fire-retardant chemical loading is used to meet the current fire retardancy requirements [5].

At the low-frequency limit, when the period of the array is small compared to the wavelength, homogenization of the transversely periodic structure in **Figure 2** with a transversely uniform medium with anisotropic permittivity and permeability can be used for the analysis. Thus, the longitudinal and transverse components of the constitutive parameters can be approximated as [36]:

$$\varepsilon_z = (1-g)\varepsilon_0 + g\varepsilon_a \tag{7}$$

$$\mu_z = (1-g)\mu_0 + g\mu_a \tag{8}$$

$$\varepsilon_t = \varepsilon_0\left[1 + g\frac{2(\varepsilon_a - \varepsilon_0)}{(1+g)\varepsilon_0 + (1-g)\varepsilon_a}\right] \tag{9}$$

$$\mu_t = \mu_0\left[1 + g\frac{2(\mu_a - \mu_0)}{(1+g)\mu_0 + (1-g)\mu_a}\right] \tag{10}$$

where the subscripts t and a respectively indicate the transverse component and parameters referring to the absorbing material. Also, the parameter g is defined based on the pyramid length L and the distance of the layer from the tip z as $g = (z/L)^2$. The accuracy of the aforesaid equations for practical materials is around 5% [36]. In another low-frequency approach, the pyramid is divided into multiple layers

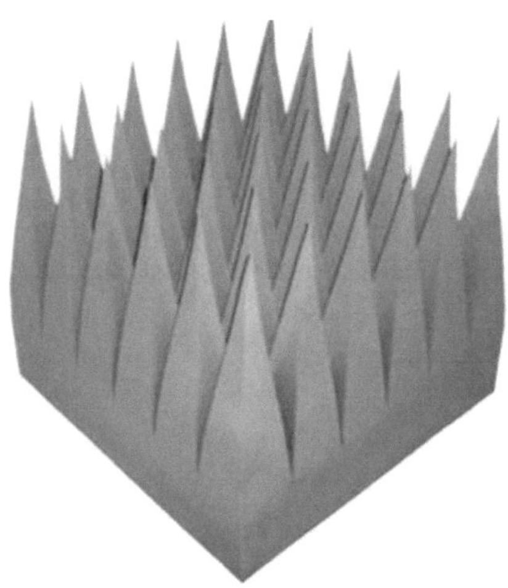

Figure 2.
The schematic view of the pyramidal absorber [24] (The photo used with permission of ETS-Lindgren).

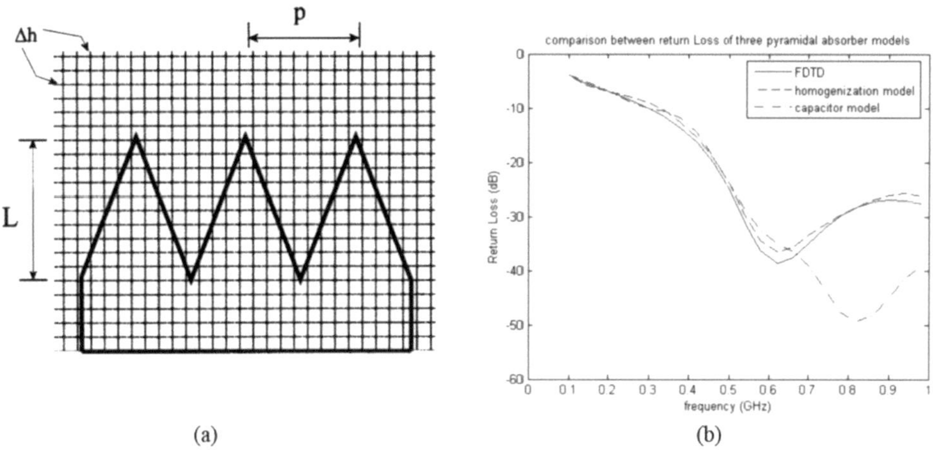

Figure 3.
(a) Spatial meshing of the pyramidal absorber for its analysis based on FDTD and (b) its return loss in comparison with the homogenization model and capacitor model ([16], reproduced courtesy of The Electromagnetics Academy).

with label i, and then, each layer is approximated with the permittivity and permeability of [37]:

$$\varepsilon_{ieq} = (1 - K_i) + \frac{K_i \varepsilon_a}{K_i + (1 - K_i)\varepsilon_a} \tag{11}$$

$$\mu_{ieq} = (1 - K_i) + \frac{K_i \mu_a}{K_i + (1 - K_i)\mu_a} \tag{12}$$

where the parameter K_i represents the normalized area of the absorber in the ith layer. The transmission-line model can be used to calculate the multiple reflection and transmissions coefficients by starting from the metal-backed bottom layer. Note that surface fraction models have been used in another homogenization approach, called the synthesized capacitance model [38].

Apart from low-frequency approximate techniques, various full-wave numerical and quasi-analytical methods such as the finite element method (FEM), the integral-equations-based method of moments (MoM), and the FDTD technique can be used for the pyramidal absorber analysis. In this regard, time-domain methods are more reasonable choices due to the wideband nature of the problem [39]. Thus, for accurate wideband modeling of pyramidal absorbers, the FDTD method can be used. The required spatial meshing for this approach is illustrated in **Figure 3(a)**, and its return loss in comparison with the homogenization model and capacitor model is included in **Figure 3(b)**, where good agreement with the former approach is attained [16].

Importantly, the bi-static or off-angle behavior of the absorbers is an important factor affecting the validity of the indoor facilities for RE/RI measurements. Therefore, polynomial approximations are proposed to predict reflected energy from pyramidal RF absorbers in different incident angles [1, 40, 41].

2.3 Hybrid absorbers

To design an absorber covering the sub-GHz and GHz frequencies, hybrid absorbers, simultaneously exploiting both technologies mentioned above, are proposed. Hybrid absorber design is not just putting the pyramidal absorbers above the

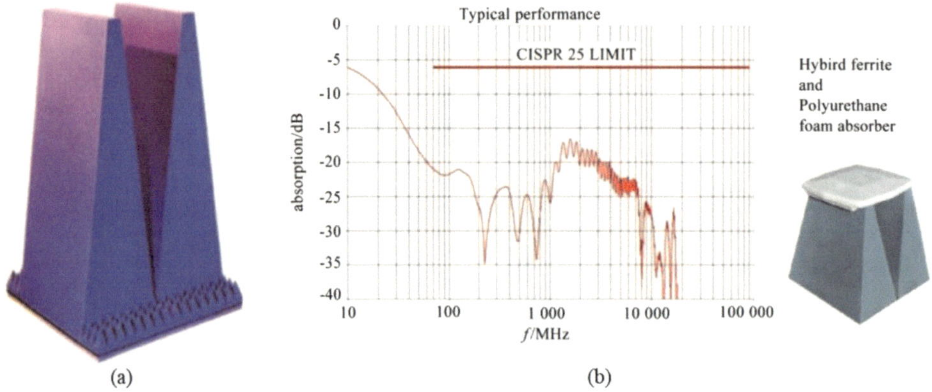

Figure 4.
(a) Hybrid absorber designed by controlling the amount of carbon loading and using pyramid's tip truncation [24] and (b) typical reflectivity of a hybrid absorber at low-frequencies [3] (The photos used with permission of ETS-Lindgren).

ferrite tiles [42]. Adjusting the carbon loading of the dielectric pyramidal absorbers, making the pyramidal absorbers hollow, or using shaped coatings of lossy paint are some solutions to improving the impedance matching of hybrid absorbers. Also, the absorption degradation due to a lower amount of carbon material is compensated by truncating the pyramid's tip, and some small-sized elements are included to improve the high-frequency performance, as in **Figure 4** [24]. Moreover, using a matching layer between the pyramidal and ferrite absorbers is another method. Fourier analysis and the mode-matching technique have been used to analyze this absorber efficiently [17]. Note that when several thousands of hybrid absorbers with complex material parameters are used in the facility design, a thin-surface impedance sheet with the same reflectivity can be used to provide time and memory efficiency instead of simulating the hybrid absorber. For this purpose, the reflectivity of the absorber must be approximated by optimizing the material parameters and thicknesses in the equivalent multilayered model [43, 44]. Modeling the absorber arrays as multilayered media through homogenization aids in analyzing the chambers using Green's function of the layered media [45].

2.4 Millimeter-wave absorbers

With the development of fifth-generation (5G) communication technology utilizing massive multiple-input multiple-output (MIMO) phased array antennas for high-speed communication, the use of millimeter-wave absorbers (30–300 GHz) with high-power handling capability is necessary for test facilities [46–48]. Convoluted (egg-carton) absorbers with the typical absorption of 50 dB at normal incidence at 30 GHz are primarily useful in millimeter-wave bands [1, 2]. The manufacturers also suggest convoluted absorber in this frequency spectrum because of tolerating wide incident angles and suggest ordering them without paint [49]. Moreover, pyramidal structures with vent holes and heat sinks are proposed, and associated power-handling capability is analyzed using multi-physics simulations [50]. Also, a millimeter-wave metamaterial absorber bandwidth is widened by embedding one of the resonators inside another in the unit cell. The device is fabricated by a standard optical photolithography process, and the measurement is done by a custom-made spectrometer, as shown in **Figure 5** [51]. Furthermore, the optical transparency requirement for some special EM protections, demanding large

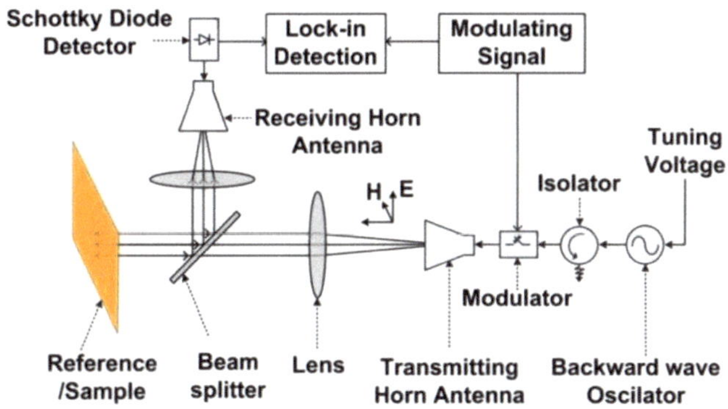

Figure 5.
Measurement setup of the millimeter-wave absorber ([51], reproduced courtesy of The Electromagnetics Academy).

light transmittance and high absorptivity, can also be satisfied [52–54]. As a final note, naturally thin and transparent graphene material has great potential in the millimeter-wave absorber design [55].

3. Conclusions

Depending on the frequency range of interest, different absorber types have been commonly used for EMC applications. For the sub-1 GHz range, the magnetic losses of unbiased ferrite tiles or grids are the main mechanism for electromagnetic absorption. Moreover, pyramidal absorbers fabricated by organic, electric, or magnetic materials have been exploited frequently in microwave frequencies. Considering the coverage of both spectrums mentioned above, efficiently designed hybrid absorbers, using both technologies simultaneously, have been proposed. They have been achieved by controlling dielectric pyramidal absorbers' carbon loading and tip truncation or by designing a matching layer between the two absorbing sections. Finally, in the millimeter-wave regime, metamaterial absorbers have been proposed for providing the measurement setups of ever-growing high-frequency applications, such as 5G communication. The one-atom-thick graphene material also has great potential for this purpose.

Acknowledgements

The author would like to thank ETS-Lindgren Company for guiding through the use of necessary sources.

Conflict of interest

The authors declare no conflict of interest.

Author details

Shiva Hayati Raad
Department of Electrical and Computer Engineering, Tarbiat Modares University, Tehran, Iran

*Address all correspondence to: shiva.hayati@modares.ac.ir

References

[1] Rodriguez V. Anechoic Range Design for Electromagnetic Measurements. Artech House; 2019

[2] Xu Q, Huang Y. Anechoic and Reverberation Chambers: Theory, Design, and Measurements. United States: Wiley-IEEE Press; 2019

[3] Rodriguez V. Automotive component EMC testing: CISPR 25, ISO 11452–2 and equivalent standards. IEEE Electromagnetic Compatibility Magazine. 2012;**1**(1):83-90

[4] Eser S, Sevgi L. Open-area test site (OATS) calibration. IEEE Antennas and Propagation Magazine. 2010;**52**(3): 204-212

[5] Hemming LH. Electromagnetic Anechoic Chambers: A Fundamental Design and Specification Guide. Wiley Interscience; 2002

[6] Demakov AV, Komnatnov ME. TEM cell for testing low-profile integrated circuits for EMC. In: 21st International Conference of Young Specialists on Micro/Nanotechnologies and Electron Devices (EDM). Chemal, Russia: IEEE; 2020. pp. 154–158

[7] Sittakul V, Hongthong S, Pasakawee S. Design and analysis of GTEM cell using the ferrite-tile and pyramid absorbers for EMC Test in National Institute of Metrology (Thailand). In: IEEE Conference on Antenna Measurements & Applications (CAMA). Chiang Mai, Thailand: IEEE; 2015. pp. 1–4

[8] Sahraei A, Aliakbarian H. On the design and fabrication of a large GTEM cell and its challenges. IEEE Electromagnetic Compatibility Magazine. 2020;**9**(1):43-50

[9] Raad SH, Atlasbaf Z. Broadband optical absorption using graphene-wrapped cross-hair/nano-rod combination. Journal of Electromagnetic Waves and Applications. 2021;**35**(3): 305-314

[10] Raad SH, Atlasbaf Z. Broadband continuous/discrete spectrum optical absorber using graphene-wrapped fractal oligomers. Optics Express. 2020; **28**(12):18049-18058

[11] Raad SH, Atlasbaf Z, Zapata-Rodríguez CJ. Broadband absorption using all-graphene grating-coupled nanoparticles on a reflector. Scientific Reports. 2020;**10**(1):1-15

[12] Dixon P. Theory and application of RF/microwave absorbers. In: York Avenue: Emerson & Cuming Microwave Products, Inc 2012

[13] Orakwue SI, Onu IP. Pyramidal microwave absorber design for anechoic chamber in the microwave frequency range of 1GHz to 10GHz. European Journal of Engineering and Technology Research. 2019;**4**(10):1-3

[14] Liu K. Analysis of the effect of ferrite tile gap on EMC chamber having ferrite absorber walls. In: Proceedings of Symposium on Electromagnetic Compatibility. Santa Clara, CA, USA: IEEE; 1996. pp. 156–161

[15] Zhu Y et al. Development of millimeter-wave em absorber with homogenization theory. SN Applied Sciences. 2019;**1**(5):398

[16] Khajehpour A, Mirtaheri SA. Analysis of pyramid EM wave absorber by FDTD method and comparing with capacitance and homogenization methods. Progress In Electromagnetics Research. 2008;**3**:123-131

[17] Fallahi A, Enayati A. Modeling pyramidal absorbers using the Fourier modal method and the mode matching

technique. IEEE Transactions on Electromagnetic Compatibility. 2016; **58**(3):820-827

[18] Xu Q, Huang Y, Zhu X, Xing L, Duxbury P, Noonan J. A hybrid FEM-GO approach to simulate the NSA in an anechoic chamber. Applied Computational Electromagnetics Society Journal. 2017;**32**(11):1035-1041

[19] Holloway CL, Delyser RR, German RF, McKenna P, Kanda M. Comparison of electromagnetic absorber used in anechoic and semi-anechoic chambers for emissions and immunity testing of digital devices. IEEE Transactions on Electromagnetic Compatibility. 1997;**39**(1):33-47

[20] http://www.ets-lindgren.com/sites/etsauthor/General_Brochures/Top%2010%20Anechoic%20Absorber%20Considerations.pdf

[21] Holloway CL, Kuester EF. A low-frequency model for wedge or pyramid absorber arrays-II: Computed and measured results. IEEE Transactions on Electromagnetic Compatibility. 1994; **36**(4):307-313

[22] Chen Z. IEEE STD 1128-1998: IEEE recommended practice for RF absorber evaluation in the range of 30 MHz to 5 GHz. In: IEEE Symposium on Electromagnetic Compatibility, Signal Integrity and Power Integrity (EMC, SI & PI), long beach CA: IEEE; 2018. pp. 1–16

[23] Tyagi SK. Development of semi-anechoic shielded chamber using ferrite-PU foam hybrids. IETE Technical Review. Taylor and francis 2006;**23**(2):105-111

[24] Wiles M, Rodriguez V. Choosing the Right Chamber for Your Test Requirements. Interference Technology USA: 2010. EMC Directory Design Guide

[25] ETS-Lindgren. Texas; http://ets-lindgren.com/datasheet/absorber/emc-absorbers/ferrite-tile-absorber/1002/100201

[26] Holloway CL, McKenna P, Johnk RT. The effects of gaps in ferrite tiles on both absorber and chamber performance. In: IEEE International Symposium on Electromagnetic Compatability. Symposium Record (Cat. No. 99CH36261), Seattle, WA, USA: IEEE; Vol. **1**. 1999. pp. 239–244

[27] Vogt A, Kolodziej H, Sowa A. An effective solution to the problem of the ferrite tile gap effect. In: IEEE EMC International Symposium. Symposium Record. International Symposium on Electromagnetic Compatibility (Cat. No. 01CH37161). Vol. **1**. Montreal, QC, Canada: IEEE; 2001. pp. 179–182

[28] Chen Z, Sha F. The simulation and analyzing of ferrite grids used in EMC anechoic chamber. In: IEEE International Symposium on Microwave, Antenna, Propagation and EMC Technologies for Wireless Communications. Beijing, China: IEEE; Vol. **1**. 2005. pp. 666–669

[29] Xiong Z, Chen Z. Homogenization modeling of periodic magnetic composite structures. In: IEEE International Symposium on Electromagnetic Compatibility & Signal/Power Integrity (EMCSI). Washington, DC, USA: IEEE; 2017. pp. 437–441

[30] Akpinar E, Erdogan Y. The important points for determining the requirements of EMC chambers. International Journal of RF and Microwave Computer-Aided Engineering. 2016;**26**(4):367-372

[31] Razavi SMJ, Khalaj-Amirhosseini M, Cheldavi A. Minimum usage of ferrite tiles in anechoic chambers. Progress In Electromagnetics Research. 2010;**19**: 367-383

[32] Clegg J et al. A Method of Reducing the Number of Ferrite Tiles in an Absorber Lined Chamber. EMC York, UK: IEEE; 1999

[33] Nornikman H, Fareq M, Soh PJ, Azremi AAH, Wee FH, Hasnain A. Parametric studies of the pyramidal microwave absorber using rice husk. Available from: http://dspace.unimap.edu.my:80/dspace/handle/123456789/33275, 2010

[34] Park M-J, Kim S-S. Design of wide bandwidth pyramidal microwave absorbers using ferrite composites with broad magnetic loss spectra. Electronic Materials Letters. 2016;**12**(5):610-614

[35] Chen Z. Common RF absorbers evaluations in W-band (75-100 GHz). In: IEEE International Symposium on Electromagnetic Compatibility & Signal/Power Integrity (EMCSI). Washington, D.C.: IEEE; 2017. pp. 1–31

[36] Kuester EF, Holloway CL. A low-frequency model for wedge or pyramid absorber arrays-I: Theory. IEEE Transactions on Electromagnetic Compatibility. 1994;**36**(4):300-306

[37] Park M-J, Choi J, Kim S-S. Wide bandwidth pyramidal absorbers of granular ferrite and carbonyl iron powders. IEEE Transactions on Magnetics. 2000;**36**(5):3272-3274

[38] Anzai H, Saikawa M, Mizumoto T, Naito Y. Analysis of the pyramid electromagnetic wave absorber-an approximated model and its application of TE wave. Electronics and Communications in Japan. 1995; 78(11):33-42

[39] Holloway CL, McKenna PM, Dalke RA, Perala RA, Devor CL. Time-domain modeling, characterization, and measurements of anechoic and semi-anechoic electromagnetic test chambers. IEEE Transactions on Electromagnetic Compatibility. 2002;**44**(1):102-118

[40] Rodriguez V, Barry E. A polynomial approximation for the prediction of reflected energy from pyramidal RF absorbers. In: AMTA Proceedings. San Diego, CA, USA: IEEE; 2016. pp. 1–6

[41] Rodriguez V. Basic rules for anechoic chamber design: Part one: RF absorber approximations. Microwave Journal. 2016;**59**(1):72-86

[42] X. Meng and W. Ying-hong, Impedance matching of compound absorber's impact on characteristic of anechoic chamber, in IEEE-APS Topical Conference on Antennas and Propagation in Wireless Communications (APWC), Cape Town, South Africa: IEEE; 2012. pp. 1278–1280.

[43] Reddy V, Kralicek P. Modeling of semi-anechoic chamber for use in automotive EMC simulations. In: IEEE MTT-S International Microwave and RF Conference (IMaRC). Hyderabad, India: IEEE; 2015. pp. 93–95

[44] Reddy V, Kralicek P, Hansen J. A novel segmentation approach for modeling of radiated emission and immunity test setups. IEEE Transactions on Electromagnetic Compatibility. 2017; **59**(6):1781-1790

[45] Chen Z, Wang Y. Efficient and accurate modeling of EMC anechoic chambers using a combined discrete complex image and raytracing method. In: 15th European Conference on Antennas and Propagation (EuCAP). Dusseldorf, Germany: IEEE; 2021. pp. 1–5

[46] Zhang Y, Zhu Y, Qi Y, Yu W, Li F, Liu L. High-power broadband honeycomb absorber for 5G millimeter wave chambers. In: IEEE International Symposium on Electromagnetic Compatibility and 2018 IEEE Asia-Pacific Symposium on Electromagnetic Compatibility (EMC/APEMC). Suntec City, Singapore: IEEE; 2018. pp. 103–105

[47] Chang K, Sun C. Millimeter-wave power-combining techniques. IEEE Transactions on Microwave Theory and Techniques. Suntec City, Singapore, IEEE; 1983;**31**:91-107

[48] Xiaoyong L, Shuyun H, Mengjun W, Yan L, Erping L. A compact ultra-wideband polarization-insensitive metamaterial absorber at 5G millimeter wave band. In: IEEE MTT-S International Conference on Numerical Electromagnetic and Multiphysics Modeling and Optimization (NEMO). Hangzhou, China: IEEE; 2020. pp. 1–4

[49] https://www.laird.com/products/microwave-absorbers/microwave-absorbing-foams/eccosorb-cv.

[50] Gong P, Liu L, Qi Y, Yu W, Li F, Zhang Y. Multiphysics analysis of millimeter-wave absorber with high-power handling capability. IEEE Transactions on Electromagnetic Compatibility. 2017; **59**(6):1748-1754

[51] Singh P, Kabiri Ameri S, Chao L, Afsar MN, Sonkusale S. Broadband millimeterwave metamaterial absorber based on embedding of dual resonators. Progress In Electromagnetics Research. 2013;**142**:625-638

[52] Xue B, Hu Y, Wu B, Chen L, Zhang W. A wideband transparent absorber for microwave and millimeter wave application. In: Sixth Asia-Pacific Conference on Antennas and Propagation (APCAP). Xi'an, China: IEEE; 2017. pp. 1–3

[53] Yin Z et al. Optically transparent and single-band metamaterial absorber based on indium-tin-oxide. International Journal of RF and Microwave Computer-Aided Engineering. 2019;**29**(2):e21536

[54] Lai S, Wu Y, Wang J, Wu W, Gu W. Optical-transparent flexible broadband absorbers based on the ITO-PET-ITO structure. Optical Materials Express. 2018;**8**(6):1585-1592

[55] Wu B et al. Experimental demonstration of a transparent graphene millimetre wave absorber with 28% fractional bandwidth at 140 GHz. Scientific Reports. 2014;**4**(1):1-7

Chapter 9

Artificial Surfaces and Media for Electromagnetic Absorption and Interference Shielding

Pai-Yen Chen, Mohamed Farhat, Zhilu Ye,
Muhammad Amin, Hakan Bagci and Danilo Erricolo

Abstract

The rapid advent of radio-frequency (RF) and microwave technologies and systems have given rise to serious electromagnetic pollution, interference and jamming for high-precision detection devices, and even threats to human health. To mitigate these negative impacts, electromagnetic interference (EMI) shielding materials and structures have been widely deployed to isolate sophisticated instruments or human settlements from potential EMI sources growing every day. We discuss recent advances in lightweight, low-profile electromagnetic absorbing media, such as metamaterials, metasurfaces, and nanomaterial-based solutions, which may provide a relatively easy solution for EMI shielding and suppressing unwanted RF and microwave noises. We present a general review of the recent progress on theories, designs, modeling techniques, fabrication, and performance comparison for these emerging EMI and electromagnetic compatibility (EMC) media.

Keywords: EMI, EMC, metamaterials, metasurfaces, perfect electromagnetic absorbers

1. Introduction

Electromagnetic absorbers have important applications in a plethora of applications, including but not limited to electromagnetic interferences and electromagnetic compatibility [1–5], stealth [6–9], camouflage [10], shielding [11–13], energy harvesting [14, 15], as well as antenna and optical measurements [2, 16–18]. Recently, the scientific interest has focused on EMC and EMI shielding that studies how to suppress noise or interference in various electronic appliances and radiative damage to humans caused by unintended EM signals. In these applications, electromagnetic absorbers play an essential role. As one of the representative examples, a high-performance and cost-effective anechoic chamber, which can provide a zero-reflection environment mimicking free space, is of paramount importance for EMC test, antenna, and scattering measurements, among many other applications. As illustrated in **Figure 1(a)**, in an anechoic chamber, the entire inner surfaces (walls, ceiling, and floor) are properly covered with absorbers to absorb waves propagating toward them and thus eliminate multipath interferences. Hence, a simple and well-defined propagation channel can be obtained between the transmitting source and

IntechOpen

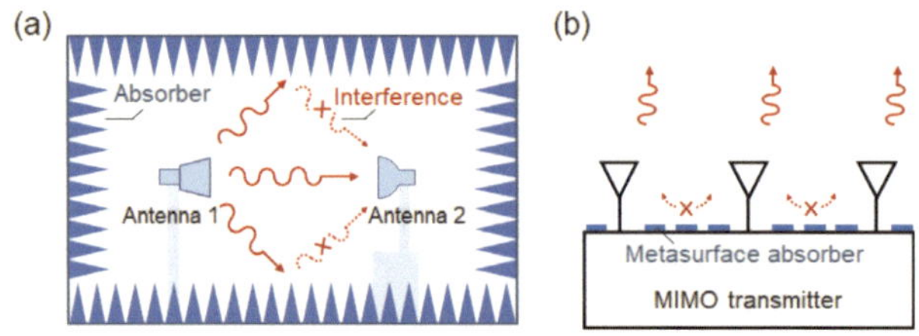

Figure 1.
Examples of practical applications of electromagnetic absorbers in EMC/EMI: (a) an anechoic chamber and (b) a high-isolation MIMO antenna system.

receiving antennas or a scattering object for radar cross-section (RCS) measurements. Nearly perfect absorption, realized with an ultrathin, lightweight, and low-cost manner, makes the artificial surfaces and media (i.e., metamaterials and metasurfaces) advantageous over the conventional electromagnetic absorbers made of natural materials. Ever since the first perfect metamaterial absorber was proposed by Landy *et al.* [19], numerous designs have been proposed for metamaterial or metasurface absorbers over a wide range of frequencies [20–25] and even with multiband operation [1, 26–28].

The applications include RCS reduction and stealth [29], EMI shielding [12, 13], sensing [30–33], terahertz imaging [34–36], energy harvesting, etc. A representative modern application of metamaterial absorbers may be the enhanced isolation in the multiple-input-multiple-output (MIMO) antenna system [37, 38], as depicted in **Figure 1(b)**. With the growing demand for miniaturization of telecommunication devices, reducing the mutual coupling or cross-talk among the antenna elements has been a challenging task in MIMO systems. A specifically designed metamaterial absorber mostly consisting of periodic resonators can help improve the isolation among antenna elements and further enhance the efficiency of a MIMO system since it can significantly absorb the unwanted interferences among antennas.

The impedance of a material is defined as $\sqrt{\mu_r\mu_0/\varepsilon_r\varepsilon_0}$, with ε_r, ε_0 (μ_r, μ_0) denoting the relative permittivity, permittivity of free space (the relative permeability, permeability of free space), respectively [39]. At microwave frequencies and even at the terahertz (THz), most of the naturally occurring materials (used, for instance, in microelectronics) exhibit very low to nonexistent magnetic resonances and thus $\mu_r \approx 1$ [40]. On the other hand, the relative permittivity can attain extremely high values, e.g., in conductors such as noble metals or graphene (in the THz regime). So, the impedance is generally much smaller than that of the surrounding ($\eta/\eta_0 \approx 1/\sqrt{\varepsilon_r} \ll 1$) [41]. This huge mismatch can cause a strong reflection of the electromagnetic signal and perturb the operation of the electrical equipment. In this context, electromagnetic compatibility is an important and active field of research that endeavors to limit undesired and unintentional scattering of microwave and THz waves to avoid, for example, unwanted behavior such as electromagnetic interference [42]. The impedance matching with the free space is hence a necessary condition to reduce much of the reflection of the incident wave. One avenue consists of a large effective relative permeability μ_r by tailoring custom RF, microwave, and THz metamaterials [43, 44]. These are artificial materials structured at the nano- or micro-scale and gained tremendous attention over the past two decades thanks to their exotic dynamic properties, generally not found in nature, such as negative optical refraction [45, 46], and electromagnetic cloaking [47, 48]. The collective oscillations of free electrons in metals, termed localized or delocalized surface plasmon polaritons (SPPs) [49], are at the origin of these unique properties, allowing a multitude of intriguing applications such as biosensors [50], optical filters [51], photodetectors [52], and nanolasers [53] to name a few.

The metallic nanostructures based on localized SPP resonance generally create strongly enhanced electromagnetic fields. The electric field confined in the junctions of metal particles, so-called electrical hot spots [49], allows for a significant improvement in emission processes and non-linearities [54], which are mediated by the electrical polarization of molecules. In particular, the magnetic activity at optical frequencies is much lower than its electrical equivalent due to the extremely low magnetic response of natural materials [40]. Accordingly, magnetic hot spots are highly desirable for enhancing the magnetic response. Many structures have been specifically designed to provide magnetic hot spots, such as diabolo antennas [55] and two parallel metal plates [56]. However, achieving simultaneous electric and magnetic hot spots in the same spatial position is rather difficult. In addition, quality factor and mode volume are critical parameters in local field enhancement engineering.

In the same vein, although the intrinsic optical loss of metals is a major limitation in the performance of these devices, it is beneficial in improving light absorption. In 2008, Landy *et al.* first proposed a perfect metamaterial absorber with near-perfect absorption by simultaneously exciting electric and magnetic resonances to achieve impedance matching with free space [19]. Soon after, important absorber designs based on different physical mechanisms have been demonstrated theoretically and experimentally in a broad spectral range, which can be classified into two categories: narrowband absorbers [57, 58] and broadband absorbers [59, 60] in terms of their absorption bandwidth. While broadband absorbers are typically used in thermophotovoltaic systems [61], perfect narrowband absorbers may be used for absorption filters [36], tailoring thermal radiation [62], and sensing [30, 32]. For the important sensing and detection applications, the experimentally performed refractive index (RI) sensor based on infrared coherent perfect absorber (CPA) shows that a narrow bandwidth and a large modulation depth are required for better performance [32, 33]. The three-layered configuration of metal–insulator–metal (MIM) employed was widely applied to the following SPP absorbers, where a thin dielectric spacer is sandwiched to allow for strong plasmonic coupling between the top resonators and the bottom metallic reflector as discussed in depth in [22, 59]. This class of CPA designs may also be intuitively treated as a single input transmission line (TL) coupled to a plasmonic resonator, where the thickness of the insulating spacer affects the radiative damping rates as well as the resonance frequency bandwidth.

2. Generalized theory for perfect electromagnetic absorbers

In this section, we will discuss the basic principles of extremely thin electromagnetic absorbers composed of an infinite two-dimensional array of electric dipoles, magnetic dipoles, or both. According to the optical theorem, when the scatterer interacts with the incident fields, the power depleted from the incident fields is the sum of the absorbed and scattered powers, i.e., $P_{ext} = P_a + P_s$. This power balance is fundamentally restricted by the energy conservation enforced by causality. The time-averaged absorbed power is the surface integral of the inward flowing flux of the Poynting vector over the surface surrounding the scatterer [63].

$$P_a = -\frac{1}{2}\,\mathrm{Re}\left[\oiint (\mathbf{E}_{inc} + \mathbf{E}_s) \times (\mathbf{H}_{inc} + \mathbf{H}_s)^* \cdot \hat{n}\, ds\right], \tag{1}$$

where $\{\mathbf{E}_{inc}, \mathbf{H}_{inc}\}$ and $\{\mathbf{E}_s, \mathbf{H}_s\}$ are the incident and scattered fields, respectively. Similarly, the time-averaged scattered power representing the re-radiated power from the scatterer is the surface integral of the outward flux of the Poynting vector of scattered fields,

$$P_s = \frac{1}{2}\,\mathrm{Re}\left[\oiint \mathbf{E}_s \times \mathbf{H}_s^* \cdot \hat{n}\,ds\right]. \tag{2}$$

Finally, the total incident power flowing through the surface S is zero

$$P_{inc} = \frac{1}{2}\,\mathrm{Re}\left[\oiint \mathbf{E}_{inc} \times \mathbf{H}_{inc}^* \cdot \hat{n}\,ds\right] \equiv 0. \tag{3}$$

From Eqs. (1)-(3), the total extracted power (power extinction) is given by the cross-terms

$$P_{ext} = -\frac{1}{2}\,\mathrm{Re}\left[\oiint \left(\mathbf{E}_i \times \mathbf{H}_s^* + \mathbf{E}_s \times \mathbf{H}_i^*\right) \cdot \hat{n}\,ds\right]. \tag{4}$$

The forward scattering sum rule is valid for any arbitrary absorbers, for which the scattered power P_s is the sum of the backward scattered (reflected) power $P_r = |R|^2 P_{inc}$ and the forward scattered power (i.e., the difference between the transmitted and incident fields) $P_f = |1-T|^2 P_{inc}$ where R and T are reflection and transmission coefficients of the compact planar absorber, respectively. The total scattered power P_s can be expressed as

$$P_s = P_r + P_f = \left(|R|^2 + |1-T|^2\right) P_{inc}. \tag{5}$$

The absorbed power P_a is the difference between the incident power and the sum of reflected and transmitted power $P_r + P_t$, given by

$$P_a = P_{inc} - P_r - P_t = \left(1 - |R|^2 - |T|^2\right) P_{inc}. \tag{6}$$

Assuming that there are no grating sidelobes, the power extinction -P_{ext} can be written as

$$P_{ext} = P_a + P_s = 2\,\mathrm{Re}\,[1-T] P_{inc}. \tag{7}$$

This can be seen as the optical theorem for infinite two-dimensional (2D) planar structures, such as metascreens or an ultrathin absorptive film. We know from Eqs. (5)-(7) that when the forward scattered power represents the total scattered power, i.e., no backward scattered (reflected) waves, the total absorption of the incident radiation can be achieved. In this case, the scattered power equals the absorbed power, $P_s = P_f = P_a$, which is somehow similar to the conjugated matching concept in the equivalent Thevenin or Norton circuit. The 100% absorption occurs only for an asymmetric system, such as a metal-back thin absorptive slab or a surface that combines both electric and magnetic dipoles [64]. For instance, electric and magnetic dipole arrays at the metascreen would radiate symmetric and anti-symmetric electric fields on either side of the metascreen. The anti-symmetry introduced by magnetic dipoles enables the complete cancelation of incident fields in the forward direction while generating no backward propagating wave, thereby achieving a perfect absorption [39]. Likewise, a metal-backed electric or magnetic dipole array, such as linear dipole or patch antenna array, also exhibits the asymmetry necessary for making a perfect absorber, which will be discussed in the following.

Consider an infinite 2D electric dipole array, separated from a PEC ground plane by a distance d; this induces an electric current sheet parallel to the ground PEC plane, as shown in **Figure 2** (a). We assume that the array period is smaller than the

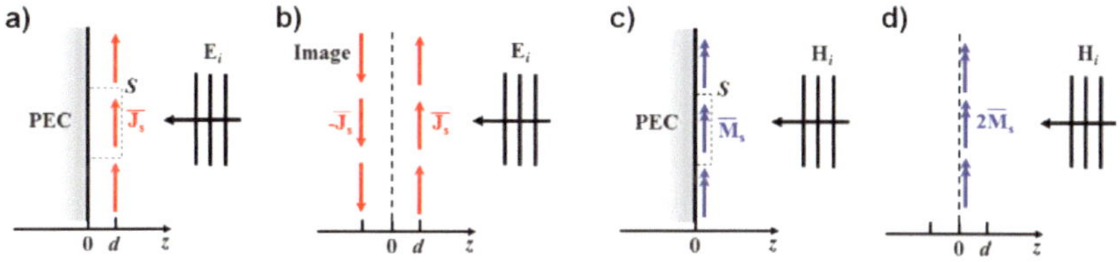

Figure 2.
(a) Electric and (c) magnetic sheet currents over a PEC surface and the corresponding images (b) and (d), respectively.

wavelength of the incident wave, and the array is excited by an x -polarized normally incident plane wave from the positive z-direction: $E_{inc} = \hat{\boldsymbol{x}} E_0 e^{jkz}$, where k is the wavenumber of the uniform background medium. For brevity, we consider an isotropic periodic grid that can be conveniently modeled by a scalar surface impedance Z_s, relating the surface-averaged electric field $\mathbf{E}_{\text{tot}} = \mathbf{E}_{\text{inc}} + \mathbf{E}_{\text{s}}$ in the array plane and the average electric surface current density of complex amplitude $\mathbf{J}_{\text{s}} = \hat{\boldsymbol{x}} J_{\text{s}} = \hat{\boldsymbol{x}} J_{\text{s}0} e^{i\varphi}$ (which is expressed in terms of the averaged tangential magnetic fields on the two sides of the array)

$$\mathbf{J}_{\text{s}} = \frac{j\omega\ \mathbf{p}}{S} = \hat{\boldsymbol{z}} \times \left(\mathbf{H}^{+} - \mathbf{H}^{-}\right) = \mathbf{E}_{\text{tot}}/Z_s, \tag{8}$$

where $\mathbf{p}$ is the electric dipole moment generated by the resonant meta-atoms, $\mathbf{E}_{\text{s}}$ is the scattered fields generated by the average electric surface current density $\mathbf{J}_{\text{s}}$ and S is the area of a unit-cell. This scattering problem can be solved using the image theory, which removes the ground plane and places an image source of $-\mathbf{J}_s$ at $z = -d$. With the principle of superposition, the total fields for $z > 0$ can be found by combining the fields from the two sources individually. Hence, the scattered electric and magnetic fields can be written as

$$E_s = \begin{cases} -\hat{\boldsymbol{x}} jJ_{\text{s}} \eta e^{-jkd} \sin kz + E_0 e^{-jkz}, & 0 < z < d \\ -\hat{\boldsymbol{x}} jJ_{\text{s}} \eta \sin kd e^{-jkz} + E_0 e^{-jkz}, & z > d \end{cases};$$
$$H_s = \begin{cases} \hat{\boldsymbol{y}} jJ_{\text{s}} e^{-jkd} \cos kz - \dfrac{E_0}{\eta} e^{jkz}, & 0 < z < d \\ -\hat{\boldsymbol{y}} jJ_{\text{s}} \sin kd e^{-jkz} + \dfrac{E_0}{\eta} e^{jkz}, & z > d, \end{cases} \tag{9}$$

where η is the intrinsic impedance of the background medium. The second terms in Eq. (9) are contributed by the reflection (backward scattering) from the conducting ground plane, whereas the first terms in Eq. (9) are produced by the induced surface current. Below the ground plane ($z > 0$), the scattering fields $\mathbf{E}_s = -\mathbf{E}_{inc}$ are always valid, regardless of the scattering environment above the ground plane. As known from Eqs. (1) and (2), the time-averaged scattered and absorbed power can be obtained from the surface integral of the inward flowing flux of the Poynting vector of the total field through the surface S enclosing a single unit-cell

$$\begin{aligned} P_s &= \frac{1}{2} \text{Re} \left[\oiint \mathbf{E}_s \times \mathbf{H}_s^* \cdot \hat{n} ds \right] \\ &= \frac{E_0^2}{\eta_0} + \frac{1}{2} J_{s0}^2 \eta_0 \sin^2(k_0 d) - E_0 J_{s0} \sin(k_0 d) \sin(\varphi), \end{aligned} \tag{10}$$

$$P_a = -\frac{1}{2}\,\mathrm{Re}\left[\oiint(\mathbf{E}_{inc}+\mathbf{E}_s)\times(\mathbf{H}_{inc}+\mathbf{H}_s)^*\cdot\hat{n}\,ds\right] \\ = J_{s0}E_0\sin(k_0d)\sin(\varphi)-\frac{1}{2}J_{s0}^2\eta_0\sin^2(k_0d). \tag{11}$$

For an arbitrary value d, the perfect absorption can be achieved when the surface electric current is tailored to be

$$J_s = \frac{E_0}{\eta\sin(kd)}\,\mathrm{e}^{\,j(\pi/2)}, \tag{12}$$

or from Eqs. (8), (9), and (12), we obtain the optimum (lossy and capacitive) surface impedance $Z_{s,opt}$ that renders the maximum absorbed power and $P_{abs} = P_{sca}$

$$Z_{s,opt} = \frac{\eta}{2}\left[2\sin^2(kd)-j\sin(2kd)\right]. \tag{13}$$

When the distance between the array plane and the ground plane $d = \lambda/4$, the optimum electric surface impedance $Z_{s,opt} = \eta$; such a result is consistent with the setting of the Salisbury screen absorber [65]. From Eq. (13), we find that a complex surface impedance is required to miniaturize an absorber with $kd \to 0$. Intuitively, a capacitive lossy surface with $\mathrm{Im}\left[Z_{s,opt}\right] = -j\eta\sin(2k_0d)/2 \simeq -j\omega\mu d$ should be exploited to compensate for the shunt inductance resulting from the grounded dielectric substrate that exhibits an input impedance $Z_{in} = j\eta\tan(kd) \approx j\omega\mu d$. An equivalent TL model sketched in **Figure 3** may be a straightforward way to describe the reflection and absorption properties of the grid array. For a plane wave normally incident on the structure, the zero reflection condition (i.e., total absorption) can be obtained as $Z_{in}^{-1} + Z_s^{-1} = \eta^{-1}$, leading to the same results of Eq. (13).

In the following, we briefly discuss how to design and synthesize the grid array that provides the required surface impedance. It is instructive to study a metamaterial surface or metasurface formed by a 2D array of meta-atoms excited by an external electric field $\mathbf{E}_{\mathrm{ext}}$ with the TE polarization (**Figure 2a**). The collective behavior of electric dipole moments induced in electric meta-atoms will result in a homogenous sheet current. The total averaged polarization related to the local field is responsible for the $\alpha_{\mathrm{e,xx}}$ component of the electric polarizability tensor $\boldsymbol{\alpha}_{\mathrm{e}}$. Considering the full coupling among the whole array of interacting electric dipoles, the dipole moment induced on the nanoparticle is given by [35, 66, 67].

$$p_{\mathrm{NxNy}} = \alpha_{\mathrm{xx}}E_{\mathrm{loc}} = \alpha_{\mathrm{e,xx}}^{-1}\left[E_{\mathrm{ext}} + \varepsilon^{-1}\sum_{(\mathrm{N_x},\mathrm{N_y})\neq(\mathrm{N'_x},\mathrm{N'_y})} G\left(r_{\mathrm{N_xN_y}} - r_{\mathrm{N'_x,N'_y}}\right)\cdot p_{\mathrm{N'_x,N'_y}}\right] \tag{14}$$

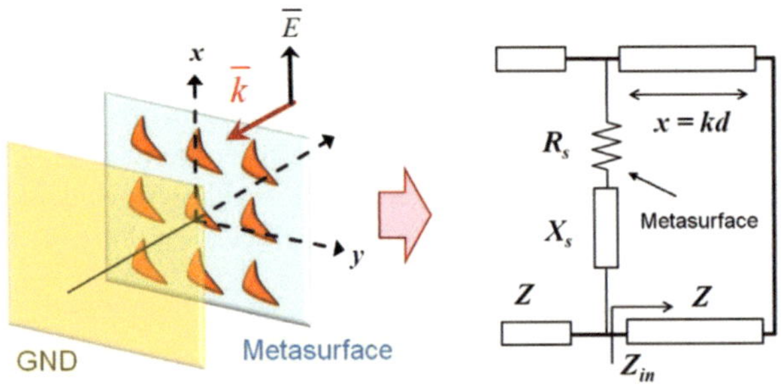

Figure 3.
Equivalent transmission line model of an ultrathin metasurface-based absorber.

where N_x and N_y are positive or negative integers, which relate the position of local dipoles and periods d_x and d_y by $x = N_x d_x$ and $y = N_y d_y$, $r_{N_x N_y}$ is the position of the dipole p^{NxNy}, and $G\left(r_{N_x N_y}, r_{N'_x N'_y}\right) = \left(\nabla\nabla + k^2 \overline{\overline{I}}\right) \frac{e^{-jk|r_{N_x N_y} - r_{N'_x N'_y}|}}{4\pi \left| r_{N_x N_y} - r_{N'_x N'_y} \right|}$ is the dyadic Green's function in a homogeneous and isotropic medium. After some algebraic manipulations, we may write (14) as:

$$p_{00} = p_{00}\hat{\boldsymbol{x}} = \frac{1}{\alpha_{e,xx}^{-1} - \beta} E_{ext} \tag{15}$$

where $\beta = \sum_{(N_x, N_y) \neq (0,0)} G\left(r_{N_x N_y}, r_{N'_x N'_y}\right) \hat{\boldsymbol{x}}\hat{\boldsymbol{x}}$ is the interaction constant that relates the fields induced by the infinite array of electric dipoles around the unit cell under consideration to the local field at the origin. The averaged surface current density on the metasurface can be expressed as:

$$J_s = J_s \hat{\boldsymbol{x}} = \frac{j\omega p_{00}}{S} = \frac{j\omega E_{ext}}{S\left(\alpha_{e,xx}^{-1} - \beta\right)} \hat{\boldsymbol{x}}, \tag{16}$$

where the area of the metasurface lattice S = $d_x d_y$. This is an accurate description of the array's electromagnetic properties, as long as the periods are small enough to ensure that only one Floquet harmonic could exist. The average surface impedance of metasurface, as the ratio of the local electric field to the surface current density, can be expressed as

$$Z_s = \frac{E_{ext}}{J_s} - \frac{\eta}{2} = \frac{S}{j\omega} \left(\alpha_{e,xx}^{-1} - \beta\right) - \frac{\eta}{2} \tag{17}$$

Under the lossless and low frequency condition, the polarizability is approximately given by $\alpha_{e,xx}^{-1} \simeq \alpha_{e0,xx}^{-1} + jk^3/(6\pi\varepsilon)$, where $\alpha_{e0,xx}$ is the static polarizability. Eq. (17) may be further manipulated by considering the physical requirements on the imaginary parts of β associated with the power balance [68].

$$\mathrm{Im}[\beta] = \frac{k^3}{6\pi\varepsilon} - \frac{\eta\omega}{2S} \tag{18}$$

So, Eq. (17) can be further reduced to

$$Z_s = \frac{S}{j\omega} \left(\alpha_{e0,xx}^{-1} - \mathrm{Re}\,[\beta]\right) - \frac{\eta}{2}. \tag{19}$$

For a subwavelength lattice, i.e., $kd_x, kd_y \ll 1$, the real part of quasi-static approximate interaction coefficient can be written as [69].

$$\mathrm{Re}\,[\beta] = \frac{1}{\varepsilon\pi d_x^3} \sum_{m=1}^{\infty} \frac{1}{m^2} - \frac{8\pi}{\varepsilon d_x^3} \sum_{m=1}^{\infty} \sum_{n=1}^{\infty} m^2 K_0 \left(mn2\pi \frac{d_y}{d_x} \right), \tag{20}$$

where $K_0(\cdot)$ is the modified Bessel function of the second kind (or Macdonald function [70]). For the square lattice $\left(d_x = d_y \ll \lambda_0\right)$, the Eq. (20) can have a compact form

$$\mathrm{Re}\,[\beta] \simeq \frac{1}{\varepsilon \mathrm{d_x^3}}\left(-\frac{\psi_2(1)}{2\pi} - 8\pi \mathrm{K_0}(2\pi)\right) \simeq \frac{0.3596}{\varepsilon \mathrm{d_x^3}}, \tag{21}$$

where ψ_n is the polygamma function [71]. Eq. (21) agrees with the low-frequency interact coefficient in Ref. [72]:

$$\begin{aligned} \beta &\simeq -\mathrm{j}\omega \frac{\eta}{4\mathrm{d_x^2}}\left(1 - \frac{1}{\mathrm{jkR_0}}\right)\mathrm{e}^{-\mathrm{jkR_0}} + \mathrm{Im}\left[\alpha_{\mathrm{e,xx}}^{-1}\right] \\ &\simeq \frac{0.3596}{\varepsilon \mathrm{d_x^3}} - \mathrm{j}\left(\frac{\eta\omega}{2\mathrm{d_x^2}} + \frac{\mathrm{k^3}}{6\pi\varepsilon}\right). \end{aligned} \tag{22}$$

where $\mathrm{R_0} \simeq \mathrm{d_x}/1.438$ and $\mathrm{Re}\,[\beta] \simeq 0.3596/\left(\varepsilon \mathrm{d_x^3}\right)$. From Eqs. (13) and (19), we find that the maximum absorption is reached when the following conditions are satisfied:

$$\mathrm{Re}\left[\alpha_{e0,\mathrm{xx}}^{-1}\right] - \mathrm{Re}\,[\beta] = \frac{\omega\eta}{2S}\sin(2kd); \tag{23}$$

$$\mathrm{Im}\left[\alpha_{e0,\mathrm{xx}}^{-1}\right] = \frac{\omega\eta}{S}\left(1 - \frac{\cos(2kd)}{2}\right). \tag{24}$$

This can be achieved by tailoring the size, shape, and material property (i.e., dispersion of the complex permittivity) of the elementary inclusion (i.e., meta-atom) in the metasurface unit-cell. For a Salisbury-type absorber with $d = \lambda/4$, Eqs. (23) and (24) becomes:

$$\mathrm{Re}\left[\alpha_{e0,\mathrm{xx}}^{-1}\right] \approx \mathrm{Re}\,[\beta], \tag{25}$$

$$\mathrm{Im}\left[\alpha_{e0,\mathrm{xx}}^{-1}\right] \approx \frac{\omega\eta}{S} = \frac{k}{\varepsilon S}. \tag{26}$$

The first condition implies an individual scatterer is at its self-resonance, which considers the cancelation of a general sense, the maximum absorption cross-section of each inclusion in densely packed arrays $\sigma_{\mathrm{sca}} = S$ is smaller than that of the same inclusion in free reactive coupling between neighboring scatters. In contrast, the second condition means that the absorption in the array is maximized; such occurs when the absorbed power is equal to the re-radiated power, i.e., $P_{abs} = P_{sca} = |E_{inc}|^2/2\eta = P_{in}$. We note that the maximum absorbed power for each lattice is $P_{abs,cell} = SP_{in}$, therefore, in space under the same illumination $\sigma_{\mathrm{sca}} = 3\lambda^2/8\pi$. It is interesting to note that inclusion in the grid can absorb as much as it is in free space when the unit cell size is: $k^2 S = 3\pi/2$. In the low-frequency region, the polarizability of a generalized inclusion is $\boldsymbol{\alpha}_e^{-1} = \boldsymbol{\alpha}_{e0}^{-1} + \frac{k^3}{6\pi\varepsilon}\mathbf{I}$. The static polarizability of a meta-atom as a function of geometry, frequency, and material property can be analytically derived or numerically retrieved. For example, the static polarizability of a spherical particle is: $\boldsymbol{\alpha}_{e0} = 4\pi a^3 \varepsilon_0 \frac{\boldsymbol{\varepsilon}_D - \varepsilon}{\boldsymbol{\varepsilon}_D + 2\varepsilon}$ [39], where a and $\boldsymbol{\varepsilon}_D$ are the radius and permittivity tensor of the spherical scatterer. For a 2D structure with negligible thickness, such as a square patch or circular disk, the static polarizability is in the form of $\boldsymbol{\alpha}_{e0} = \alpha_{e,xx}\hat{\boldsymbol{x}}\hat{\boldsymbol{x}} + \alpha_{e,yy}\hat{\boldsymbol{y}}\hat{\boldsymbol{y}}$. For example, for a conducting circular disk $\alpha_{e0,xx} = \alpha_{e0,yy} = 16\pi a^3/3\varepsilon$ where a is its radius. In the extreme case, the polarizability of a 1D conducting rod with a deeply subwavelength radius a and height h, aligned parallel to the electric field of the incident wave, has only the $\alpha_{e,xx}$ component [39]. For modern meta-atom designs such as electric or magnetic split-ring resonators [73–77], the resonant electric/magnetic polarizability can be fitted to a generalized Lorentzian dispersion model

$$\alpha_{e/m}(\omega) = \frac{A\omega^2}{\omega_0^2 - \omega^2 + j\omega\Gamma}, \tag{27}$$

where ω_0 is the resonant frequency of electric/magnetic polarizability, Γ is the damping factor, and A is the amplitude.

Figures 4 and **5** show metasurface absorbers based on 2D arrays of plasmonic disks [78] and electric ring resonators [79], respectively. It is clearly seen that the ultrathin and low-profile metasurface absorbers can significantly absorb the incident wave in the frequency band of interest. We note that these electric resonant inclusions can be equivalent to a 2D array of electric dipoles discussed above.

It is also possible to exploit a metasurface constituted by magnetic or magneto-electric meta-atoms [9, 19, 22, 23] to build a perfect electromagnetic absorber. For example, **Figure 2(c)** and **(d)** consider a magnetic dipole array over a ground plane and the image. When a plane wave normally incident on the structure, the magnetic current density is induced: $\mathbf{M}_s = j\omega\mathbf{m}_{00}/S$. In a generalized scenario, the metasurface is excited by normally incident plane wave with fields $(\mathbf{E}_{ext}, \mathbf{H}_{ext})$, and an individual meta-atoms is excited by the local field $(\mathbf{E}_{loc}, \mathbf{H}_{loc})$:

$$\begin{aligned} \mathbf{p}_{00} &= \boldsymbol{\alpha}_e \mathbf{E}_{loc} + \boldsymbol{\alpha}_{em} \mathbf{E}_{loc}; \\ \mathbf{m}_{00} &= \boldsymbol{\alpha}_{me} \mathbf{E}_{loc} + \boldsymbol{\alpha}_m \mathbf{E}_{loc}, \end{aligned} \tag{28}$$

and

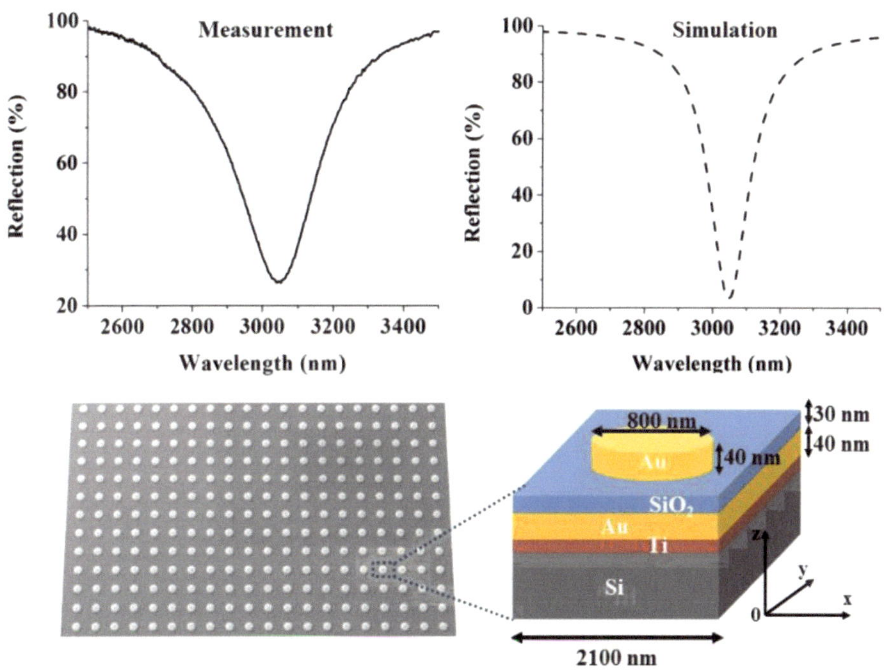

Figure 4.
Comparison between measurement and simulation. Above: Reflection spectra of the metamaterial absorber consisting of periodic arrays of Au cylinders/disks deposited on a stack of SiO_2/Au/Ti/Si layers measured by FTIR spectrometer and calculated by FDTD simulation. Bottom: SEM image of the absorber and a relevant unit cell needed for simulation.

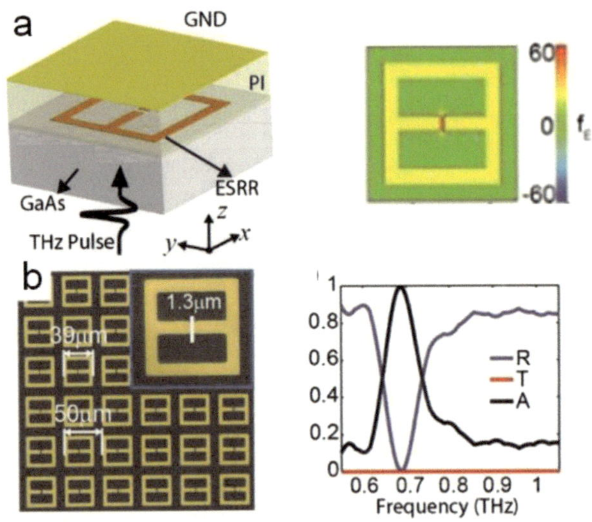

Figure 5.
(a) Terahertz metasurface absorber based on electric resonant inclusions, which displays a strong electric field localization in the capacitive gap region. (b) Transmission, reflection, and absorption against frequency for the metasurface in (a).

$$\begin{aligned} \mathbf{E}_{loc} &= \mathbf{E}_{ext} + \mathbf{p}_{00} \cdot \boldsymbol{\beta}; \\ \mathbf{H}_{loc} &= \mathbf{H}_{ext} + \mathbf{m}_{00} \cdot \boldsymbol{\beta}', \end{aligned} \tag{29}$$

where $\boldsymbol{\alpha}$ is the dyadic polarizability and the interaction coefficient are

$$\boldsymbol{\beta} = \beta \mathbf{I} + j\left(\frac{\eta\omega}{2S} + \frac{k^3}{6\pi\varepsilon}\right)\mathbf{I} \quad \text{and} \quad \boldsymbol{\beta}' = \frac{\boldsymbol{\beta}}{\eta^2}. \tag{30}$$

The total averaged fields on the metasurface can be expressed as

$$\begin{pmatrix} \mathbf{E}_{tot} \\ \mathbf{H}_{tot} \end{pmatrix} = \begin{pmatrix} \mathbf{Z}_e & \mathbf{Z}_{em} \\ \mathbf{Z}_{me} & \mathbf{Z}_m \end{pmatrix} \cdot \begin{pmatrix} \mathbf{J} \\ \mathbf{M} \end{pmatrix}, \tag{31}$$

where the surface impedances are given by

$$\begin{aligned} \mathbf{Z}_e &= -j\frac{S}{\omega}\left[\left(\boldsymbol{\alpha}_e - \boldsymbol{\alpha}_{em} \cdot \boldsymbol{\alpha}_{mm}^{-1} \cdot \boldsymbol{\alpha}_{me}\right)^{-1} - \boldsymbol{\beta}_e - j\frac{\eta\omega}{2S}\mathbf{I}\right]; \\ \mathbf{Z}_m &= -j\frac{S}{\omega}\left[\left(\boldsymbol{\alpha}_m - \boldsymbol{\alpha}_{me} \cdot \boldsymbol{\alpha}_{ee}^{-1} \cdot \boldsymbol{\alpha}_{em}\right)^{-1} - \boldsymbol{\beta}_m - j\frac{\omega}{2\eta S}\mathbf{I}\right]; \\ \mathbf{Z}_e &= -j\frac{S}{\omega}\left[\left(\boldsymbol{\alpha}_e - \boldsymbol{\alpha}_{em} \cdot \boldsymbol{\alpha}_{mm}^{-1} \cdot \boldsymbol{\alpha}_{me}\right)^{-1} - \boldsymbol{\alpha}_{em} \cdot \boldsymbol{\alpha}_{mm}^{-1}\right]; \\ \mathbf{Z}_e &= -j\frac{S}{\omega}\left[\left(\boldsymbol{\alpha}_m - \boldsymbol{\alpha}_{me} \cdot \boldsymbol{\alpha}_{ee}^{-1} \cdot \boldsymbol{\alpha}_{em}\right)^{-1} - \boldsymbol{\alpha}_{me} \cdot \boldsymbol{\alpha}_{ee}^{-1}\right]. \end{aligned} \tag{32}$$

If the reader is interested in learning details about anisotropic magnetoelectric metasurfaces, please see Refs. [80–82].

Figure 6 shows an ultrathin electromagnetic absorber based on resonant magnetic structures. Under the excitation of a plane wave, a magnetic dipole array can be induced, and at resonance, the incident power can be absorbed, analogous to the

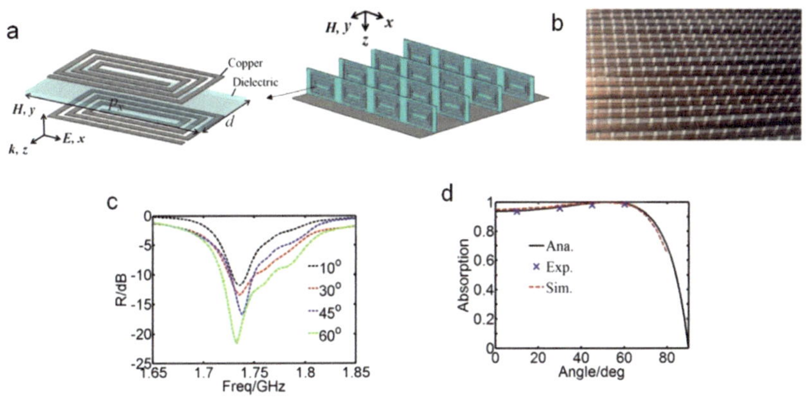

Figure 6.
(a) Sketch and (b) photo of an ultrathin perfect absorber based on magnetic resonant structures and its measured (c) reflection and (d) absorption.

function of the electric dipole array discussed above. The experimental results reported an absorption efficiency above 93% at 1.74 GHz at illumination angles up to 60 degrees [83]. Moreover, this absorber is 98% lighter than traditional microwave absorbers made of natural materials working at the same frequencies. As an alternative explanation, this structure can be seen as a metal-backed magnetic-near-zero (MNZ) metamaterial slab [84]. For an ultrathin metal-backed MNZ absorber, the permeability $\mu = (\mu' - j\mu'')\mu_0$ required for perfect absorption can be derived as: $\mu'\mu_0 \approx 0$ and $\mu''\mu_0 \approx 1/k_0t$, where k_0 and μ_0 are the wavenumber and the permeability of free space, and t is the thickness of the metamaterial slab. This magnetic metamaterial absorber can achieve zero backward scattering ($R = 0$) and a large forward scattering ($T = 0$), i.e., $P_f = P_{inc}$ and $P_{ext} = P_a + P_s = 2P_{inc}$.

On the contrary, if symmetric designs are considered (e.g., a suspended resistive film with optimal sheet impedance $Z_s = \eta_0/2$ or a suspending epsilon-near-zero (ENZ) thin-slab with permittivity $\varepsilon = -j2/k_0t$), only 50% of incident radiation can be absorbed. Due to the symmetric geometry and the negligible thickness of a resistive sheet/film, the induced surface current would equally re-radiate fields to

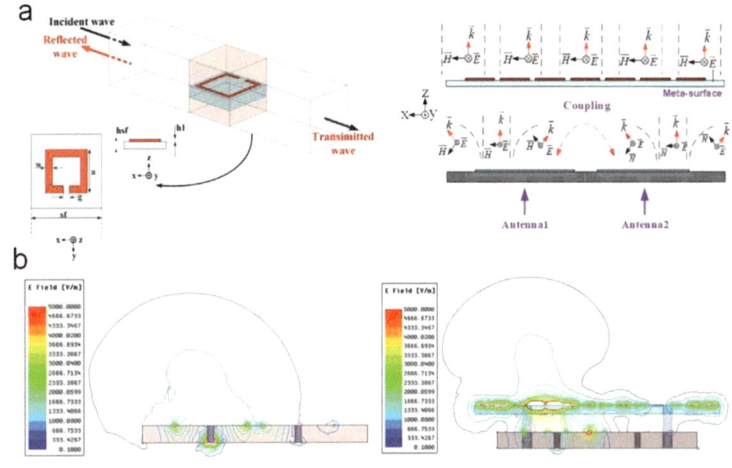

Figure 7.
(a) Magnetic metasurface for absorbing lateral radiation from the microstrip patch antenna. (b) Mutual coupling in a two-element MIMO system (left) without and (right) with the magnetic metasurface.

either side of the absorptive sheet, i.e., $P_r = P_f = |R|^2 P_{inc}$ leads to a maximum absorbed power $P_a = P_s = 0.5P_{inc}$ and thus $P_{ext} = P_{inc}$. Flat magnetic resonant structures have also been used to improve the isolation of MIMO antennas that are densely packed in a limited area. **Figure** 7 shows a two-element array of rectangular patch antennas covered by the magnetic metasurface constituted by split-ring resonators [85]. From **Figure** 7, it is clearly seen that at the magnetic resonance, the resonators can effectively absorb the lateral radiation causing mutual coupling between antennas, thus effectively reducing the envelope correlation coefficient (ECC) of the MIMO system.

3. Hyperbolic metamaterial absorbers

Hyperbolic metamaterials (HMMs) (see **Figure 8**) are known for their isofrequency contour and broadband singularity in the density of photonic states, which have led to many new photonic and optical applications, including the substrate for molding spontaneous emissions into a directional beam and the "rainbow trapping" structure for broadband light absorption [25, 86–88]. The effective medium theory can describe the effective permittivity of such an artificial anisotropic medium as:

$$\bar{\bar{\varepsilon}}_{\mathrm{eff}} = \varepsilon_0 \left(\varepsilon_{\parallel} \hat{x}\hat{x} + \varepsilon_{\parallel} \hat{y}\hat{y} + \varepsilon_{\perp} \hat{z}\hat{z} \right), \tag{33}$$

$$\varepsilon_{\perp} = \left(\sum_{i=1}^{N} \varepsilon_i^{-1} \rho_i \right)^{-1}, \tag{34}$$

$$\varepsilon_{\parallel} = \sum_{i=1}^{N} \varepsilon_i \rho_i, \tag{35}$$

where ε_i and ρ_i are relative permittivity and volume fraction of i-*th* constituent materials, and ε_0 is the permittivity of free space. The isofrequency relation for the transverse-magnetic (TM) wave propagating in a uniaxially anisotropic medium, as schematically shown in **Figure 8**, can be expressed as

$$\frac{k_y^2 + k_z^2}{\varepsilon_{\perp}} + \frac{k_x^2}{\varepsilon_{\parallel}} = \frac{\omega^2}{c^2}. \tag{36}$$

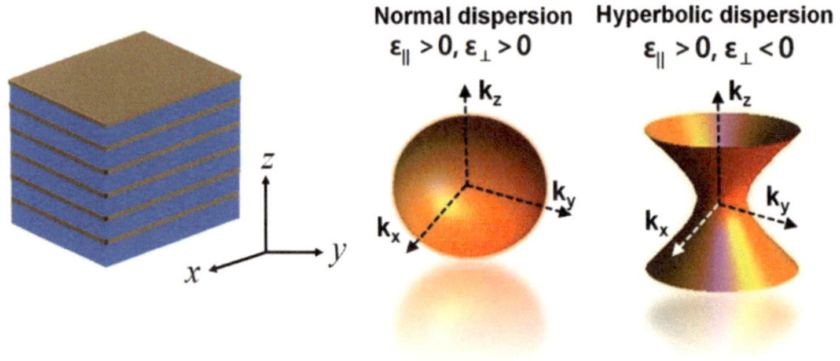

Figure 8.
Left: Schematics of HMM (left); right: Isofrequency surfaces of extraordinary waves in normal dielectrics and HMMs (right).

Therefore, for an anisotropic medium with extreme material properties, i.e., $\mathrm{Re}\left[\varepsilon_{\perp}\right] \cdot \mathrm{Re}\left[\varepsilon_{\parallel}\right] < 0$, the spherical isofrequency surface of vacuum distorts to an ellipsoid, as shown in **Figure 8**.

Figure 9a considers a trenched HMM slab with a periodicity P and air slots with width W, illuminated by a TM-polarized plane wave. The trenched HMM substrate is, in some sense, similar to a periodic array of air/HMM/air waveguides, for which a guided mode propagating along the waveguide axis can exhibit a near-zero group velocity, i.e., $v_g = \partial\omega/\partial\beta \approx 0$, at certain frequencies, where ω is the angular frequency and β is the modal propagation constant. With the periodic boundary conditions, the dispersion relation can be derived using the transverse resonance method [88–91] as:

$$\tan\left[\sqrt{k_0^2\varepsilon_{\perp} - \left(\varepsilon_{\perp}/\varepsilon_{\parallel}\right)\beta^2}\,\frac{P-W}{2}\right] - \varepsilon_{\perp}\frac{\sqrt{\beta^2 - k_0^2},}{\sqrt{k_0^2\varepsilon_{\perp} - \left(\varepsilon_{\perp}/\varepsilon_{\parallel}\right)\beta^2}}\tanh\left[\sqrt{\beta^2 - k_0^2},\frac{W}{2}\right] = 0 \tag{37}$$

where β is the modal propagation constant and k_0 is the free space wavenumber. Provided the periodicity of each MIM unit-cell is subwavelength ($\mathrm{P} < <\lambda_0$), all diffraction orders, except for the zero*th* mode, are evanescent. The scattering and absorption properties of this HMM-based device can be solved using the TL model (TLM) and the transfer matrix method (TMM), as shown in **Figure 9**. For the TM-wave illumination normally incident upon the HMM substrate, the characteristic impedance per unit length can be expressed as $\mathrm{Z}_0 = \mathrm{V}_0/\mathrm{I}_0 = \int_0^{\mathrm{P}/2}\mathrm{E_x dx}/\mathrm{H_y} = \eta_0\mathrm{P}/2$, which is defined as the ratio between the voltage across one period $V_0 = \int_0^{P/2}E_x dx = |\mathbf{E}_0|d\cos\theta$ and the current per unit length $I_0 = H_y = \varepsilon_0 c|\mathbf{E}_0|$, where ε_0 and c are permittivity and speed of light in free space. Similarly, the characteristic

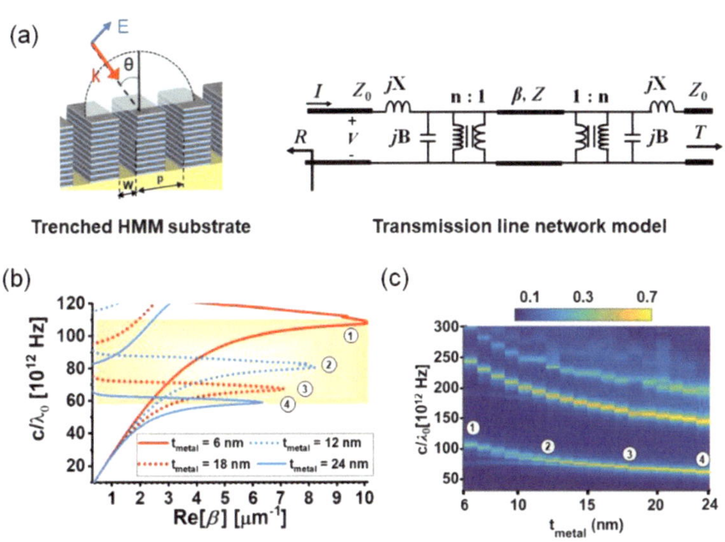

Figure 9.
(a) Schematics of a slow-light structure based on the structured HMM that can be described using a transmission line network model. (b) Dispersion diagram for a waveguide array made by carving an HMM substrate made of repeated Ag-NbO-Cu lattices with P = 300 nm and W = 100 nm. Labels in (a) and (b) indicate the slow-wave modes and the corresponding absorption peaks. (c) Calculated contours of absorptance for the HMM in (b) constituted by 12 MIM heterojunctions, varying the metal thickness (t_{metal}) and the operating wavelength; [89].

impedance of HMM $Z = (\beta/\omega\varepsilon_0) \times (W/2)$. At the interface between air and metasurface, there exists a shunt surface susceptance B imposed by the discontinuous fields [92].

$$B = \frac{2P}{\lambda_0}\left(\log\left[\frac{1-\alpha^2}{4\alpha}\left(\frac{1+\alpha}{1-\alpha}\right)^{\frac{\alpha+\alpha^{-1}}{2}}\right] + 2\frac{C_1+C_2+2C_3}{C_1C_2-C_3^2} + \left(\frac{P}{4\lambda_0}\right)^2\left(\frac{1-\alpha}{1+\alpha}\right)^{4\alpha}\left(\frac{5\alpha^2-1}{1-\alpha^2}+\frac{4}{3}\frac{\alpha^2C_3}{C_1}\right)^2\right), \tag{38}$$

$$\begin{aligned} C_1 &= \left(\frac{1+\alpha}{1-\alpha}\right)^{2\alpha}\frac{1+\sqrt{1-(P/\lambda_0)^2}}{1-\sqrt{1-(P/\lambda_0)^2}} - \frac{1+3\alpha^2}{1-\alpha^2}, \\ C_2 &= \left(\frac{1+\alpha}{1-\alpha}\right)^{2/\alpha}\frac{1+\sqrt{1-(W/\lambda_0)^2}}{1-\sqrt{1-(W/\lambda_0)^2}} + \frac{3+\alpha^2}{1-\alpha^2}, \\ C_3 &= \left(\frac{4\alpha}{1-\alpha^2}\right)^2, \end{aligned} \tag{39}$$

and $\alpha = W/P$. The reflection/transmission spectrum and absorption peaks can be calculated based on the TLM mode shown in **Figure 9**.

Figure 9b shows the calculated dispersion diagram of a periodically trenched HMM slab constituted by stacked silver (Ag) and copper (Cu) thin-films with thickness t_{metal}, separated by 1 nm-thick niobium oxide (Nb_2O_5) insulating layers, of which the periodicity and air gap size are 300 nm and 150 nm, respectively. The realistic material properties extracted from experiments are used [89]. It can be seen from **Figure 9b** that near-zero group velocity can be achieved at certain wavelengths, resulting in the slow light effect and light trapping and absorption in the lossy (dissipative plasmon loss) and anisotropic HMM region. **Figure 9c** presents the associated contour plot of absorptance as a function of metal thickness t_{metal} [nm] and wavelength λ [μm]. We find excellent agreement between the wavelengths of maximum absorption in **Figure 9c** and the near-zero group velocity points in the dispersion diagram plotted in **Figure 9b**. We note that total HMM size is still subwavelength since the large value of $\text{Re}\,[\beta]$ suggests that the incident light can be effectively absorbed in a long-wavelength range.

Additionally, the absorption spectrum can be readily tailored by varying the volume fraction of metal, which determines the permittivity tensor elements of HMM. Although a linear HMM-based absorber can exhibit a high and angle-independent optical absorption, its bandwidth is limited around the slow-wave modes. This limitation can be mitigated by exploiting a tapered-HMM, as shown in **Figure 10a**, which has been proposed to realize wideband photodetection and solar energy harvesting applications [88–91]. Using tapered geometry, the bandwidth is expected to increase due to the superposition of multiple slow-wave modes. **Figure 10b** shows the contour plot of absorptance as a function of the photon energy and the angle of incidence. It is evidently seen that a broadband (1 eV to 1.6 eV) and wide-angle (0° to grazing angle) optical absorption can be obtained with the tapered HMM substrate backed by a metallic mirror (50 nm Ag thin-film). The capability to effectively trap photons over a wide range of photon energies and illumination angles is essential for building efficient hot-electron energy harvesters or photodetectors. Finally, we note that the general limitation on the maximum bandwidth of a ground-backed absorber should always obey the passivity and causality [93], following the physical bound $\left|\int_0^\infty \ln|r(\hbar\omega)|d\hbar\omega\right| \leq 2\pi^2\sum_i \mu_{s,i}t_i/\mu_0$, where $r(\hbar\omega)$

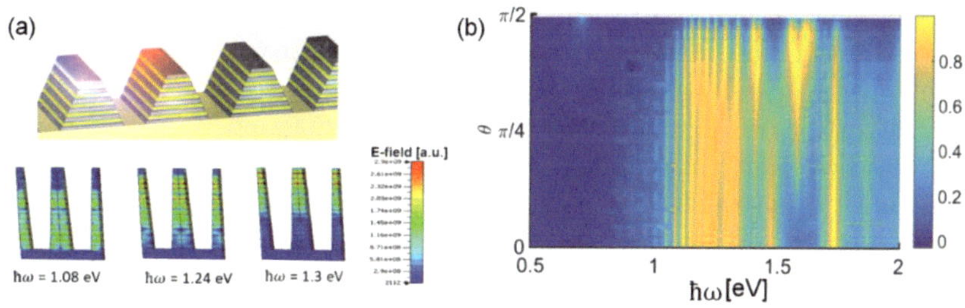

Figure 10.
(a) Top: Schematics of the HMM with a periodic array of tapered trenches, which allows one to tailor the broadband of light trapping; bottom: Snapshots of electric field distributions inside the tapered HMM at photon energies of 260 THz, 300 THz, and 320 THz (left to right). (b) Contour of absorptance as a function of photon energy and incident angle for the tapered-HMM [88].

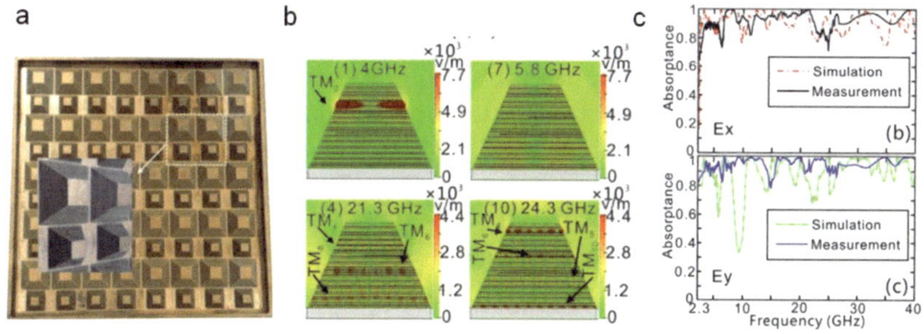

Figure 11.
(a) Photograph of a microwave HMM absorber. (b) Simulated electric field distributions at different frequencies. (c) Measured absorption spectra for different polarizations.

is the reflection coefficient as a function of photon energy, $\mu_{s,i}$ and t_i are the static permeability and the thickness of the i-*th* layer of the multilayered absorptive slab. For an effective non-magnetic medium, we can define Rozanov's limiting factor as $\eta_{Rozanov} = \left|\int_0^\infty \ln|r(\hbar\omega)| d\hbar\omega\right| / 2\pi^2 t$, where t is the total thickness of the absorber. $\eta_{Rozanov}$ must be less than unity and a larger value corresponds to a wider absorption bandwidth. For the tapered HMM in **Figure 10b**, the calculated $\eta_{Rozanov}$ can reach ~40%, which, although not optimized yet, is moderately high for the polarized light.

Similarly, an ultra-broadband HMM absorber can be realized in the RF and microwave regions by integrating two different-sized tapered HMM waveguides, as shown in **Figure 11a** [94], each of which has wide but different absorption bands, leading to a broadband slow-light response. By properly selecting the geometrical parameters for each waveguide in the HMM, multiple absorption bands can be achieved with different waveguides, as shown in **Figure 11b**. Such an approach can effectively widen the total absorption bandwidth. Experimental results in Ref. [94] validate the theoretical results, showing a very large absorption bandwidth ranging from 2.3 GHz to 40 GHz.

4. Future trends: reconfigurable absorbers based on plasmonic, graphene, and beyond

Electromagnetic wave absorbers are mainly utilized as boundary structures to prevent the scattering of electromagnetic fields. They can be divided into two main categories based on their operating bandwidth, i.e., resonant and broadband absorbers [95, 96]. Resonant absorbers typically depend on the designed material assemblies interrelating with the incident waves around certain resonance frequencies [96]. On the other hand, the wideband absorbers rely on material damping characteristics that are largely independent of electromagnetic frequencies and typically made of lossy dispersive materials [96]. Such broadband absorbing materials along with structural transition are commonly used in anechoic chambers to effectively emulate a non-reflecting unbounded medium suitable for testing radiating antenna elements [95]. Lately, there is an emphasis in the scientific community to design efficient plasmonic metamaterial-based absorbers. Light matter interaction in subwavelength metamaterial structures allowed various other applications including perfect lenses [45, 97], chiral surfaces [98, 99], transformational surfaces [100, 101], optical cloaking [100, 102, 103], spatial light switching [104, 105] and IR camouflage and microwave antennas because of certain useful characteristics [96]. In addition to these applications, the perfect metamaterial absorber (PMA) is designed as a tool to efficiently absorb electromagnetic waves utilizing plasmonic resonator elements embedded within its assembly.

Generally, metamaterial absorbers are composed of a patterned metal film over a continuous thin metallic film with a dielectric substrate sandwiched between them [96]. It should be noted that the definition of total absorptivity considers reflected electric field vector, i.e., $\vec{\mathbf{R}} = R_{ii}\hat{\boldsymbol{a}}_i + R_{ji}\hat{\boldsymbol{a}}_j$, that considers both co- ($R_{ii} = E_i^{ref}/E_i^{inc}$) and cross ($R_{ji} = E_j^{ref}/E_i^{inc}$) polarized components of reflected fields [21]. Here, subscript $\{i,j\} = \{x,y\}$refer to x or y coordinates. Oftentimes, the plasmonic metasurface absorbers inhibits only co- polarized reflection (i.e., $R_{ii} = 0$) and converts incident fields to cross polarized reflected fields components (R_{ji}) [21, 106], as shown in **Figure 12**. Such metasurface is designed with certain structural anisotropy that allows the flow of electric currents that suppresses co- polarization states (R_{ii}) for the reflected fields. As a result, such plasmonic metasurface converts the polarization of incident electromagnetic fields to orthogonal reflected field component (R_{ji}). Special cases of circular or elliptical polarized reflected fields may also be attained by specific vector combination of co- (R_{ii}) and cross (R_{ji}) polarized components [107, 108]. In the case of a full absorber as shown in **Figure 12(a)**, the normalized perfect absorption $A(\omega)$ due to the plasmonic metasurface can be found from the amplitude of reflected fields i.e., $A(\omega) = 1 - |R(\omega)|^2 - |T(\omega)|^2$. Here, $|R(\omega)|$ and $|T(\omega)|$ are amplitudes of reflected and transmitted fields, respectively. The ideal back reflector typically supports negligible transmission i.e., $|T(\omega)| = 0$.

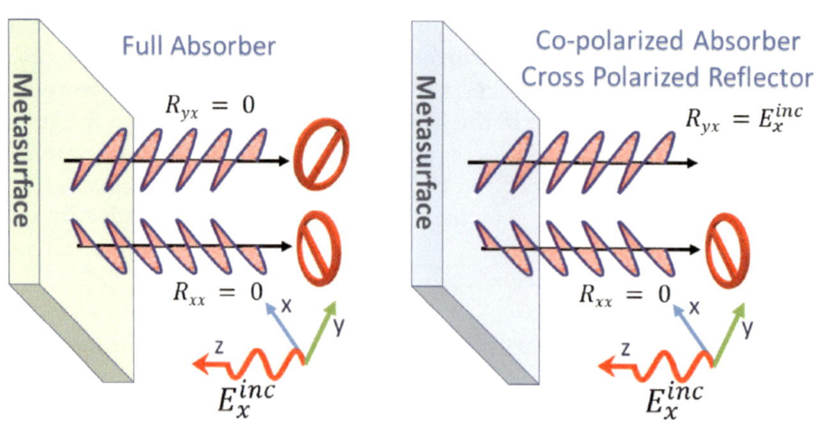

Figure 12.
(Left) full absorber to inhibit any scattering from the metasurface. (Right) cross polarizer to suppress the co-polarized component ($R_{xx} = 0$) while not affecting cross-polarized reflection component (R_{yx}).

The response of PMA is often described as an effective medium and characterized by properties such as complex electric permittivity (ε) and magnetic permeability (μ) [109]. Although much of the work on effective medium properties of metamaterials has been focused on the real part of ε and μ, as it contributes to negative refractive index properties, it is equally important to reduce losses represented by the imaginary part of ε and μ for practical applications related to wave propagation within negative refractive index medium [110, 111]. In contrast, the metamaterial absorbers rely on high material losses (large value of imaginary parts of ε and μ) and impedance matching with the background medium. Landy et al. provided the first experimental demonstration of near perfect metasurface absorption at GHz frequencies. The working principle of PMA relies on impedance matching between effective medium forming metasurface with dielectric and background medium to the free space background, rejecting the reflection and therefore efficiently absorbing the incident EM wave [19, 20]. Ever since, the design of PMAs has attracted significant attention ranging from microwave to optical frequencies [112, 113]. Apart from resonant absorption characteristics, the plasmonic effects also provide tremendous near field enhancement to improve the efficiency of solar cells [114], support for sensing [31], and enhanced thermal emission and photo-detection [115].

Recently, the electro-optic tunability of resonant absorption spectrum at THz frequencies is made possible by varying the frequency dependent conductivity of plasmonic materials through the use of graphene metasurface [22, 59]. The frequency dependent optical properties of graphene in the THz frequency range are controlled by its optoelectronic properties. The permittivity (ε_g) of graphene is function of the surface conductivity (σ_g).

$$\varepsilon_g(\omega) = 1 + j\frac{\sigma_g(\omega)}{\varepsilon_0 \omega \Delta}. \tag{40}$$

Here, ε_0 is the vacuum permittivity, and Δ = 1 nm is the monoatomic thickness of graphene.

The surface conductivity of graphene (σ_g) can be deduced from the Kubo formula, at low THz frequencies (inter-band conductivity can be neglected in this regime),

$$\sigma_g(\omega) = j\frac{e_0^2 k_B T}{\pi\hbar(\hbar\omega + j\Gamma)}\left\{\frac{\mu_c}{k_B T} + 2\log\left[1 + e^{-\mu_c/(k_B T)}\right]\right\}. \tag{41}$$

Here, μ_c[eV] is the chemical potential of graphene, $\Gamma = -\left(e_0 \hbar v_f^2\right)/(\mu\mu_c)$ is the damping coefficient, v_f [m/s] is the Fermi velocity, and μ [cm^2/Vs] is the electron mobility, e_0 is the electronic unit-charge, T is temperature, k_B is the Boltzmann constant, and $\hbar$ is the reduced Plank's constant. Therefore, the surface conductivity σ_g and permittivity of graphene ε_g can be directly controlled by varying μ_c. The chemical potential can be controlled by doping graphene or by applying an electrostatic bias to the graphene sheet.

The real-time tunability of Graphene Surface Plasmons (GSPs) is a distinct feature offered by graphene metasurface when compared to traditional noble metals. In addition, GSPs offer other benefits such as tighter mode volume confinement and lower intrinsic losses compared to conventional surface plasmon materials. The unique electro-optic tunability of graphene conductivity within the THz band makes it an attractive candidate for plasmonic metamaterial applications. Therefore, graphene enabled the research in surface plasmons to be redirected toward reconfigurable THz wave optics applications, including GSPP waveguides [116], modulators [24], THz cloaks [117], THz antennas [118], Fourier optics [119], photonic crystal nano-cavities [120], and biochemical sensors [31, 121]. The reconfigurable response of resonant absorption can offer additional functionalities, including wave modulation, polarization conversion, and sensing.

Here, we discuss a few selected applications of reconfigurable THz metasurface absorbers, as shown in **Figure 13**. **Figure 13(a)** shows graphene micro-ribbon metasurface design capable of efficiently absorbing THz radiation [22]. The chemical potential (μ_c) can be used to control graphene conductivity. As a result, the reconfigurable response with near perfect absorption can be utilized for THz wave modulation applications. **Figure 13(b)** shows a design of multilayer graphene

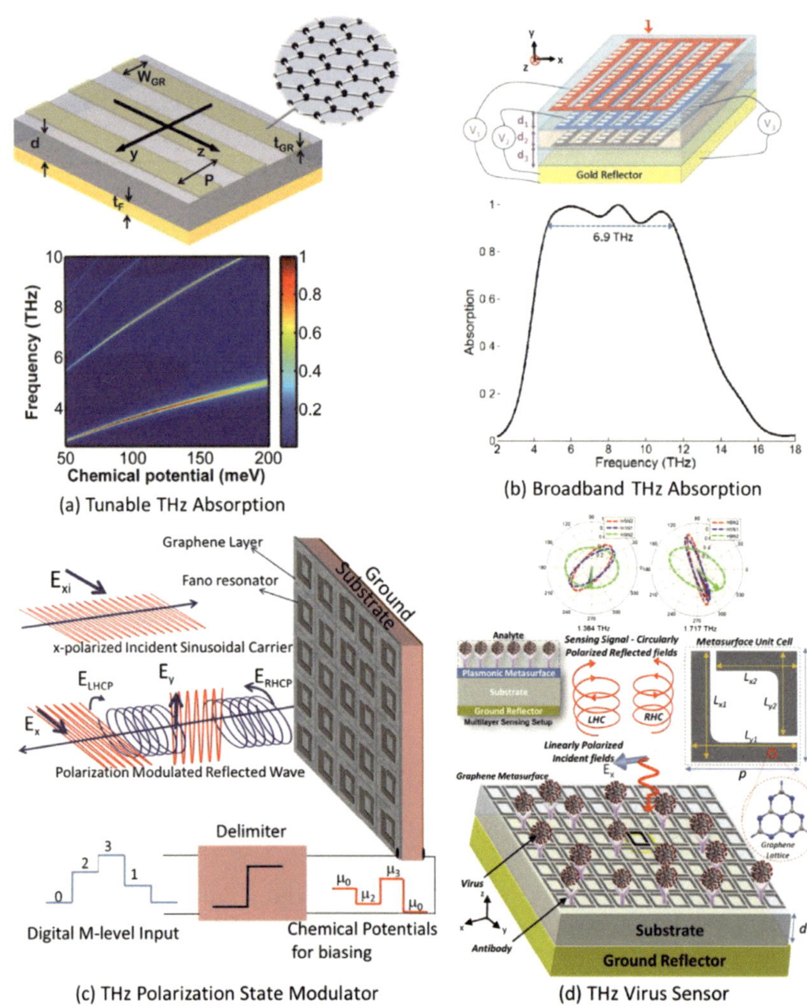

Figure 13.
(a) (Top) Schematic illustration of graphene micro-ribbon metasurface. (Bottom) Tunable absorption characteristics due to variation in graphene chemical potential (μ_c) [22]. (b) (Top) Multilayer graphene metasurface biased at different levels to efficiently absorb incident radiation. (Bottom) Wideband absorption spectrum for THz waves for the normal incident condition [59]. (c) Description of polarization state modulation of a digital M-level signal by chiral graphene metasurface. The 4-level digital stream is fed to a limiter that converts it into the required chemical potential and consequently produces desired polarization state for the reflected field [98]. (d) The chiral biosensor is constructed by graphene metasurface supporting chiral reflection characteristics to test ligand-antigen bindings on the surface. The resonance frequency supports a distinct contour path traversed by locus tip of electric field vector in time for three different strains of influenza viruses H1N1, H5N2, and H9N2 [31].

metasurface that operates at THz frequencies [59]. The graphene layers are designed to generate quadrupolar localized surface plasmons that destructively interfere with the dipole mode. The patterned graphene layers are biased to operate at different chemical potential levels and backed up with dielectric substrates stacked on top of each other. Full-wave electromagnetic simulations demonstrate that the absorption spectrum is not only tunable but can be optimized to a large bandwidth of operation, *i.e.*, 6.9 THz bandwidth is obtained for over 90% normalized absorption. **Figure 13(c)** shows a schematic illustration of the design of graphene metasurface supporting polarization state modulation with high spectral efficiency [98]. The structural chirality of metasurface is utilized to generate chiral reflection along with highly dispersive Fano resonance. Several polarization states, including two orthogonal linearly polarized, right-and left-handed circular polarized reflections, are demonstrated for a narrow electro-optic tuning range of chemical potentials between 500 and 700 meV. By exploiting these properties, highly efficient modulation stages of modern communication systems can be designed. **Figure 13(d)** shows a polarization-state sensing setup to distinguish closely resembling optical properties of biomolecules such as viruses [31]. The measurement consists of a plasmonic metasurface with chiral unit cells, and the polarization properties of reflected fields can determine the optical characteristics of the analyte. It is shown that the proposed sensor can distinguish three closely resembling influenza virus strains i.e., H1N1, H5N2, and H9N2 based on the variation of the reflected polarization states.

5. Conclusion

In this chapter, we have reviewed the most recent advances in the field of metamaterial and metasurface-based electromagnetic perfect absorbers. We have first provided a thorough theoretical investigation describing the material and geometrical conditions that may lead to a near-perfect absorption of light. Based on the well-known optical theorem, this analysis gives a power balance and clarifies the design process in the microwaves and terahertz. Next, we have discussed a peculiar and interesting class of perfect absorbers that are hyperbolic metamaterial absorbers. These devices exploit a particular dispersion of hyperbolic media and lead to robust and tunable absorbers. We finally have discussed the newly proposed graphene plasmonics based absorbers, which exploit the high conductivity and tunable optical properties of this 2D material to build some of the most appealing and versatile absorbers, with applications spanning energy harvesting, biosensing, or light polarization manipulation. This chapter can be helpful to theorists and experimentalists alike, working on the design of novel absorbers of light or even other types of waves.

Author details

Pai-Yen Chen[1*], Mohamed Farhat[2], Zhilu Ye[1], Muhammad Amin[3], Hakan Bagci[1] and Danilo Erricolo[2]

1 Department of Electrical and Computer Engineering, University of Illinois Chicago, Chicago, IL, USA

2 Computer, Electrical, and Mathematical Science and Engineering (CEMSE) Division, King Abdullah University of Science and Technology (KAUST), Thuwal, Saudi Arabia

3 College of Engineering, Taibah University, Madinah, Saudi Arabia

*Address all correspondence to: pychen@uic.edu

References

[1] Mishra N, Kumari K, Chaudhary RK. An ultra-thin polarization independent quad-band microwave absorber-based on compact metamaterial structures for EMI/EMC applications. Int. J. Microw. Wireless Technol. 2018;**10**:422

[2] Holloway CL, DeLyser RR, German RF, McKenna P, Kanda M. Comparison of electromagnetic absorber used in anechoic and semi-anechoic chambers for emissions and immunity testing of digital devices. IEEE Transactions on Electromagnetic Compatibility. 1997;**39**:33

[3] Crawford ML, Workman JL, Thomas CL. Expanding the Bandwidth of TEM Cells for EMC Measurements. IEEE Transactions on Electromagnetic Compatibility. 1978;**EMC-20**:368

[4] Groh C, Karst JP, Koch M, Garbe H. TEM waveguides for EMC measurements. IEEE Transactions on Electromagnetic Compatibility. 1999;**41**:440

[5] Mishra SR, Pavlasek TJF. Design of Absorber-Lined Chambers for EMC Measurements Using a Geometrical Optics Approach. IEEE Transactions on Electromagnetic Compatibility. 1984;**EMC-26**:111

[6] L. Zhao, H. Liu, Z. He, and S. Dong, All-metal frequency-selective absorber/emitter for laser stealth and infrared stealth, Appl. Opt., AO **57**, 1757 (2018).

[7] Panwar R, Puthucheri S, Singh D, Agarwala V. Design of Ferrite–Graphene-Based Thin Broadband Radar Wave Absorber for Stealth Application. IEEE Transactions on Magnetics. 2015;**51**:1

[8] Chakradhary VK, Baskey HB, Roshan R, Pathik A, Akhtar MJ. Design of Frequency Selective Surface-Based Hybrid Nanocomposite Absorber for Stealth Applications. IEEE Transactions on Microwave Theory and Techniques. 2018;**66**:4737

[9] Kim J, Han K, Hahn JW. Selective dual-band metamaterial perfect absorber for infrared stealth technology. Scientific Reports. 2017;7:6740

[10] Kim T, Bae J, Lee N, Cho HH. Hierarchical Metamaterials for Multispectral Camouflage of Infrared and Microwaves. Adv. Funct. Mater. 2019;**29**:1807319

[11] Sabah C, Dincer F, Karaaslan M, Unal E, Akgol O, Demirel E. Perfect metamaterial absorber with polarization and incident angle independencies based on ring and cross-wire resonators for shielding and a sensor application. Optics Communications. 2014;**322**:137

[12] Gholampoor M, Movassagh-Alanagh F, Salimkhani H. Fabrication of nano-Fe3O4 3D structure on carbon fibers as a microwave absorber and EMI shielding composite by modified EPD method. Solid State Sciences. 2017;**64**:51

[13] Micheli D, Apollo C, Pastore R, Bueno Morles R, Laurenzi S, Marchetti M. Nanostructured composite materials for electromagnetic interference shielding applications. Acta Astronautica. 2011;**69**:747

[14] Farhat M, Chen P-Y, Bagci H, Amra C, Guenneau S, Alù A. Thermal invisibility based on scattering cancellation and mantle cloaking. Scientific Reports. 2015;**5**:9876

[15] Dincer F. Electromagnetic energy harvesting application based on tunable perfect metamaterial absorber. Journal of Electromagnetic Waves and Applications. 2015;**29**:2444

[16] Chung BK, Chuah HT. Design and construction of a multipurpose

wideband anechoic chamber. IEEE Antennas and Propagation Magazine. 2003;**45**:41

[17] S. F. Gregson, J. Dupuy, C. G. Parini, A. C. Newell, and G. E. Hindman, in 2011 Loughborough Antennas Propagation Conference (2011), pp. 1–4.

[18] Qu Y, Li Q, Gong H, Du K, Bai S, Zhao D, et al. Spatially and Spectrally Resolved Narrowband Optical Absorber Based on 2D Grating Nanostructures on Metallic Films. Advanced Optical Materials. 2016;**4**:480

[19] Landy NI, Sajuyigbe S, Mock JJ, Smith DR, Padilla WJ. Perfect Metamaterial Absorber. Phys. Rev. Lett. 2008;**100**:207402

[20] Tao H, Landy NI, Bingham CM, Zhang X, Averitt RD, Padilla WJ. A metamaterial absorber for the terahertz regime: Design, fabrication and characterization, Opt. Express, OE. 2008;**16**(7181)

[21] B.-Y. Wang, S.-B. Liu, B.-R. Bian, Z.-W. Mao, X.-C. Liu, B. Ma, and L. Chen, A novel ultrathin and broadband microwave metamaterial absorber, Journal of Applied Physics **116**, 094504 (2014).

[22] Alaee R, Farhat M, Rockstuhl C, Lederer F. A perfect absorber made of a graphene micro-ribbon metamaterial, Opt. Express, OE. 2012;**20**(28017)

[23] Cao T, Wei C, Simpson RE, Zhang L, Cryan MJ. Broadband Polarization-Independent Perfect Absorber Using a Phase-Change Metamaterial at Visible Frequencies. Scientific Reports. 2014;**4**:3955

[24] Zhang Y, Feng Y, Zhu B, Zhao J, Jiang T. Graphene based tunable metamaterial absorber and polarization modulation in terahertz frequency, Opt. Express, OE. 2014;**22**(22743)

[25] Zhou J, Kaplan AF, Chen L, Guo LJ. Experiment and Theory of the Broadband Absorption by a Tapered Hyperbolic Metamaterial Array. ACS Photonics. 2014;**1**:618

[26] D. Chaurasiya, S. Ghosh, S. Bhattacharyya, and K. V. Srivastava, in 2014 IEEE International Microwave and RF Conference (IMaRC) (2014), pp. 96–99.

[27] Singh AK, Abegaonkar MP, Koul SK. A Triple Band Polarization Insensitive Ultrathin Metamaterial Absorber for S- C- and X-Bands. Progress In Electromagnetics Research M. 2019;**77**: 187

[28] W. Li, X. Zhou, Y. Ying, X. Qiao, F. Qin, Q. Li, and S. Che, Polarization-insensitive wide-angle multiband metamaterial absorber with a double-layer modified electric ring resonator array, AIP Advances **5**, 067151 (2015).

[29] Liu T, Cao X, Gao J, Zheng Q, Li W, Yang H. RCS Reduction of Waveguide Slot Antenna With Metamaterial Absorber. IEEE Transactions on Antennas and Propagation. 2013;**61**: 1479

[30] Luo S, Zhao J, Zuo D, Wang X. Perfect narrow band absorber for sensing applications, Opt. Express, OE. 2016;**24**(9288)

[31] Amin M, Siddiqui O, Abutarboush H, Farhat M, Ramzan R. A THz graphene metasurface for polarization selective virus sensing. Carbon. 2021;**176**:580

[32] Liu N, Mesch M, Weiss T, Hentschel M, Giessen H. Infrared Perfect Absorber and Its Application As Plasmonic Sensor. Nano Lett. 2010;**10**: 2342

[33] Zhang L, Farhat M, Salama KN. Spectrometer-Free Graphene

Plasmonics Based Refractive Index Sensor. Sensors. 2020;**20**:2347

[34] Landy NI, Bingham CM, Tyler T, Jokerst N, Smith DR, Padilla WJ. Design, theory, and measurement of a polarization-insensitive absorber for terahertz imaging. Phys. Rev. B. 2009; **79**:125104

[35] Chen P-Y, Alù A. Subwavelength Imaging Using Phase-Conjugating Nonlinear Nanoantenna Arrays. Nano Lett. 2011;**11**:5514

[36] Wesemann L, Panchenko E, Singh K, Della Gaspera E, Gómez DE, Davis TJ, et al. Selective near-perfect absorbing mirror as a spatial frequency filter for optical image processing. APL Photonics. 2019;**4**:100801

[37] Sharawi MS, Numan AB, Khan MU, Aloi DN. A Dual-Element Dual-Band MIMO Antenna System With Enhanced Isolation for Mobile Terminals. IEEE Antennas and Wireless Propagation Letters. 2012;**11**:1006

[38] Zhang S, Ying Z, Xiong J, He S. Ultrawideband MIMO/Diversity Antennas With a Tree-Like Structure to Enhance Wideband Isolation. IEEE Antennas and Wireless Propagation Letters. 2009;**8**:1279

[39] Jackson JD. Classical Electrodynamics. 3rd ed. New York: Wiley; 1999

[40] Pendry JB, Holden AJ, Robbins DJ, Stewart WJ. Magnetism from conductors and enhanced nonlinear phenomena. IEEE Transactions on Microwave Theory and Techniques. 1999;**47**:2075

[41] D. M. Pozar, Microwave Engineering (John Wiley & Sons, 2011).

[42] H. W. Ott, Electromagnetic Compatibility Engineering (John Wiley & Sons, 2011).

[43] J. D. Baena, R. Marqués, F. Medina, and J. Martel, Artificial magnetic metamaterial design by using spiral resonators, Phys. Rev. B **69**, 014402 (2004).

[44] Bilotti F, Toscano A, Vegni L, Aydin K, Alici KB, Ozbay E. Equivalent-Circuit Models for the Design of Metamaterials Based on Artificial Magnetic Inclusions. IEEE Transactions on Microwave Theory and Techniques. 2007;**55**:2865

[45] Pendry JB. Negative Refraction Makes a Perfect Lens. Phys. Rev. Lett. 2000;**85**:3966

[46] Shelby RA, Smith DR, Schultz S. Experimental Verification of a Negative Index of Refraction. Science. 2001;**292**: 77

[47] Schurig D, Mock JJ, Justice BJ, Cummer SA, Pendry JB, Starr AF, et al. Metamaterial Electromagnetic Cloak at Microwave Frequencies. Science. 2006; **314**:977

[48] M. Farhat, P.-Y. Chen, S. Guenneau, and S. Enoch, Transformation Wave Physics: Electromagnetics, Elastodynamics, and Thermodynamics (CRC Press, 2016).

[49] S. A. Maier, Plasmonics: Fundamentals and Applications (Springer Science & Business Media, 2007).

[50] Brolo AG. Plasmonics for future biosensors. Nature Photonics. 2012;**6**:709

[51] Xiao S, Liu L, Qiu M. Resonator channel drop filters in a plasmon-polaritons metal, Opt. Express, OE. 2006;**14**(2932)

[52] Liu Y, Cheng R, Liao L, Zhou H, Bai J, Liu G, et al. Plasmon resonance enhanced multicolour photodetection by graphene. Nature Communications. 2011;**2**:579

[53] Ma R-M, Oulton RF, Sorger VJ, Bartal G, Zhang X. Room-temperature sub-diffraction-limited plasmon laser by total internal reflection. Nature Materials. 2011;**10**:110

[54] C. Argyropoulos, P.-Y. Chen, G. D'Aguanno, N. Engheta, and A. Alù, Boosting optical nonlinearities in ε-near-zero plasmonic channels, Phys. Rev. B **85**, 045129 (2012).

[55] Grosjean T, Mivelle M, Baida FI, Burr GW, Fischer UC. Diabolo Nanoantenna for Enhancing and Confining the Magnetic Optical Field. Nano Lett. 2011;**11**:1009

[56] Kildal P-S, Alfonso E, Valero-Nogueira A, Rajo-Iglesias E. Local Metamaterial-Based Waveguides in Gaps Between Parallel Metal Plates. IEEE Antennas and Wireless Propagation Letters. 2009;**8**:84

[57] Teperik TV, García de Abajo FJ, Borisov AG, Abdelsalam M, Bartlett PN, Sugawara Y, et al. Omnidirectional absorption in nanostructured metal surfaces. Nature Photonics. 2008; **2**:299

[58] Y. D. Chong, L. Ge, H. Cao, and A. D. Stone, Coherent Perfect Absorbers: Time-Reversed Lasers, Phys. Rev. Lett. **105**, 053901 (2010).

[59] Amin M, Farhat M, Bağcı H. An ultra-broadband multilayered graphene absorber, Opt. Express, OE. 2013;**21** (29938)

[60] Aalizadeh M, Khavasi A, Butun B, Ozbay E. Large-Area, Cost-Effective, Ultra-Broadband Perfect Absorber Utilizing Manganese in Metal-Insulator-Metal Structure. Scientific Reports. 2018;**8**:9162

[61] Rephaeli E, Fan S. Absorber and emitter for solar thermo-photovoltaic systems to achieve efficiency exceeding the Shockley-Queisser limit, Opt. Express, OE. 2009;**17**(15145)

[62] M. Diem, T. Koschny, and C. M. Soukoulis, Wide-angle perfect absorber/thermal emitter in the terahertz regime, Phys. Rev. B **79**, 033101 (2009).

[63] M. Kerker, The Scattering of Light and Other Electromagnetic Radiation: Physical Chemistry: A Series of Monographs (Academic Press, 2013).

[64] Kwon D-H, Pozar DM. Optimal Characteristics of an Arbitrary Receive Antenna. IEEE Transactions on Antennas and Propagation. 2009;**57**:3720

[65] Fante RL, McCormack MT. Reflection properties of the Salisbury screen. IEEE Transactions on Antennas and Propagation. 1988;**36**:1443

[66] Chen P-Y, Argyropoulos C, D'Aguanno G, Alù A. Enhanced Second-Harmonic Generation by Metasurface Nanomixer and Nanocavity. ACS Photonics. 2015;**2**:1000

[67] A. Alù, First-principles homogenization theory for periodic metamaterials, Phys. Rev. B **84**, 075153 (2011).

[68] S. Tretyakov, Analytical Modeling in Applied Electromagnetics (Artech House, 2003).

[69] P. A. Belov and C. R. Simovski, Homogenization of electromagnetic crystals formed by uniaxial resonant scatterers, Phys. Rev. E **72**, 026615 (2005).

[70] G. N. Watson, A Treatise on the Theory of Bessel Functions (Cambridge University Press, 1995).

[71] M. Abramowitz and I. A. Stegun, Handbook of Mathematical Functions with Formulas, Graphs, and Mathematical Tables (U.S. Government Printing Office, 1964).

[72] Simovski CR, Zouhdi S, Yatsenko VV. Electromagnetic interaction in dipole grids and prospective high-impedance surfaces. Radio Science. 2005;**40**:1

[73] Smith DR, Padilla WJ, Vier DC, Nemat-Nasser SC, Schultz S. Composite Medium with Simultaneously Negative Permeability and Permittivity. Phys. Rev. Lett. 2000;**84**:4184

[74] D. R. Smith, J. Gollub, J. J. Mock, W. J. Padilla, and D. Schurig, Calculation and measurement of bianisotropy in a split ring resonator metamaterial, Journal of Applied Physics **100**, 024507 (2006).

[75] T. Driscoll, G. O. Andreev, D. N. Basov, S. Palit, S. Y. Cho, N. M. Jokerst, and D. R. Smith, Tuned permeability in terahertz split-ring resonators for devices and sensors, Appl. Phys. Lett. **91**, 062511 (2007).

[76] A. K. Azad, A. J. Taylor, E. Smirnova, and J. F. O'Hara, Characterization and analysis of terahertz metamaterials based on rectangular split-ring resonators, Appl. Phys. Lett. **92**, 011119 (2008).

[77] Rockstuhl C, Zentgraf T, Guo H, Liu N, Etrich C, Loa I, et al. Resonances of split-ring resonator metamaterials in the near infrared. Appl. Phys. B. 2006; **84**:219

[78] Le KQ, Bai J, Ngo QM, Chen P-Y. Fabrication and Numerical Characterization of Infrared Metamaterial Absorbers for Refractometric Biosensors. Journal of Elec Materi. 2017;**46**:668

[79] X. Zhao, J. Zhang, K. Fan, G. Duan, G. D. Metcalfe, M. Wraback, X. Zhang, and R. D. Averitt, Nonlinear terahertz metamaterial perfect absorbers using GaAs [Invited], Photon. Res., PRJ **4**, A16 (2016).

[80] Asadchy VS, Díaz-Rubio A, Tretyakov SA. Bianisotropic metasurfaces: physics and applications. Nanophotonics. 2018;**7**:1069

[81] Liang F, Hanson GW, Yakovlev AB, Lovat G, Burghignoli P, Araneo R, et al. Dyadic Green's Functions for Dipole Excitation of Homogenized Metasurfaces. IEEE Transactions on Antennas and Propagation. 2016;**64**:167

[82] Yatsenko VV, Maslovski SI, Tretyakov SA, Prosvirnin SL, Zouhdi S. Plane-wave reflection from double arrays of small magnetoelectric scatterers. IEEE Transactions on Antennas and Propagation. 2003;**51**:2

[83] Zhong S, He S. Ultrathin and lightweight microwave absorbers made of mu-near-zero metamaterials. Scientific Reports. 2013;**3**:2083

[84] Chen P-Y, Farhat M, Bağcı H. Graphene metascreen for designing compact infrared absorbers with enhanced bandwidth. Nanotechnology. 2015;**26**:164002

[85] Wang Z, Zhao L, Cai Y, Zheng S, Yin Y. A Meta-Surface Antenna Array Decoupling (MAAD) Method for Mutual Coupling Reduction in a MIMO Antenna System. Scientific Reports. 2018;**8**:3152

[86] Poddubny A, Iorsh I, Belov P, Kivshar Y. Hyperbolic metamaterials. Nature Photonics. 2013;**7**:948

[87] M. Y. Shalaginov, V. V. Vorobyov, J. Liu, M. Ferrera, A. V. Akimov, A. Lagutchev, A. N. Smolyaninov, V. V. Klimov, J. Irudayaraj, A. V. Kildishev, A. Boltasseva, and V. M. Shalaev, Enhancement of single-photon emission from nitrogen-vacancy centers with TiN/(Al,Sc)N hyperbolic metamaterial, Laser & Photonics Reviews **9**, 120 (2015).

[88] Sakhdari M, Hajizadegan M, Farhat M, Chen P-Y. Efficient, broadband and wide-angle hot-electron

transduction using metal-semiconductor hyperbolic metamaterials. Nano Energy. 2016;**26**:371

[89] P.-Y. Chen, M. Hajizadegan, M. Sakhdari, and A. Alù, Giant Photoresponsivity of Midinfrared Hyperbolic Metamaterials in the Photon-Assisted-Tunneling Regime, Phys. Rev. Applied **5**, 041001 (2016).

[90] T. Guo, L. Zhu, P.-Y. Chen, and C. Argyropoulos, Tunable terahertz amplification based on photoexcited active graphene hyperbolic metamaterials [Invited], Opt. Mater. Express, OME **8**, 3941 (2018).

[91] M. Hajizadegan, M. Sakhdari, and P.-Y. Chen, in Micro- and Nanotechnology Sensors, Systems, and Applications X (International Society for Optics and Photonics, 2018), p. 106390M.

[92] N. Marcuvitz and M. I. of T. R. Laboratory, Waveguide Handbook (IET, 1951).

[93] Rozanov KN. Ultimate thickness to bandwidth ratio of radar absorbers. IEEE Transactions on Antennas and Propagation. 2000;**48**:1230

[94] Yin X, Long C, Li J, Zhu H, Chen L, Guan J, et al. Ultra-wideband microwave absorber by connecting multiple absorption bands of two different-sized hyperbolic metamaterial waveguide arrays. Scientific Reports. 2015;**5**:15367

[95] E. F. Knott, J. F. Schaeffer, and M. T. Tulley, Radar Cross Section (SciTech Publishing, 2004).

[96] Watts CM, Liu X, Padilla WJ. Metamaterial Electromagnetic Wave Absorbers. Advanced Materials. 2012; **24**:OP98

[97] Amin M, Siddiqui O, Farhat M, Khelif A. A perfect Fresnel acoustic reflector implemented by a Fano-resonant metascreen. Journal of Applied Physics. 2018;**123**:144502

[98] M. Amin, O. Siddiqui, and M. Farhat, Polarization-State Modulation in Fano Resonant Graphene Metasurface Reflector, Journal of Lightwave Technology 1 (2021).

[99] Ouyang L, Wang W, Rosenmann D, Czaplewski DA, Gao J, Yang X. Near-infrared chiral plasmonic metasurface absorbers. Opt. Express. 2018;**26**:31484

[100] M. Amin, O. Siddiqui, W. Orfali, M. Farhat, and A. Khelif, Resonant Beam Steering and Carpet Cloaking Using an Acoustic Transformational Metascreen, Phys. Rev. Applied **10**, 064030 (2018).

[101] Martini E, Mencagli M, Maci S. Metasurface transformation for surface wave control, Philosophical Transactions of the Royal Society A: Mathematical. Physical and Engineering Sciences. 2015;**373**:20140355

[102] Chen P-Y, Soric J, Padooru YR, Bernety HM, Yakovlev AB, Alù A. Nanostructured graphene metasurface for tunable terahertz cloaking, New J. Phys. 2013;**15**:123029

[103] L. Hsu, T. Lepetit, and B. Kanté, Extremely Thin Dielectric Metasurface for Carpet Cloaking, ArXiv:1503.08486 [Physics] (2015).

[104] Buchnev O, Podoliak N, Kaczmarek M, Zheludev NI, Fedotov VA. Electrically Controlled Nanostructured Metasurface Loaded with Liquid Crystal: Toward Multifunctional Photonic Switch. Advanced Optical Materials. 2015;**3**:674

[105] Komar A, Paniagua-Domínguez R, Miroshnichenko A, Yu YF, Kivshar YS, Kuznetsov AI, et al. Dynamic Beam Switching by Liquid Crystal Tunable Dielectric Metasurfaces. ACS Photonics. 2018;**5**:1742

[106] Artiga X, Bresciani D, Legay H, Perruisseau-Carrier J. Polarimetric Control of Reflective Metasurfaces. IEEE Antennas and Wireless Propagation Letters. 2012;**11**:1489

[107] Wu X, Meng Y, Wang L, Tian J, Dai S, Wen W. Anisotropic metasurface with near-unity circular polarization conversion. Appl. Phys. Lett. 2016;**108**: 183502

[108] Zhu HL, Cheung SW, Chung KL, Yuk TI. Linear-to-Circular Polarization Conversion Using Metasurface. IEEE Transactions on Antennas and Propagation. 2013;**61**:4615

[109] Koschny T, Kafesaki M, Economou EN, Soukoulis CM. Effective Medium Theory of Left-Handed Materials. Phys. Rev. Lett. 2004;**93**: 107402

[110] R. Liu, T. J. Cui, D. Huang, B. Zhao, and D. R. Smith, Description and explanation of electromagnetic behaviors in artificial metamaterials based on effective medium theory, Phys. Rev. E **76**, 026606 (2007).

[111] C. R. Simovski, On electromagnetic characterization and homogenization of nanostructured metamaterials, J. Opt. **13**, 013001 (2010).

[112] M. A. El-Aasser, Design optimization of nanostrip metamaterial perfect absorbers, JNP **8**, 083085 (2014).

[113] Hedayati MK, Javaherirahim M, Mozooni B, Abdelaziz R, Tavassolizadeh A, Chakravadhanula VSK, et al. Design of a Perfect Black Absorber at Visible Frequencies Using Plasmonic Metamaterials. Advanced Materials. 2011; **23**:5410

[114] Liu D, Wang L, Cui Q, Guo LJ. Planar Metasurfaces Enable High-Efficiency Colored Perovskite Solar Cells. Advanced Science. 2018;**5**:1800836

[115] D. B. Durham, S. R. Loria, F. Riminucci, K. Kanellopulos, X. Shen, F. Ciabattini, A. Mostacci, P. Musumeci, A. M. Minor, S. Cabrini, and D. Filippetto, in Plasmonics: Design, Materials, Fabrication, Characterization, and Applications XVIII (International Society for Optics and Photonics, 2020), p. 1146222.

[116] Locatelli A, Capobianco A-D, Midrio M, Boscolo S, De Angelis C. Graphene-assisted control of coupling between optical waveguides. Opt. Express. 2012;**20**:28479

[117] Farhat M, Rockstuhl C, Bağcı H. A 3D tunable and multi-frequency graphene plasmonic cloak, Opt. Express, OE. 2013;**21**(12592)

[118] Filter R, Farhat M, Steglich M, Alaee R, Rockstuhl C, Lederer F. Tunable graphene antennas for selective enhancement of THz-emission, Opt. Express, OE. 2013;**21**(3737)

[119] A. Vakil and N. Engheta, Fourier optics on graphene, Phys. Rev. B **85**, 075434 (2012).

[120] Gan X, Mak KF, Gao Y, You Y, Hatami F, Hone J, et al. Strong Enhancement of Light–Matter Interaction in Graphene Coupled to a Photonic Crystal Nanocavity. Nano Lett. 2012;**12**:5626

[121] Amin M, Farhat M, Bağcı H. A dynamically reconfigurable Fano metamaterial through graphene tuning for switching and sensing applications. Scientific Reports. 2013;**3**:2105

Section 4

Measurements and Emission

Chapter 10

EMI Pre-Compliance Measurements Reveal Sources of Interference

Bystrík Dolník

Abstract

The chapter focuses on the electromagnetic compatibility of a prototype of electrical equipment such as street lighting with various LED controllers, LED information boards, and audio equipment. The requirements for the harmonic content of the input currents of the conducted emission power lines are used as a basis for the analysis of compliance with the EMC standards. The results obtained from the experiments indicated that some commercially produced voltage drivers are not compatible with the requirements for the harmonic current content of the input line. The problem is caused by two factors: a bad design by the manufacturer or the wrong LED driver design concerning the rated load. EMI radiation measurements indicate the need to precisely design all functional blocks placed on a PCB with suitable grounding and shielding techniques. This chapter is intended for engineers and researchers working in the development of electrical equipment as well as the general public interested in EMC issues.

Keywords: EMI, coupling, LED driver, LED information board, audio device

1. Introduction

Today, no one doubts that the production and the use of large amounts of electrical equipment and electronics have brought an important requirement: electromagnetic compatibility (EMC). It turns out that the requirements for an electrical product to comply with EMC have two aspects. The first aspect follows naturally to ensure that the product complies with EMC regulations. The second aspect, still influenced by environmental experts whose efforts are to monitor the long-term effects of electromagnetic fields generated by technical devices in the environment on living organisms. It is likely that with the growing awareness of the population, in the future, more attention will pay to the effects of electromagnetic fields acting simultaneously from multiple electrical devices on the environment.

According to the International Electrotechnical Vocabulary (IEV) IEV 161-01- 07, EMC is the ability of an equipment or system to function satisfactorily in its electromagnetic environment without introducing intolerable electromagnetic disturbances to anything in that environment [1]. Electrical equipment and electrical systems are always exposed to internal and external electromagnetic disturbances. However, it should be noted that each electrical device is at the same time more or less a source of electromagnetic interference (EMI) [2–6].

IntechOpen

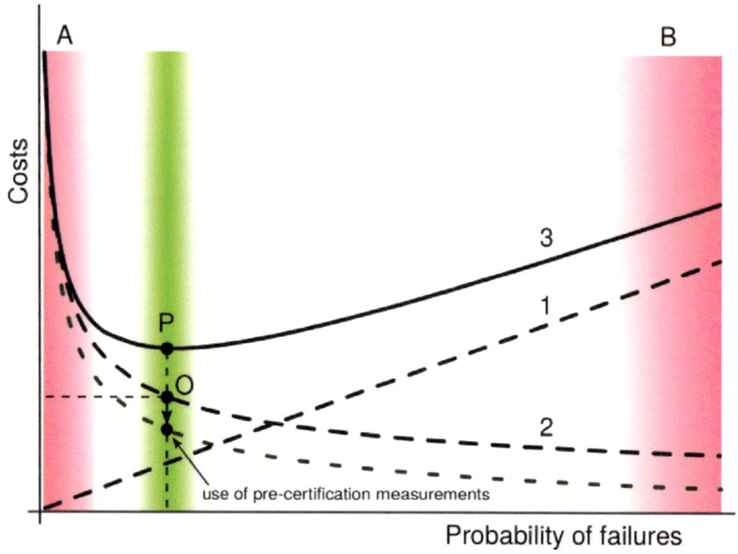

Figure 1.
Optimal financial costs for EMC provision: Point O; 1 – Operating costs, 2 – Costs of ensuring EMC compliance, 3 – Total costs; pre-compliance measurements shift the point O to lower values.

If the EMI level of the equipment under test[1] (EUT) is below the permitted EMC level of the standard, the EMC compatibility goal of the product is achieved. Otherwise, further measures are needed to reduce the EMI level from the EUT. The role of the engineers is to adjust the internal layout of the functional blocks in the EUT and connect (or disconnect) some components to the EUT to eliminate EMI, especially according to their professional experience. Subsequently, the results of the modifications are verified by re-sending the EUT to the EMC test facility. The process of making changes to reduce EMI and then repeat testing in the EMC test room can be time-consuming and costly until the EUT passes the EMC test. Every manufacturer of electrical equipment wants to gain a foothold in the market and sell the product. The necessary condition for selling products on the marketplace is to obtain an EMC certificate of conformity. The manufacturer must therefore take into account the costs to ensure the EMC conformity of the product.

In recent years much effort has been made to reduce the cost of EMC compliance and speed up the EMC compliance process. The EMC pre-compliance methodology in the test room or at the open site of the product manufacturer is proposed in the early 1990s. In the initial phase, EMC preliminary conformity measurements were performed in a product manufacturer's test facility with standard EMC test equipment and standard test methods. Later, more flexible EMC compliance measurements were developed and introduced an alternative test apparatus or test methodology to reduce compliance costs [7].

The selection of specific permissible levels and the related reserves for the devices are not prescribed. It is within the competence of the device manufacturer. If the manufacturer chooses too large EMC reserves, then unnecessarily high financial costs for interference are incurred. Conversely, if the EMC reserve is too small, there is a greater risk that the equipment will not pass EMC certification. Subsequent measures must take to achieve EMC.

---

[1] In some professional and scientific sources such as journals, standards, and books, DUT (Device Under Test) is used instead of EUT.

The graph in **Figure 1** represents a typical course of financial costs for the production and operation of technical equipment depending on the probability of failures. It is clear from the graph that as the number of failures increases, additional operating costs increase. Conversely, the failure probability by failing to comply with EMC principles is inversely proportional to the amount of investment in providing EMC equipment.

Therefore, the total operating costs of the equipment are the sum of the two mentioned components, which have the minimum marked as point P in **Figure 1**. The part of the EMC compliance investment of the equipment must choose (point O) so that the total costs are equal to the minimum.

The optimal costs of providing EMC should be about 2–10% of the development and production costs. Under certain conditions, the total cost of providing EMC is less than 1%, which can be considered a success [8]. The area marked with the letter "A" corresponds to too high a cost of providing EMC at the planning stage. The area marked with the letter "B" corresponds to the conditions under which the costs of ensuring EMC are neglect at the planning stage of the equipment production, which results in a high rate of failures.

Pre-certification EMC pre-compliance measurements significantly reduce the cost of developing new electrical equipment and shorten the time to market. Thus, point O in the graph in **Figure 1** defining the optimal cost for EMC provision shifts towards lower values.

2. EMC pre-compliance measurements

Data from EMC testing laboratories report that 85% of products submitted for final EMC compliance testing fail for the first time. However, using pre-compliance EMC measurements, it is possible to reverse these statistics so that products pass EMC tests successfully [9].

Performing pre-compliance measurements significantly increases the probability of a successful first transition to complete EMI compliance testing, saves time and money. Large companies developing products for medical, automotive, military, and other industrial applications perform pre-compliance measurements as part of a standard procedure. Small companies and startups are starting to follow this way as well, as they can benefit from investing in setting up pre-compliance measurements.

Pre-compliance EMC measurements can define as the ability to perform internal testing of EMC products before going to the final certified testing laboratory. In doing so, pre-compliance measurements can be performed by the manufacturer either internally or in agreement with an experienced development laboratory.

Of course, there are even more reasons to do pre-compliance measurements, which are as follows:

1. Ensure that newly developed equipment has sufficient immunity to maintain the specified parameters in the presence of EMI.

2. Early identification of potential EMC issues. It allows detection problems with EMC of developing products before testing in a certified laboratory using custom testing.

3. Reduction of EMC related costs. Products are more likely to be certified for the first time, avoiding additional development time and costs.

4. Accelerates time to market by reducing delays.

5. Immediate feedback on the impact of design changes.

6. An immediate inspection of the modifications made to the product results in the acquisition of experience in the design of other products.

7. Better communication between test and development groups provides fast real-time feedback from internal sources.

Figure 2 shows very clearly an example of the process of product development using pre-compliance measurements. Whether the product passes the individual stages of development or not is determined by the results of not only pre-compliance measurements. Although pre-compliance tests may seem to have only advantages, they have a higher measurement uncertainty than compliance EMC tests due to their simplification. However, this disadvantage is many times compensated by savings not only in money but especially in time during product development.

More complex electrical equipment contains several functional blocks, such as power supply circuits, switching power supply, input/output (I/O) circuits, sensitive analog circuits, digital circuits, high-frequency (HF) circuits, converters, peripherals, control circuits, I/O connectors, and filters, reference ground plane. These circuits must arrange in such a way that they do not interfere with each other and that they are at the same time sufficiently immune to external electromagnetic fields. **Figure 3** shows an example of the optimal arrangement of circuits on a PCB in compliance with EMC principles.

The precise design of a printed circuit board (PCB), and the selection of quality components, play a significant role in the product prototype design. The PCB designer must have good experience to eliminate interference generation by precise PCB design. Currently, CAD software can use in PCB design, which provides the ability to select different tracing strategies. In the design of PCBs, the generation of Joule losses and methods of loss dissipation must also take into account.

Some researchers are working on methods for diagnosing and managing enhanced EMC for a specific area of EMC, such as mobile phones. The authors in [10] constructed a knowledge graph that presents interference and sensitive units based on mathematical rules. EMC diagnostic and management reports are generated by searching knowledge graphs with an extracted entity and parameter information. The authors performed an experiment and found that the proposed intelligent method of EMC diagnostics and control significantly increased the efficiency of the calculation, saved storage space, and increased the accuracy of identification.

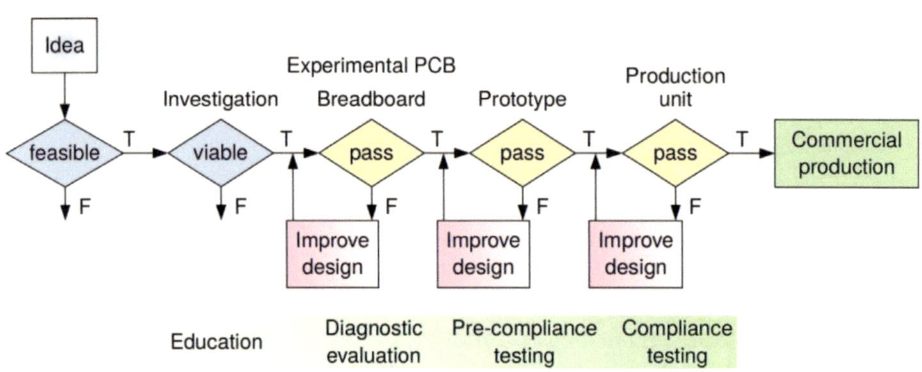

Figure 2.
Product development cycle; the meaning of capital letters is: T – true, F – false.

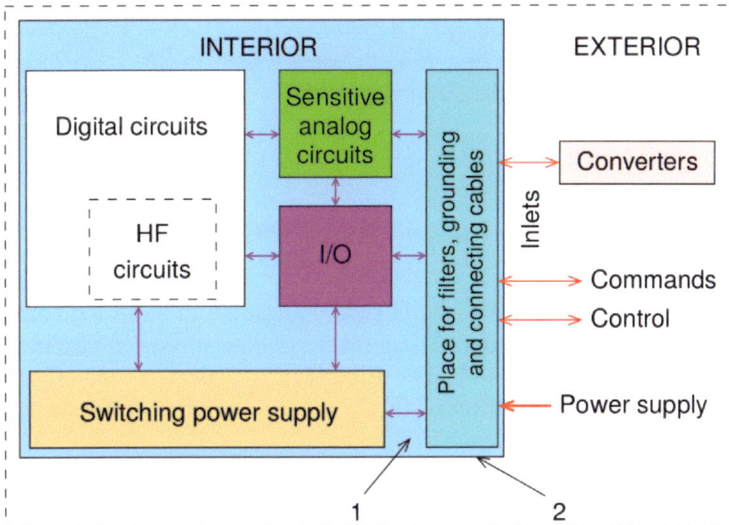

Figure 3.
Circuit layout on the PCB in compliance with EMC principles: 1 – The necessary connections must filter, 2 – The reference ground plane located below all elements and conductive surfaces.

2.1 Solving EMI problems during pre-compliance measurements

When solving EMI problems during pre-compliance measurements, engineers face several challenges. Some EMI problems can be detected and eliminated relatively quickly and easily; some are more complicated and time-consuming. Therefore, engineers must address the technical side of the problem along with the requirement to minimize costs. Remember that each metal wire in an electrical circuit acts simultaneously as an EMI source and an EMI receiver. Thus, any interference can affect the electrical equipment either by metallic conductors or by electromagnetic waves. A distinction exists between EMI by conduction (conducted Interference) and EMI by radiation (radiated interference).

Radiated interference is due to radio noise or unwanted signals over the air, not through a physical medium. Conducted interference is due to conducted radio noise or unwanted signals entering a victim by direct coupling (via cables).

According to the IEV 161-03-28, a radiated disturbance is an electromagnetic disturbance for which the energy is transferred through space in the form of electromagnetic waves; IEV 161-03-27 conducted disturbance is an electromagnetic disturbance for which the energy is transferred via one or more conductors. IEV 161-01-05 defines electromagnetic disturbance as an electromagnetic phenomenon that can degrade the performance of a device, equipment or system, or adversely affect living or inert matter [1]. Strategies based on theoretical knowledge of circuit technology, micro-electronics, and professional experience are used to detect EMI sources. The EMI problems that need to address are closely related to the electromagnetic couplings between the source and the victim. The procedure can be as follows:

1. all EMI sources must detect on the PCB,

2. check interference levels in all metallic conductors,

3. short-distance EMI verification of the prototype (1 m or 3 m).

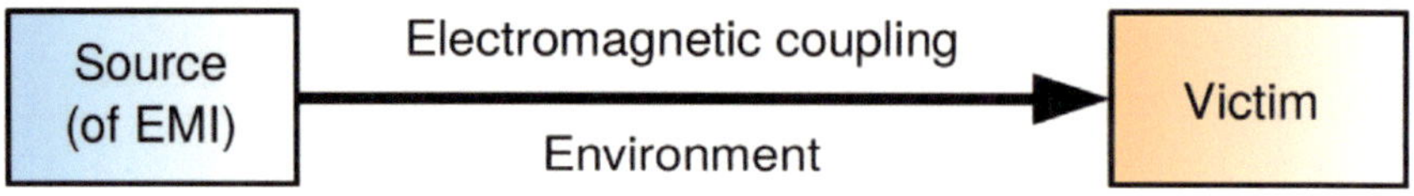

Figure 4.
EMC base chain.

2.1.1 Detect all EMI sources on the PCB

Use near-field probes (either E- or H-field). These small hand-held antennas are mainly used in the development and diagnostics of electronic devices, monitoring the radiation of components and blocks directly inside the developed device, for the most accurate search for the source of the interfering signal at the area of interest or for detecting electromagnetic leaks in shields.

Measurements using a near-field probe are not regulated by any standards. It is only a matter of determining the relative degree of interfering radiation in a given place or a given circuit. EMI sources generally include fast-edge digital signals. When the product contains a shielded enclosure, check the area around the joined parts and near the holes.

2.1.2 Check interference levels in all metallic conductors

Use a current probe to measure high-frequency currents flowing through a wire quickly. When the laboratory is better equipped, a line impedance stabilization network[2] (LISN) connection can use. Remember that cables are the most likely structure emitting radio-frequency (RF) energy. During the measurement, move the probe back and forth along the cable to maximize the highest currents. Be careful not to affect the measurements with the user's own body (hand movement).

2.1.3 Verify short-distance EMI

Use a short-distance antenna to find out which of the harmonic contents radiates. Then compare these harmonics with indoor measurements and EMI measurements in cables to determine the most likely energy sources that are coupling to and radiate through the wires or slots.

2.2 Measures used to eliminate EMI

The technical measures used to reduce EMI in any EMC chain are called interference suppressors. Interference suppressors include filters, surge protection devices (SPDs), shielding, and grounding. Through these measures, the electromagnetic coupling between the source and the victim is eliminated. **Figure 4** shows the relationship between the EMI source and the victim via coupling.

SPDs eliminate the destructive effect of current pulses and high-energy voltage pulses. EMC grounding diverts interfering signals from the victim, creates a reference ground plane on PCBs, grounds the cable shield and metal enclosures for safety purposes. The shielding attenuates EMI propagated by radiation.

[2] The LISN is a low pass filter located between the AC or DC power source and the DUT to create a known impedance and provide a port for measuring HF noise. The main function of the LISN is to provide an accurate impedance to the EUT power port. It also isolates unwanted radio-frequency signals from a power source.

In EMC, the couplings are classified as follows:

- galvanic,
- inductive,
- capacitive,
- and electromagnetic coupling.

Knowing the principle of how coupling works using equivalent models, a procedure can make to reduce coupling [11]. **Figure 5** shows simplified equivalent circuits for galvanic and inductive coupling between the source and the victim. The magnitude of the interfering voltage on the series impedance according to **Figure 5a** can calculate as follows

$$u_{\mathrm{EMI}}(t) = R \cdot i(t) + L\,\frac{\mathrm{d}i(t)}{\mathrm{d}t} \tag{1}$$

where R is the resistance of the conductor, L is the self-inductance of the conductor, and $i(t)$ is the disturbing current. According to Eq. (1), several measures can perform to eliminate the interfering voltage. For example, reduce the voltage drop across resistor R and inductance L. The resistance R can reduce by shortening the length of the conductors, reducing the number of bends, or increasing the cross-section of the conductors. These resistivity reduction measures follow from Eq. (2)

$$R = \rho \cdot \frac{l}{S} \tag{2}$$

where l is the length of the conductor, S is the conductor cross-section, and ϱ is the specific resistivity of the conductor. The voltage drop across the reactance component is done by reducing the inductance L or reducing the time changes $(\mathrm{d}i(t)/\mathrm{d}t)$

Figure 5.
Equivalent circuit: Galvanic coupling (a) and inductive coupling (b); R – Resistance, L – Self-inductance, M_{12} – Mutual inductance between wires, i – Disturbing current, u_{EMI} – Disturbing voltage, H – Magnetic field strength.

of the interfering current. The magnitude of the interfering voltage across the inductive coupling can calculate as follows

$$u_{\mathrm{EMI}}(t) = M_{12}\frac{\mathrm{d}i(t)}{\mathrm{d}t} \quad (3)$$

where $i(t)$ is the disturbing current and M_{12} is the mutual inductance between wires. The mutual inductance between the conductors is bound to the magnetic flux as follows

$$\Phi_{12} = B_1 \cdot S_2 \quad (4)$$

where S_2 is the area bounding the victim's circumference, and B_1 is the magnetic flux density generated by the EMI source. Based on (Eqs. (3) and (4)), the following measures can use to reduce the inductive coupling: reducing the mutual inductance by shortening the length of the parallel conductors, increasing the mutual distance between the circuits, orthogonal distribution of the circuits, and others. Similarly, an analysis of equivalent circuits for capacitive coupling and electromagnetic coupling can perform.

3. Precompliance measurements examples

In the following section, we will show three examples of pre-compliance measurements. The first case will concern the detection of EMI sources in a prototype of an LED street light. The second case will matter the detection of EMI sources from the LED information board. Finally, the third case points to unexpected changes in EMI radiation during the development of audio equipment.

3.1 Prototype of a LED street light

The best choice to achieve the highest system efficiency is a switched-mode power supply. Correcting the harmonic distortion of the input line current is one of the main goals in designing the street light power supply. For the LED street light EN 61000-3-2, Class C applies at full load [12, 13]. EN 55015 is a product family standard (largely based on CISPR 15). Key EMC standards include EN 61000-3-2 for limits on harmonic current emissions, EN 61000-3-3 for limits on voltage changes, voltage fluctuations, flicker, and EN 61547 for immunity requirements.

The pre-compliance measurements realized on the prototype of LED street light focus on the power line conducted emissions and harmonic current emissions. The power line conducted emissions measured on the first 60 W LED driver shows **Figure 6**. It is clear from **Figure 6** that the level of emission of the LED driver is below the limits specified in the standard EN 55015. Therefore, a peak detector in a frequency range from 9 kHz to 30 MHz is used.

Table 1 lists harmonic currents generated by the 60 W LED driver. The power factor (PF) is 0.444, and the corresponding limit for the 3rd harmonics is 13.32%. The third column of **Table 1** lists the multiples of limits. The corresponding average total harmonic distortion (THD) of the current generated by the 60 W LED driver is equal to 181.33% the maximal THD is 181.73%.

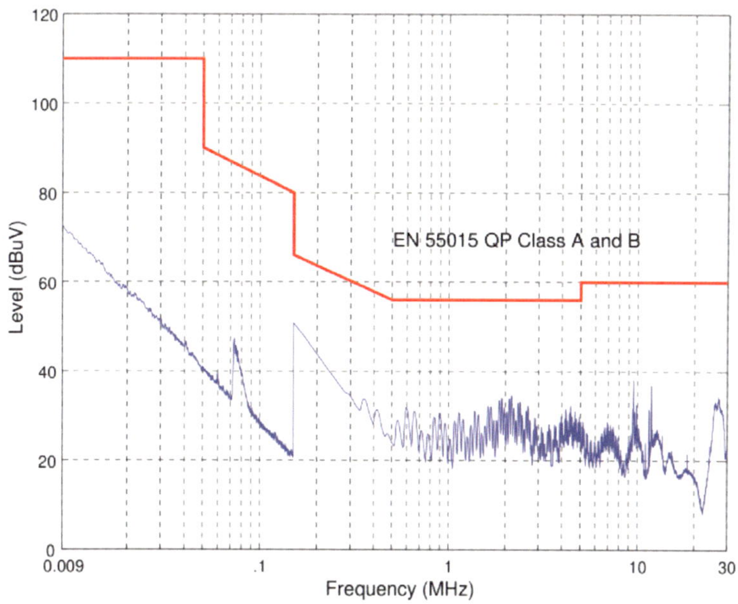

Figure 6.
Limits and measured EMI level of the conducted emissions on the first 60 W LED driver between the phase conductor and earth.

Harmonic order	Harmonic current	Multiple of limit[a]
n	(%)	
2	2.46	1.23
3	89.71	6.74
5	84.12	8.41
7	76.39	10.91
9	66.89	13.38
11	56.38	18.79

[a]*Calculated.*

Table 1.
Measured harmonic current emissions of the 60 W LED driver – Class C, power factor PF = 0.444.

It is clear from **Table 1** that the harmonic currents generated by the 60 W LED driver far exceed the limits set by the standard. This fact is confirmed by the time course of the terminal voltage and the current flowing into the first 60 W LED driver shown in **Figure 7**. Measured values of the odd harmonics extremely exceed the specified limits, thus resulting in the high value of THD of the supply current and low power factor.

Figure 7 shows the time course of the terminal voltage and current flowing into the 60 W LED driver and the THD of the current through the 60 W LED driver and corresponding PF measured for the first 2500 minutes of measurement. It is clear that the current time course in **Figure 7** is periodic but does not have a sinusoidal shape. The deformation of the current waveform is due to harmonic currents emission from the LED driver.

For comparison, **Table 2** lists the harmonic currents generated by the 50 W LED driver. Measured levels of the conducted emissions on the 50 W LED driver

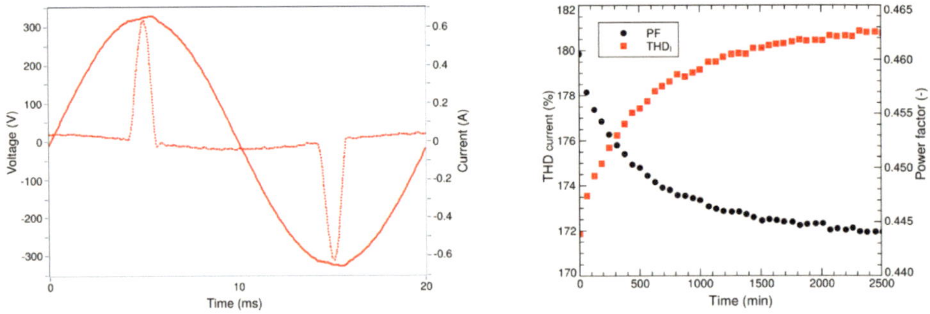

Figure 7.
Time course of the terminal voltage and current flowing into the 60 W LED driver during one period (left), and corresponding THD and PF measured in phase conductor during 2500 minutes (right).

Harmonic order	Harmonic current	Multiple of limit[a]
n	(%)	
2	0.13	0.065
3	7.30	0.246
5	4.83	0.483
7	4.13	0.590
9	3.13	0.626
11	2.35	0.783

[a]*Calculated.*

Table 2.
Measured harmonic current emissions of the 50 W LED driver – Class C, power factor PF = 0.991.

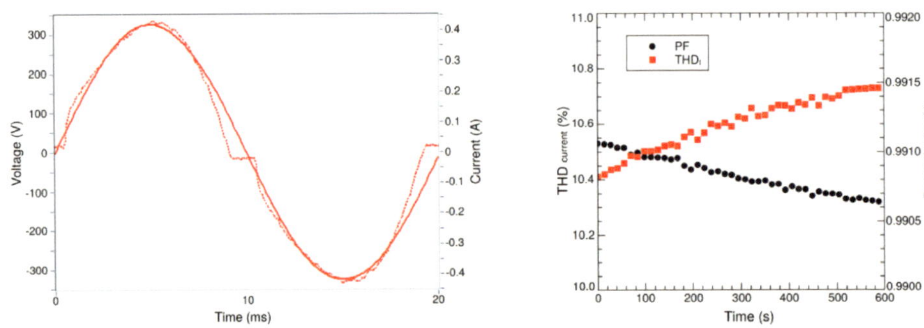

Figure 8.
Time course of the terminal voltage and current flowing into the 50 W LED driver during one period (left), and corresponding THD and PF measured in phase conductor during 600 seconds (right).

between the phase conductor and earth need not indicate as they are below the specified limits. It is clear from **Table 2** that the 50 W driver meets the requirements for harmonic current emissions with a sufficient margin. The corresponding average THD of the current generated by the 50 W LED driver is equal to 10.73%. The maximal THD is 10.77%.

Figure 8 shows the time course of the terminal voltage and current flowing into the 50 W LED driver and the trend over time of the THD of the current through the 50 W LED driver and corresponding PF measured for the first 600 seconds of measurement.

By comparing the measured data, it can conclude that all LED drivers comply with the standard EN 55015 with a sufficient margin and are not sources of EMI. However, as for the harmonic current emissions, the measurements have shown the lack of some drivers. The reason lies in permissible limits according to the standard EN 61000-3-2 were exceeded. Further, the THD of the supply current is high, and the measured PF is low. Therefore, harmonic current emissions are due to either improper circuit design of the LED driver or wrong LED driver design concerning the rated load of the LED street light.

3.2 LED information board

Measures to mitigate the galvanic coupling through AC mains wires and DC power supply of the LED modules to the LED information board are investigated. The LED information board enclosure is a galvanized steel with a protection rating of IP65 and anti-reflective protective glass. The LED information board supplies from a single-phase AC low-voltage network converted by DC-to-DC converters with a maximum current of 20 ampers. A total of three DC-to-DC converters are in the LED information board. For receiving and sending data in several ways, an optical converter, router, and ADSL modem are installed. A mini PC controls data processing and process control [14]. The results show that the EMI emissions in horizontal polarization are higher in comparison to vertical polarization. The following measures were sequentially used to reduce the EMI emissions:

- twisted power wires,
- chokes on wires (straight or twisted),
- suppression capacitors,
- impedance matching resistors,
- data wires on PCB with SMD LC filters,
- twisted data wires.

Figure 9 shows the interference level with three different EMI mitigation measures. Pre-compliance testing shows that the most critical frequency range is 30 to 130 MHz. From the initial measurements without EMI measures, the results

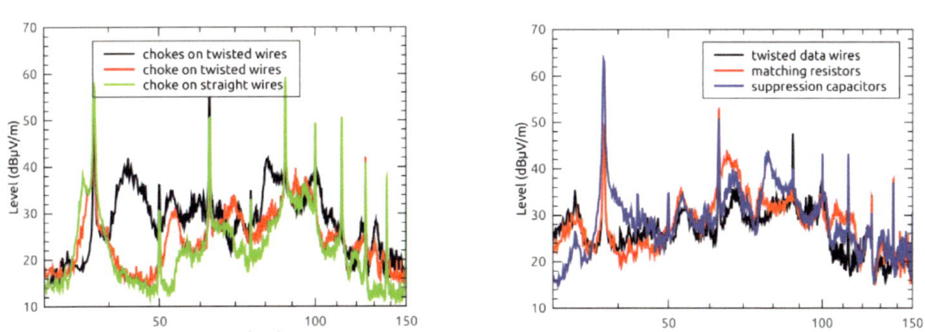

Figure 9.
Frequency characteristic of EMI level for different EMI mitigation measures.

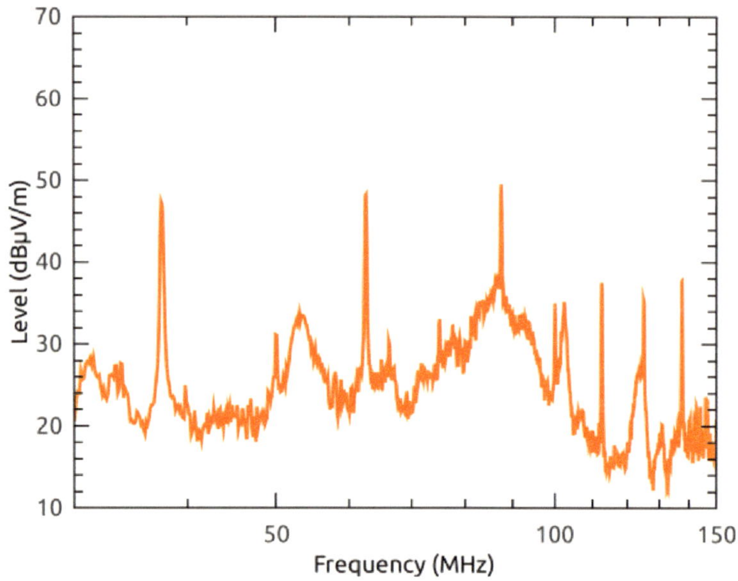

Figure 10.
Frequency characteristic of EMI level for combined EMI mitigation measures.

show that the maximum EMI radiation is at 37.5 and 87.5 MHz. Other sources of EMI emissions are at 62.5, 100, 112.5, and 125 MHz. It is evident from the list that frequencies are multiples of 12.5 MHz.

The most optimal EMI reduction reach by the simultaneous use of several EMI reduction measures. An example of the measured EMI emissions levels for such a case shows **Figure 10**, where chokes on twisted pairs from AC-to-DC converters at both ends, impedance matching resistors, and data wires on PCB with SMD LC filters are used. Other variants used in testing had only a slight effect, a few tenths of dB. To compare the effects of EMI measures with each other, the vertical scales in **Figures 9** and **10** are the same.

3.3 Audio equipment

Sound equipment is still popular, and home theaters provide quality surround sound. Retro designs with small dimensions are popular. In the past, analog amplifiers have been used in high-performance audio devices to emphasize sound quality. In recent years, digital amplifiers have improved sound quality and become popular in home audio. Today, audio devices have implemented multiple communication and I/O technologies to provide comfort. This is associated with EMI problems in the development of new audio equipment.

The EMI sources are the display unit, switching power supply, Bluetooth, USB port, S/PDIF connection. Due to a large number of connections, the cabinet of the audio device contains many holes. Holes are places where the generated interfering electromagnetic fields radiate out into space. **Figure 11** shows a selection of pre-compliance measurements on audio devices in different operating conditions.

Figure 11a shows the EMI for an audio device (amplifier with compact disc (CD) drive) with a metal cover, Bluetooth active, CD playback on, the maximum power delivered to the rated load, and the connected S/PDIF cable. It is clear from the measured data that the device emits a discrete spectrum from 80 to 700 MHz, 930 to 1000 MHz, and a significant continuous spectrum between 220 and 300 MHz.

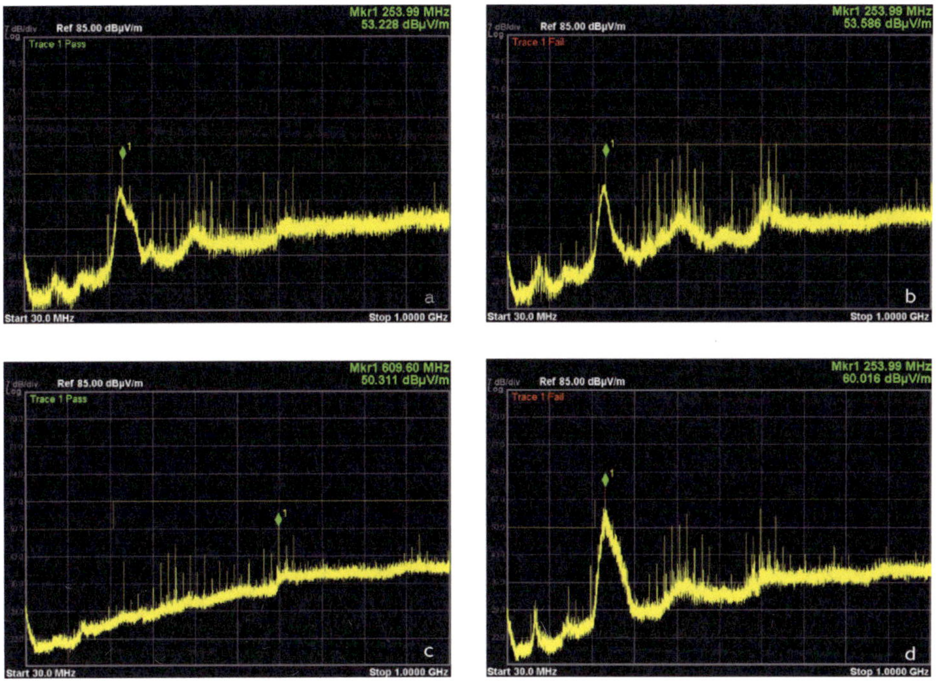

Figure 11.
EMI pre-compliance measurements on audio devices in different operating conditions.

Figure 11b shows the EMIs for the same audio device under the same conditions with a change in the use of the choke in the power conductors inside the cabinet of the audio device. Surprisingly, this measure has the opposite effect. Instead of the expected reduction of EMI field radiation, emissions are increased by 6 to 8 dB in the band 340 to 680 MHz. In addition, the use of a choke affected the width of the continuous spectrum. As a result, the continuous spectrum is narrower by shifting the upper limit from 300 to 270 MHz.

Figure 11c shows EMIs for an audio device (amplifier with CD drive) with a cabinet in which some parts are wood. In this case, the audio device is powered by a battery pack. The CD drive is empty without a CD inserted, and the audio cables are not connected. The aim is to determine the basic level of EMI radiation with the possibility of subsequently monitoring the effect of connecting cables to audio inputs and outputs and a CD drive with a CD inserted. The measured EMI in **Figure 11c** shows only the discrete spectrum of EMI field radiation. **Figure 11d** shows the EMIs for the same audio device with a CD inserted in the CD drive, and the device operates in "play" mode. S/PDIF cables are connected as well. The measured EMI emissions in **Figure 11d** show the continuous emission spectrum of EMI fields from 220 to 300 MHz. The same CD drive and CD drive control are used, as in the first measured audio device (see **Figure 11a** and **b**). Thus, a component generating continuous EMI emissions in an audio device is classified.

4. Summary and conclusions

Electromagnetic smog is a problem for technical equipment and living organisms. The manufacturer of electrical equipment wants to gain a foothold in the market and sell the product. The successful sales of the product on the market require an EMC certificate of conformity, which increases the development and production costs of the product. To reduce equipment development costs, shorten time to market, and obtain a risk-free EMC certificate, pre-compliance measurements are needed. In addition, shortening the time to market accelerates the professional experience of engineers.

This chapter has provided three examples of EMI measurements using pre-compliance measurements. The first case concerns the generation of unwanted harmonic currents and the detection of EMI sources by conduction in a prototype of street LED lighting. The second case concerns EMI sources from the LED information board and how some measures reduce EMI interaction. Finally, the third case points to unexpected changes in EMI radiation during the development of audio equipment.

Acknowledgements

This work was supported by the Cultural and Educational Grant Agency of the Ministry of Education, Science, Research and Sport of the Slovak Republic (KEGA) under project No. 008TUKE-4/2019.

Conflict of interest

The author declares no conflict of interest.

Abbreviations

CAD	Computer aided design
CD	Compact Disc
DUT	Device Under Test
EMC	Electromagnetic Compatibility
EMI	Electromagnetic Interference
EUT	Equipment under test
IEV	International Electrotechnical Vocabulary
HF	High-Frequency
I/O	Input/Output
LISN	Line Impedance Stabilization Network
PCB	Printed Circuit Board
PF	Power Factor
RF	Radio-Frequency
SPD	Surge Protection Device
THD	Total Harmonic Distorsion

Author details

Bystrík Dolník[†]
Technical University of Košice, Košice, Slovakia

*Address all correspondence to: bystrik.dolnik@tuke.sk

† These authors have contributed equally.

References

[1] IEC. International Electrotechnical Commission;. Accessed: 2021-06-05. https://www.electropedia.org/iev/iev.nsf/index?openform&part=161.

[2] Prudenzi A, Grasselli U, and Lamedica R. IEC Std. 61000-3-2 harmonic current emission limits in practical systems: need of considering loading level and attenuation effects. In: Power Engineering Society Summer Meeting, 2001. vol. 1. IEEE; 2001. p. 277-282.

[3] Schlabbach J. Harmonic current emission of photovoltaic installations under system conditions. In: Electricity Market, 2008. EEM 2008. 5th International Conference on European. IEEE; 2008. p. 1-5.

[4] Dugan RC, McGranaghan MF, Santoso S, and Beaty HW. Electrical Power Systems Quality. 2nd ed. McGraw-Hill; 2002.

[5] Uddin S, Shareef H, Mohamed A, and Hannan MA. An analysis of harmonics from LED lamps. In: cccc2012 Asia-Pacific Symposium on Electromagnetic Compatibility, APEMC 2012 - Proceedings; 2012. p. 837-840.

[6] Medved' D, and Kanálik M. Measurement of power quality on the COTEK S1500-124 Inverter's terminals in case of linear load supplying. Universal Journal of Electrical and Electronic Engineering. 2014;2(4):178-182.

[7] Videnka R, and Svačina J. Introduction to EMC pre-compliance testing. In: MIKON 2008 - 17th International Conference on Microwaves, Radar and Wireless Communications. IEEE; 2008. p. 1-4.

[8] Schwab AJ. Elektromagnitnaja sovmestimost' (Electromagnetic compatibility). 2nd ed. Energoatomizdat; 1998.

[9] Ott HW. Electromagnetic Compatibility Engineering. 1st ed. John Wiley & Sons, Inc.; 2009.

[10] Shi D, Wang N, Zhang F, and Fang W. Intelligent Electromagnetic Compatibility Diagnosis and Management With Collective Knowledge Graphs and Machine Learning. IEEE Transactions on Electromagnetic Compatibility. 2021;63(2):443-453.

[11] Perotti M, and Fiori F. A Closed Loop Delay Compensation Technique to Mitigate the Common Mode Conducted Emissions of Bipolar PWM Switched Circuits. IEEE Transactions on Power Electronics. 2021;36(5):5450-5459.

[12] Li Y, Lim JW, and Kim HJ. A low-cost digital PWM-controlled LED driver with PFC and low light flicker. Journal of Electrical Engineering and Technology. 2015 11;10:2334-2342.

[13] Chen N, and Chung HSh. An LED Lamp Driver Compatible with Low- and High-Frequency Sources. IEEE Transactions on Power Electronics. 2013;28(5):2551-2568.

[14] Dolník B, Mészáros A, Pavlík M, Zbojovský J, Špes M, and Čonka Z. On the udesirable EMI of Large LED information board; 2017. p. 335-338. Available from: https://www.scopus.com/inward/record.uri?eid=2-s2.0-85029690105&partnerID=40&md5=6f66831c96ea064913cb6d81b7219bfd.

Chapter 11

Power Converters Electromagnetic Emissions with Methods to Measure, Compare and Reduce Noise Fields

Debasish Nath

Abstract

Power electronic converters find widespread applications in the present times. However, they have been known to be rich sources of electromagnetic emissions. The converters are required to operate in compliance with their electromagnetic environment, for which they must adhere to emission limits imposed by the standards. The power converters are duly tested, and their emission levels are measured and checked for compliance. Therefore, understanding the emission mechanism is necessary to design a power converter that will satisfy the EMC requirements. In addition to the switches, parasitic elements, which are invariably present in every circuit, play a crucial role, especially in the higher frequency range, in degrading the EMC performance of converters. The present chapter takes up these aspects and discusses the reasons behind emissions and the roles played by switch characteristics and parasitic elements. Demonstrations through simulations and computations are presented all along with the discussions, and inferences are accordingly drawn. Finally, the standard specifications for emission measurements and analysis through measuring EMI receiver are briefly introduced. A few popular methods to reduce emissions are demonstrated. The final improvement in the EMI performance is shown with the help of EMI receiver output, in accordance with the CISPR standards.

Keywords: Power Converters, EMC, EMI, EMI receiver

1. Introduction

In electrical power systems, a number of potential sources capable of producing electromagnetic disturbances or noise are present. These include natural phenomena like lightning or electrical corona and electronic, especially power electronics circuits and devices. These sources either introduce or draw from the supply terminals, non-sinusoidal currents typically characterized by the rapid rise and fall times, or equivalently, high-frequency components (in MHz or GHz range) in their frequency-domain characteristics. Such currents can influence the operation of other circuits or devices present in the electrical systems in two possible ways:

1. By directly reaching the circuits through closed electrical paths

IntechOpen

2. By producing electromagnetic fields of sufficient strength, which may couple with the other circuits directly through the air.

The former is known as conducted emission and has a frequency range of interest between 150 kHz to 30 MHz, whereas the latter is referred to as radiated emission, with the frequency range of interest being >30 MHz. With the proliferation of power electronic devices in everyday life, there is a potential increase for devices to interfere with each other. Applications of power electronics in the present times are abundant. This includes everyday household use in laptop and mobile battery chargers, Uninterrupted Power Supply (UPS), induction cooking, etc. Furthermore, in industrial applications such as control of motor drives or railway traction systems, power transmission systems such as Flexible Alternating Current Transmission System (FACTS) and High Voltage Direct Current (HVDC) transmission of electric power, utilization of power electronic systems can be found. In addition, with power electronics finding increased applications in renewable energy systems in recent times, the electromagnetic compatibility of power converter systems is a significant concern.

In view of the above, the present chapter will focus on understanding the origin of the electromagnetic noise from power converters, some methods to reduce emissions, and measurement procedures followed to quantify the noise.

2. Origin of electromagnetic noise from power converters

Power electronic converters employ semiconductor switching devices, which are operated in a manner required to produce the desired output. For example, consider the DC-DC Buck converter, the basic topology of the same has been shown in **Figure 1**. The buck converter is used to step down the input DC voltage to a lower level at the output. A controlled semiconductor switch, typically a MOSFET, is periodically turned ON and OFF to achieve the purpose.

During every switching action, the currents through certain elements or voltages at certain circuit nodes undergo rapid transition. For example, assuming every element, including the switches shown in **Figure 1**, to be ideal, the input current waveform I_{sw} for an ideal buck converter is a periodic pulse train, as shown in **Figure 2(a)**. The current undergoes rapid transitions to zero when the controlled switch is turned OFF and similarly rises rapidly (to the load current) when the switch is turned ON [1].

Figure 2(a) shows the current waveform for a switching frequency of *100 kHz*, duty cycle D of 0.4, a rise time of *25 ns,* and fall time of *100 ns* of the PWM pulse (control signal to switch). The FFT of the current is shown in **Figure 2(b)**. It can be observed that even though the switching frequency is only *100 kHz,* the current

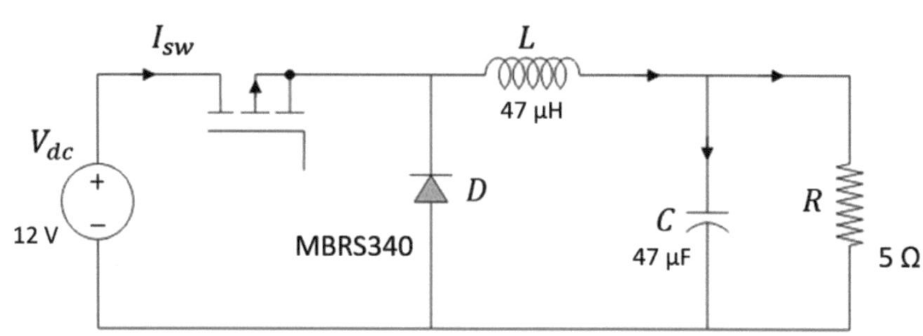

Figure 1.
Buck converter topology.

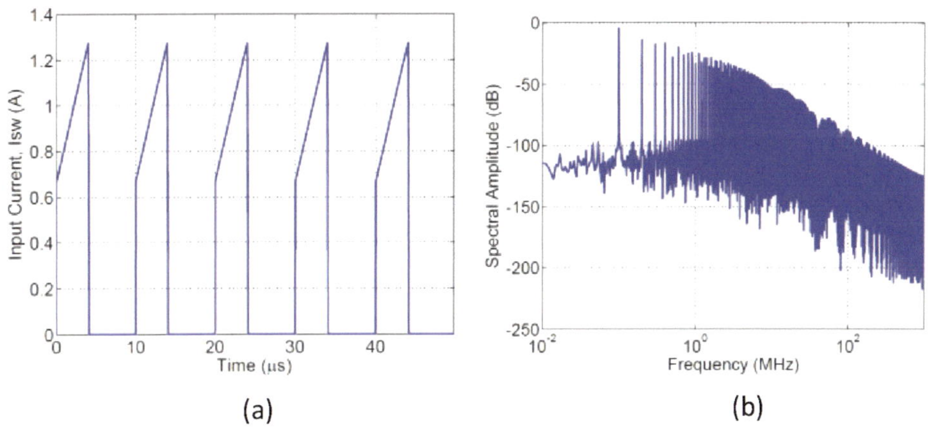

Figure 2.
Input current to the ideal Buck converter. (a) Time domain waveform (b) frequency Spectrum of the currents.

waveform has significant spectral amplitude even in the MHz range (high-frequency components).

Typically, the DC input V_{dc} to the converter is obtained by rectifying the *220 V or 110 V* (50 or 60 *Hz*) AC supply. Therefore, the non-sinusoidal (pulse train) current drawn by the buck converter is reflected on the AC supply side, and the high-frequency components can flow through the electrical paths reaching various other neighboring devices and circuits connected to the supply. This can result in electromagnetic interference, and since the coupling of the electromagnetic noise takes place through closed electrical paths, this forms an example of *conducted emission* (CE), introduced earlier.

In practice, the entire circuitry contains numerous metallic or conducting paths and connections. Although, for example, the input DC supply (and even the load) is connected to the converter through a cable harness, the converter layout could be realized in a Printed Circuit Board (PCB) in which the PCB traces provide the connections. In addition, there are various metallic connectors and wires present in the entire circuit. When the high-frequency currents flow through these conducting elements, some of them (depending on the dimensions and frequency or wavelength) behave like radiating antennas. In such situations, the electromagnetic fields may couple directly through the air between two different circuits. This is the case of *Radiated Emission* (RE) which has earlier been introduced.

The above analysis considering the ideal buck converter topology clearly demonstrates that the switching transients that take place in a power electronic converter can lead to electromagnetic interference through conducted and radiated emissions. Although the buck converter topology has been used as an example for demonstration, it is clear that since the switching transients take place in all power electronic converters, electromagnetic noise is produced by all power electronic converters.

Although the ideal buck converter is a good example to consider for a first analysis, the situation is much more complicated in practical circuits, chiefly due to the presence of parasitic elements. Some of these aspects are discussed in the following section.

3. Effect of non-linearities and parasitic elements

The power semiconductor switches are inherently non-linear in their operational characteristics. These non-linearities lead to additional system dynamics during the switching transients. This results in worsening of the EMI performance of practical converters compared to analysis carried out for ideal converters.

To demonstrate the above, the buck converter shown in **Figure 1** is simulated again. This time, however, the freewheeling diode is no longer assumed to be ideal. Instead, the well-known quasi-static model (diode models used in LTSpice), presented in **Figure 3**, is employed to obtain the terminal characteristics of the diode [2]. The equations of the model have also been presented below [2]. For more advanced diode models, the reader may refer to recent literature [3, 4].

The model provides the current i_d through the diode and the voltage across its terminals v_d. To computei_d, the saturation current I_s, emission coefficient N and the threshold voltage V_T need to be known. The model also includes a series resistor R_s and a non-linear junction capacitor C_J. The junction capacitance is a function of the voltage v_d, and can be computed by knowing the zero-bias junction capacitance C_{J0}, junction grading coefficient M and the junction potential V_J. For the diode considered, MBRS340, the values as obtained from the LTSpice library are; $I_s = 22.6\ \mu A$, $N = 1.094$, $R_s = .042\ \Omega$, $C_{J0} = 480\ pF$, $M = 0.61$. V_T is taken to be $26\ mV$ and $V_J = 1\ V$, the default value.

The buck converter is simulated with the above diode model, keeping the switching frequency, duty cycle, and the load same as that of the ideal converter analysis. The simulated input current waveform obtained is shown in **Figure 4(a)**.

There is a sharp spike observed in the current waveform at the rising edge in the present case. This is not present in the ideal converter input current shown in **Figure 2(a)**. **Figure 4(b)** shows the rising edge of both the input current and current through the diode are plotted together. The sudden spike in the diode current is required to charge the junction capacitance when the controlled switch (MOSFET) is turned on and the diode is, turned off. This current is drawn directly from the DC supply V_{dc} and therefore reflects as a spike in the input current waveform. The effect of this spike on the emission performance can be clearly understood by comparing the frequency domain characteristics of the input currents in the ideal and practical buck converter cases. For example, the FFTs of the two currents (shown in **Figures 2(a)** and **4(a)**) are shown in **Figure 5**.

From **Figure 5**, it is clear that for the practical converter, the spectral amplitude of the high-frequency components has increased by many times. The difference is even a few tens of dB in the higher frequency range. Therefore, it is clearly seen that the nonidealities of the power converter switches can play an extremely important role towards the magnitude of the electromagnetic emissions from the power converter circuit.

The situation in practical converters is further complicated by the parasitic elements which are inadvertently present in the circuits. To demonstrate this, a parasitic inductance of *1 nH* is considered on the converter input side, as shown in **Figure 6**. Such inductances arise due to the connecting cables or wires, PCB traces, contacts, and connectors, etc., and are unavoidable in the practical circuits.

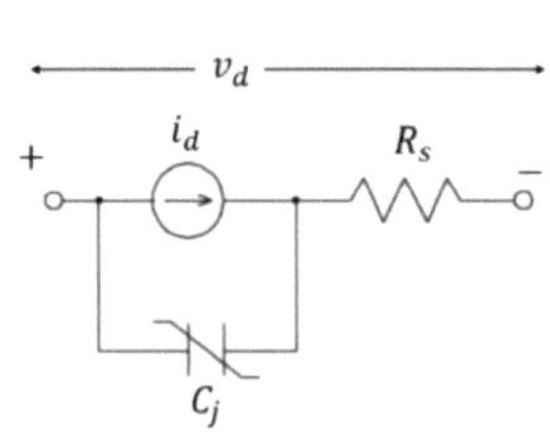

$$i_d = I_s\left(e^{v_d/NV_T} - 1\right)$$

$$C_J = C_{J0}\left[1 - \left(\frac{v_d}{V_J}\right)\right]^{-M}$$

Figure 3.
Model of freewheeling diode.

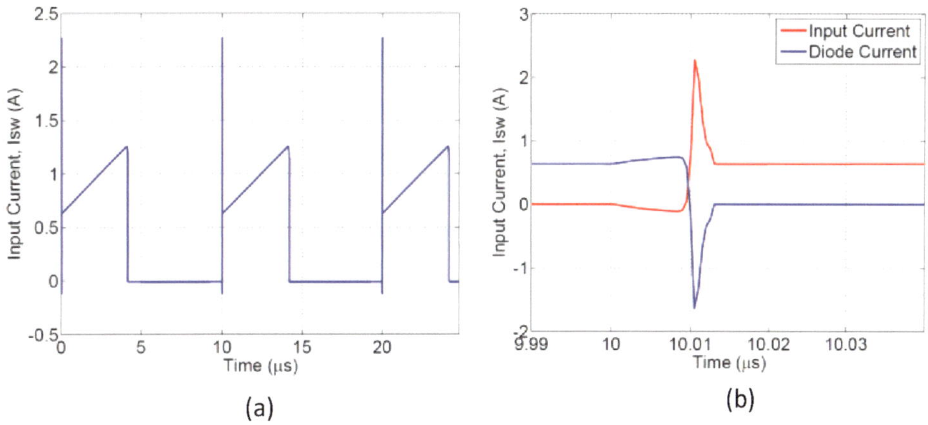

Figure 4.
Simulated currents for the practical Buck converter. (a) Input current and (b) input current and the freewheeling diode current.

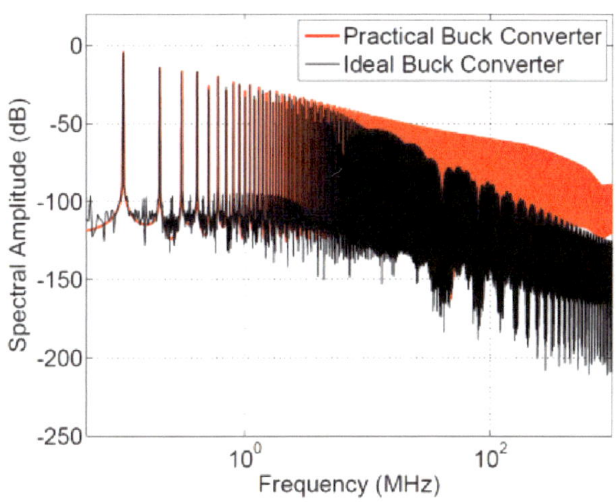

Figure 5.
Frequency domain characteristics of the input currents to ideal and practical buck converters.

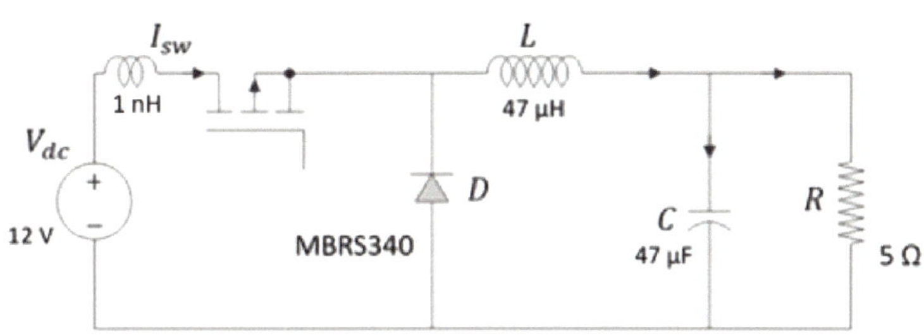

Figure 6.
Buck converter with input side parasitic inductance.

Keeping everything else unchanged, the above circuit is simulated, and the input current is shown in **Figure 7(a)**. It is observed that the input current contains a damped, high-frequency oscillatory response at the rising (and falling) portions. Such oscillations are known as ringing and are due to the L-C resonance between the parasitic inductance and the diode junction capacitance. The frequency-domain

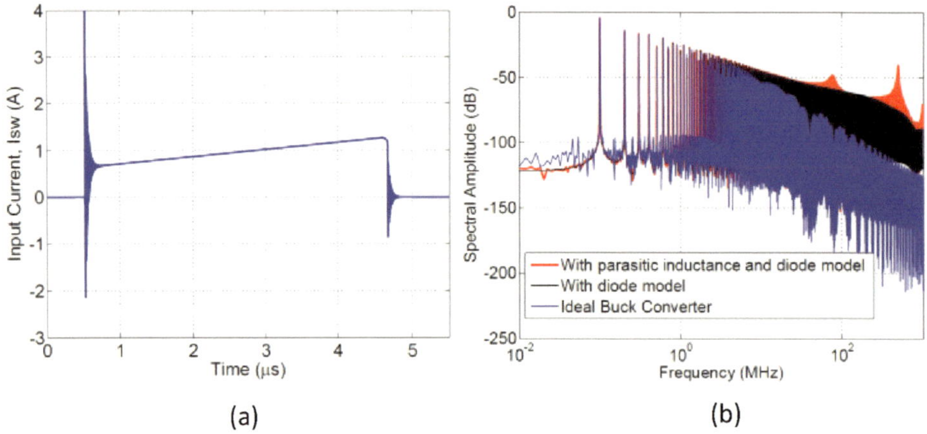

Figure 7.
Buck converter with input side parasitic inductance and diode effects considered. (a) Temporal variation of input current and (b) comparison of frequency domain characteristics.

characteristics are shown in **Figure 7(b)** along with the earlier two cases for comparison. The spectral amplitude of the current with parasitic inductance (plus diode) is the highest. In addition, resonant peaks are clearly observed in the high-frequency end of the spectrum. Therefore, the parasitic elements increase the high-frequency components of the input current and, correspondingly, the emissions from the power converter. These are, therefore, serious considerations in assessing the EMI performance of the converter.

In practice, it is often necessary to reduce the emission from power converters in order for the design to be EMC compliant. Damping out the ringing shown in **Figure 7(a)** is extremely important in this regard.

The discussion so far has been carried out towards understanding the origin of the electromagnetic noise (emissions) from power converters. In addition, the effects of parasitic elements and switching device features have also been examined. It is quite clear that electromagnetic noise from power converters is unavoidable due to their intrinsic switching operations. However, the design must also be EMC compliant. With this view, the discussion will next be directed towards methods to improve the EMI performance of power converters.

4. Methods to improve EMI performance of power converters

4.1 With *RC* snubber

In the previous section, it has been shown that the ringing in the current (and voltage) waveforms degrades the EMI performance of the circuit. Therefore, damping out the ringing will improve the performance. A simple method to achieve the same is the use of an *RC* snubber. Considering the case of a buck converter, the snubber is connected across the freewheeling diode (snubber is to be connected between switching node and ground), as shown in **Figure 8**. The values of the snubber elements R_s and C_s are computed as follows:

$C_s \approx 3$ times the diode junction capacitance. Since the junction capacitance is a varying quantity (with junction voltage), the maximum value corresponding to zero bias value, i.e., C_{J0} is considered for calculation. Therefore, $C_s \approx 3 \times 480\ pF$. This results in a value of *1.44 nF,* and so a value of *2 nF* is selected.

$R_s \approx 2\omega_0 L_p$ where, ω_0 is the angular frequency corresponding to the ringing oscillations to be damped and L_p is the parasitic inductance, *1 nH* at present.

The ringing oscillations in the current waveform shown in **Figure 7(a)** have a frequency of ≈500 *MHz*. With these values, R_s is computed to be *6.28 Ω*, and therefore, the value of *10 Ω* is selected.

With the snubber elements included in the buck converter circuit as shown in **Figure 8**, the input current is computed and is shown in **Figure 9(a)**. Also shown in the Figure is the current without snubber (shown earlier in **Figure 7(a)**) for comparison.

Clearly, the snubber results in damping out the ringing at the rising edge of the input current waveform. The same inference can also be drawn from the frequency domain characteristics shown in **Figure 9(b)**. The snubber has clearly damped out the resonant peak due to the high-frequency ringing. With snubber included in the circuit, the frequency spectrum plot clearly shows a reduction in the spectral amplitude of the high-frequency components, and hence an improvement in the emission performance.

It is also important to check the power loss in the snubber resistor and its impact on the efficiency of the circuit. The power loss is given by [5], $f_{sw} C_f V^2$ where, f_{sw} is the switching frequency (100 kHz), $V$is the peak voltage which appears across the diode terminals (approximately equal to the input voltage V_{dc}).

The power loss is calculated and found to be *28.8 mW*. The output power is $\frac{(0.4\times 12)^2}{5} = 4.6\ W$. Therefore, the power loss in the snubber is $\frac{28.2\times 10^{-3}}{4.6} \times 100 =$ 0.625% of the power delivered at the output and can be neglected for all practical purposes.

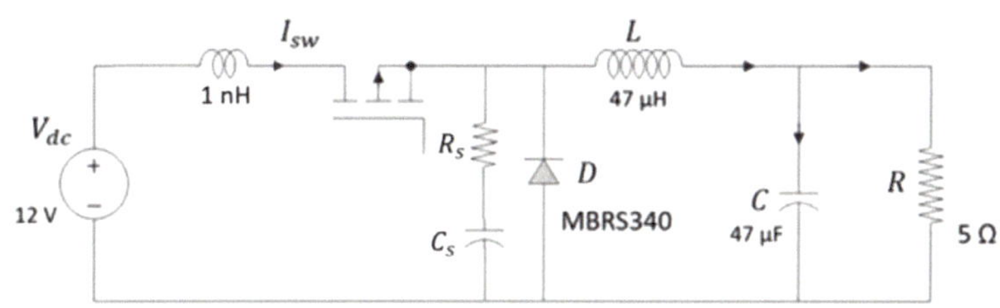

Figure 8.
Buck converter with input side RC *snubber.*

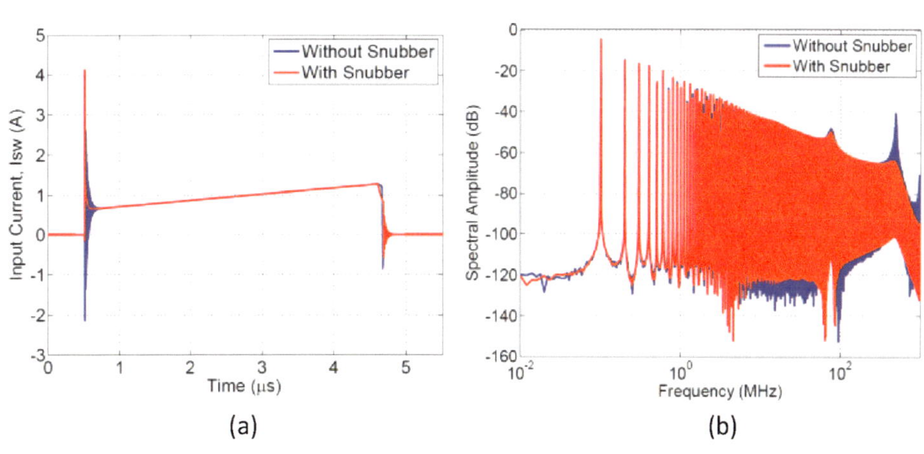

Figure 9.
Input currents in buck converters with and without snubber. (a) Temporal variation of input current and (b) comparison of frequency domain characteristics.

Although the snubber improves the EMI performance of the converter, designing the snubber elements can prove to be difficult since it is difficult to estimate the parasitic inductance accurately and switch capacitance and the maximum voltage at the switching node (due to overshoot). In addition, the snubber damps out or attenuates the high-frequency oscillations, which take place at (or around) a certain frequency. If all high-frequency components (above a certain cutoff or threshold frequency) could be attenuated, further improvement in the EMI performance can be obtained. This can obviously be achieved by using a low pass filter. This is discussed in the following.

4.2 With damped *LC* filter

One popular method to reduce emissions from power converter and improve their EMI performance is using filters. Considering the example of the buck converter shown in **Figure 6**, a damped *LC* filter is added on the input side, as shown in **Figure 10**.

The filter elements are the inductance L_f, capacitance C_f and the damping resistor R_f. The value of the damping resistor is selected from the condition $R_f << R/D^2$ [6]. Selecting $R_f = 1\,\Omega$ easily satisfies the criteria. The values of L_f and C_f are selected to be *100 μH* and *100 pF,* respectively. With these values, the resonant frequency is ≈5 *MHz*. This is sufficiently away from the switching frequency of *100 kHz* and therefore does not slow down the converter's response. However, it provides sufficient attenuation to the high-frequency components and, therefore, is expected to improve EMI performance. To validate the same, the buck converter with the input side filter is simulated, and the input current I_{sw} is obtained. This is shown in **Figure 11 (a)**, and the input current without the filter is shown in **Figure 7(a)** for a practical converter. The frequency spectrum of the currents with and without filters is shown in **Figure 11(b)**.

In the time domain current waveform, the effect of the damped *LC* filter in reducing the oscillation is clearly noticeable. The amplitude of the overshoot is also observed to have reduced. The attenuation provided to the high-frequency components can be seen from the frequency response. The reduction in the amplitude at some of the high-frequency components is around *20–30 dB.* This is a significant improvement in the performance of the converter operation from the EMI/EMC point of view. Also, it is worthwhile to note that the damped *LC* filter attenuates both the resonant peaks present in the frequency characteristics. This can be compared with the method employing snubber elements (**Figure 9(b)**), where only one peak has been attenuated.

However, it is important to mention here that the *LC* filter changes the converter dynamics, often leading to a significant degradation in the transient response. Moreover, for converters operated under closed-loop control, the addition of the filter may even make the control system unstable [7]. Therefore, the

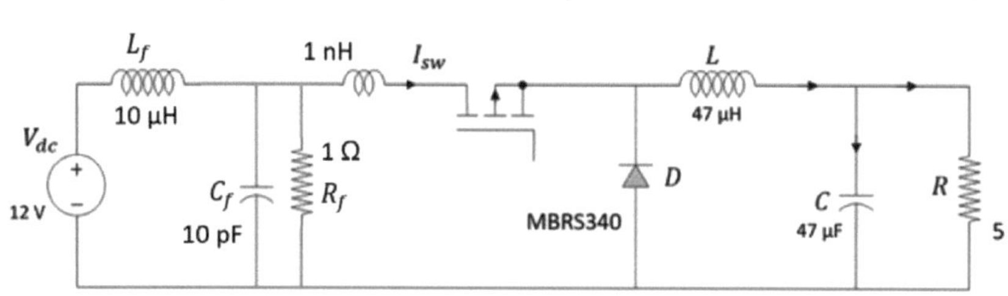

Figure 10.
Buck converter with input side damped LC filter.

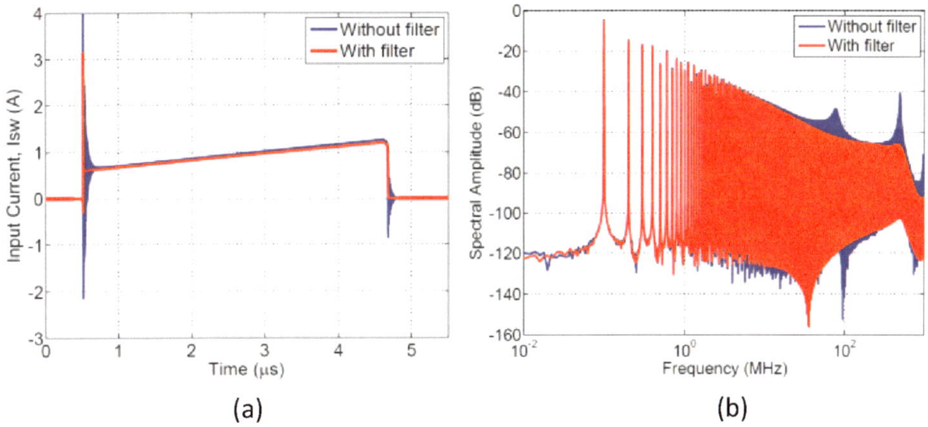

Figure 11.
Input currents in buck converters with and without input side filter. (a) Temporal variation of input current and (b) comparison of frequency domain characteristics.

design of the filter is not a straightforward task, and the converter performance, closed-loop response, and EMI specifications must all be considered in satisfying all the requirements.

So far, the mechanism behind emissions from power converters, along with some of the important factors responsible for degrading the EMI performance, has been discussed. In addition, some popular methods to improve the EMI performance have also been discussed. In the following sections, the measurement of the emissions and analysis of the same through standard receivers will be briefly discussed.

5. Measurements of electromagnetic emissions

Measurements of emissions are performed to test for EMC compliance of any product. Therefore, the measurements must be carried out so that the results are easy to correlate across different measurement sites or EMC labs. To understand this, assume that the noise current injected by a certain product into the power supply needs to be measured for compliance. This could be conveniently carried out by using a current probe. However, the noise current injected into the supply depends on the equivalent impedance measured between the supply's terminals. This impedance varies over frequency and will be different at different places, time of the day, seasons, etc. As a result, the noise current will also vary, and therefore the measurement results cannot be correlated.

In order to eliminate the above uncertainties, a Line Impedance Stabilization Network (LISN) is employed for the measurement of Conducted Emissions (150 kHz- 30 MHz) [8]. The block diagram of the measurement setup is shown in **Figure 12**.

The LISN is connected between the supply lines and the product, usually referred to as Device under Test (DuT) or Equipment under Test (EuT). LISN has the following two purposes [8]:

1. To offer a constant impedance (of 50 Ω), looking from the terminals of the product, over the entire CE frequency range (150 kHz – 30 MHz).

2. To block the noise that may enter from the supply terminals and otherwise be wrongly attributed to the DuT.

Therefore, with **Figure 12**, only the noise current from the DuT flows through the LISN to the supply terminals. Since the impedance is fixed, a linearly related equivalent noise voltage drop is produced at the LISN, which is then fed to a measuring receiver (EMI receiver). The receiver's measured noise is then analyzed, which will be briefly discussed in the following section.

In the case of radiated emissions, the emissions or noise is coupled directly through the air. Hence an antenna is used to capture the emissions, which are then fed to the measuring receiver for analysis. The block diagram is shown in **Figure 13**.

In the measurement of radiated emissions, typically, over the frequency range of 30–200 MHz, the biconical antenna is used, from 200 MHz – 1 GHz, a log-periodic antenna is employed, and beyond 1 GHz, horn type antennas are used. Discussion on antennas is beyond the scope of the present chapter. However, more detailed discussions are available in the literature [8].

The emission or noise measured by the antenna is sometimes passed through a pre-amplifier stage before feeding to the EMI receiver. The noise emitted by the devices or equipment must be within limits set by the standards. In the United States, the Federal Communications Commission (FCC) limits are required to be adhered to whereas, in most of the European countries, the limits specified by the International Special Committee on Radio Interference (CISPR) are followed. In order to test whether the device or equipment emits the noise, lies within the limit, the measured noise (by LISN or Antenna) is analyzed by a measuring receiver or EMI receiver. The standards also stipulate the analysis procedure to be followed by

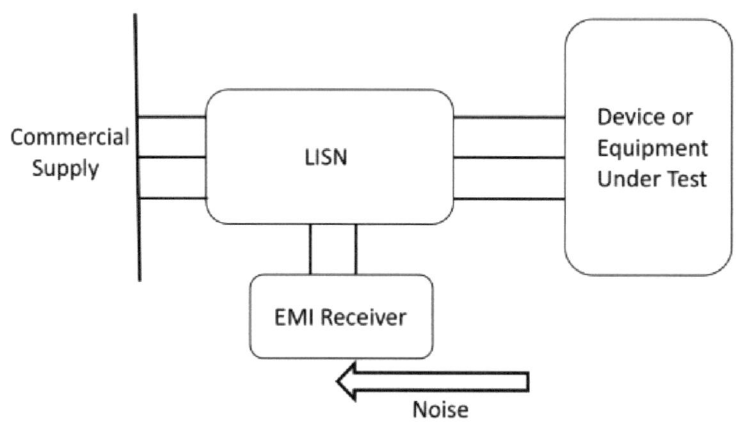

Figure 12.
Block diagram of conducted emission measurement setup.

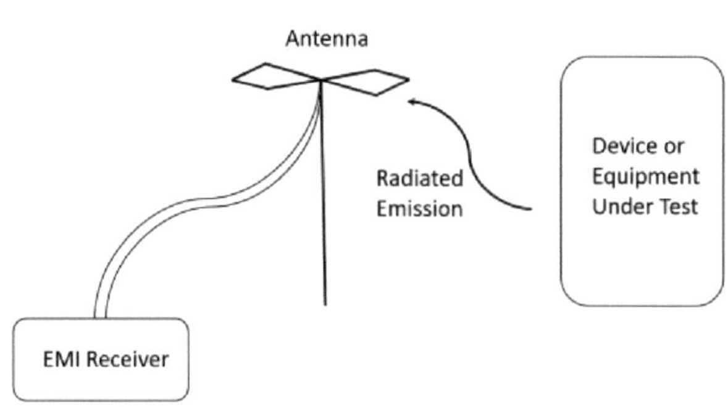

Figure 13.
Block diagram of radiated emission measurement setup.

such a receiver. The receiver, according to CISPR 16–1-1 standards, is briefly discussed in the following.

5.1 EMI receiver according to CISPR 16–1-1

As specified by the CISPR 16–1-1 standard, a measuring EMI receiver must be employed to obtain the noise produced by the DuT. The receiver is coupled to different devices (LISN for CE or different antennas for RE) depending upon the type of emission to be tested. The input impedance of the receiver must be 50 Ω or as close as possible. Otherwise, impedance mismatch might lead to standing waves which will introduce error in the results. The receiver can be operated in different modes, such as Peak, Quasi-Peak, Average, or RMS. The modes are selected based on the type of signal to be measured [9].

In the following, the peak detector mode is briefly discussed since it can be used for different types of input signals and indicates the worst-case scenario. In other words, if the emission of equipment is above the stipulated limit in the peak detector mode, it will fail all the other modes as well. Discussion and modeling of the receiver in quasi-peak, average and RMS modes can be found in literature [10, 11].

The receiver frequency range is divided into a number of different bands. Each band has its range as well as bandwidth for the Intermediate Filter (IF). **Table 1** specifies the important requirements for the receiver in peak detector mode.

CISPR Band	A	B	C	D	E
Frequency Range	9 kHz – 150 kHz	150 kHz – 30 MHz	30 MHz – 300 MHz	300 MHz – 1000 MHz	1 GHz – 18 GHz
Bandwidth B6 (6 dB)	200 Hz	9 kHz	120 kHz	120 kHz	1 MHz

Table 1.
Requirements of CISPR 16–1-1 EMI receiver in peak detector mode.

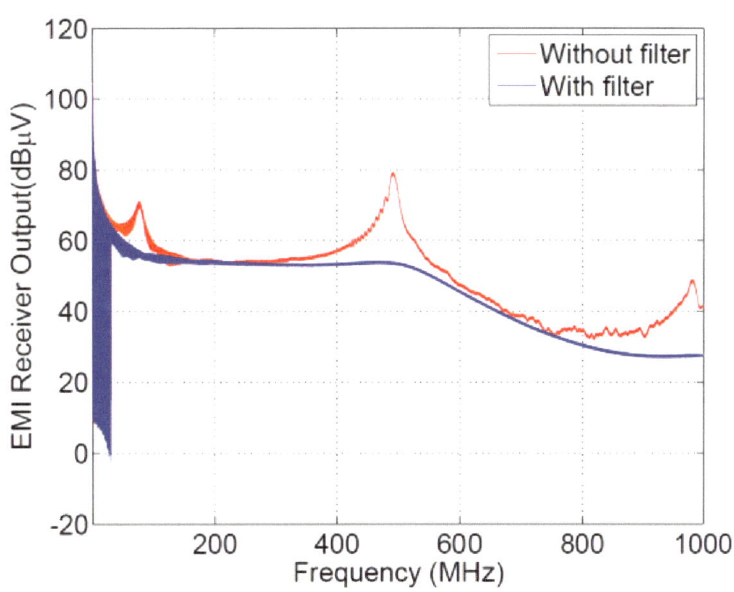

Figure 14.
EMI receiver output for the input current to the buck converter.

The input time-domain signal to the EMI receiver is typically transformed to the frequency domain through the FFT algorithm. Thereafter, the windowing operation is performed with the IF filter, determined by the band specified in **Table 1**. An extremely short time constant for charging and a very long time for discharging is employed for the peak detector mode. For the other modes, the time constants are specified in the standard. The output of the peak detector stage is obtained across the frequency range of the EMI receiver to produce the emission result for the DuT [6].

As an example, consider the input currents to the buck converter shown in **Figures 7(a)** and **11(a)**, i.e., without and with the damped input filter, respectively. The noise currents are analyzed with the CISPR 16–1-1 EMI receiver (numerically implemented). The results are presented in **Figure 14**, up to 1 GHz.

The improvement in the emission due to the buck converter with the damped *LC* filter is clearly observed in the EMI receiver output. The output of the EMI receiver can be compared with the limits set by the standards to test if the equipment is compliant with the EMC standards or not. Since the standardized procedure is followed, the results can be correlated across different sites.

6. Conclusion

In this chapter, the emission from power electronic converters has been discussed. Beginning with a discussion on the origin of the emission, which was attributed to the rapid switching transients taking place during the operation of the switches, the effect of the device characteristics as well as circuit parasitic elements were analyzed. At each step, through simulations and computations, the effects were demonstrated to degrade the converter's EMI performance compared to an ideal buck converter performance. Thereafter, a couple of methods employed to improve the EMI performance were discussed and demonstrated.

The importance of stipulating a standard procedure for measuring and analyzing the emissions was briefly discussed. The different coupling devices used in measuring emissions over different frequency ranges were introduced along with the measuring EMI receiver according to CISPR 16–1-1 standard. The results of the buck converter currents were analyzed with a numerically modeled EMI receiver as per the standard, and it was shown that the input filter dramatically improves the EMI performance of the power converter. Finally, since the switching transient is common to all power electronic converters, the discussions hold good for power converters in general.

Conflict of interest

"The author declares no conflict of interest."

Thanks

The author would like to thank his mother, Mrs. Jyotsna Nath, for everything.

Author details

Debasish Nath
Indian Institute of Science, SimYog Technology Pvt. Ltd., Bangalore, India

*Address all correspondence to: debasishnath100@gmail.com

References

[1] Mohan N, Undeland TM, Robbins WP. Power electronics : converters, applications, and design: John Wiley & sons 2003.

[2] Halkias CC, Millman J, Jit S. Electronic devices and circuits: Mcgraw Hill, 1967

[3] Zaikin Dl. Basic Diode SPICE model extension and a software characterization tool for reverse recovery simulation: IEEE International Conference on Industrial Technology, pp. 941-945, 2015

[4] Wang J, Liang S, Deng L, Yin X, Shen ZJ. An improved SPICE model of SiC BJT incorporating surface recombination effect: IEEE Transactions on Power Electronics, pp. 6794-6802, 2018

[5] Zach Z. RC snubber design in EZBUCK circuit: Alpha Omega Semiconductor, Inc. 2008.

[6] Erickson RW, Maksimovic D. Fundamentals of power electronics: Springer Science & Business Media; 2007

[7] Chu Y, Wang S, Wang Q, Goswami R. Modeling and stability analysis of active/hybrid common-mode EMI filters for DC/DC power converters: IEEE Energy Conversion Congress and Exposition, pp. 247-254, 2015

[8] Paul CR. Introduction to electromagnetic compatibility: John Wiley & Sons; 2006.

[9] Schwarzbeck D. The EMI receiver according to CISPR 16-1-1: Schwarzbeck Mess-Elektronik.

[10] Li C, Zhang L, Dong T, Wang T, Chen H. An EMI receiver model with consideration of the intermediate frequency filter: International Symposium on Electromagnetic Compatibility, pp. 233-236, 2016

[11] Li H, See KY. Conversion factors between common detectors in EMI measurement for impulse and Gaussian Noises: IEEE Transactions on Electromagnetic Compatibility, pp. 657-663, 2013

Section 5

Modelling and Computational Methods

Chapter 12

Return Stroke Process Simulation Using TCS Model

Fridolin Heidler

Abstract

The Traveling Current Source (TCS) model describes the electrical processes during the lightning return stroke phase. The TCS model assumes that the lightning current is injected at the top of the increasing return stroke channel represented by a transmission line. The electric and magnetic field is calculated based on the spatial and temporal distribution of the lightning current along the return stroke channel. It is shown that the main characteristics of the measured electric and magnetic fields can be reproduced with the TCS model. These are the Initial Peak of the electric and magnetic fields for near intermediate and far distances, the Ramp (up to the maximum) of the near electric field, the Hump of the near magnetic field after the initial peak, and the Zero Crossing of the far distant electric and magnetic fields. The fundamentals of the model are presented, and the model is extended to consider the current reflections occurring at the ground and the upper end of the return stroke channel. To this end, the ground reflection factor ρ and the top reflection factor R are introduced. Due to the increasing return stroke channel, the top reflection factor is a function of the return stroke velocity. The total current is composed of the source current according to the TCS model and the reflected currents. It is shown that the ground reflection causes significant variation in the waveform of the channel-base current and the electric and magnetic fields.

Keywords: Return Stroke, Lightning, Electric Field, Magnetic Field, Simulation, TCS model, Ground Reflection, Channel Top Reflection

1. Introduction

The threat of lightning can be classified into two separate groups, given by the direct and the indirect effects. The direct effects include physical losses due to the hot lightning channel and the high lightning current. Typical direct effects are mechanical damage, fire ignition, and the life-threatening hazard by lightning impact to persons. The basic protection measures against this threat are installing air termination systems, down conductor systems, and grounding systems [1, 2].

In contrast, the indirect effects are caused by nearby lightning events. Typical indirect effects are over-voltages which affect the electric and electronic systems and devices. The over-voltages are caused by partial currents that enter the structure and the coupling effects due to the high electric and magnetic fields radiated by lightning.

Meanwhile, the economic losses caused by the indirect effects are much higher compared to the direct effects [3]. This is attributed to the widespread use of electrical and electronic systems and devices in private buildings and industrial

IntechOpen

facilities. Countermeasures require the integration of lightning protection into the rules of electromagnetic compatibility (EMC) [4].

The lightning current may contain several components, from which the so-called return stroke current represents the highest threat. The return stroke current is a short current pulse, which lasts some tens to some hundreds of microseconds and may have an amplitude up to more than 100 kA (For example, see [5]). The currents generate electric and magnetic fields, which may be so intense that they couple over-voltages of several kilo-volts into installations inside buildings.

Examining these over-voltages requires simulation models that consider the return stroke process, including the electric and magnetic fields. To this end, return stroke models were developed which calculate the electric and magnetic field from the spatial and temporal distribution of the lightning current along the return stroke channel [6–10]. In these models, current reflections at the ground are commonly ignored. For this reason, the so-called traveling current (TCS)-model [11, 12] was developed, which considers the current reflections at the striking point.

One key task of the EMC is to evaluate the maximum threat which the electrical equipment and systems have to withstand. In the case of lightning, the electric and magnetic fields are highest if the orientation of the lightning channel is perpendicular to the earth's surface. For this reason, the lightning channel is considered with vertical orientation.

2. Physical background on TCS model

Most of the observed cloud-to-earth flashes are of negative polarity. For this reason, the TCS model is presented for the negative return stroke. The return stroke phase involves two periods, the initial connecting leader period, followed by the second period when the downward leader channel is discharged.

2.1 Connecting leader period

The negative cloud-to-ground lightning starts with processes in which charges are separated and rearranged inside the thundercloud. Due to these processes, negative charges are accumulated, and the center of the negative charge is built up in the lower part of the thundercloud. When the accumulated charge exceeds a critical value, a negative leader is formed, propagating from the negative charge center towards the ground.

The hot core of the leader is surrounded by negative charges, which also move down. When the downward propagating leader comes close to the ground, the electric field increases due to the charge approach. Then, a connecting leader starts from the ground as soon as the electric field exceeds a critical value.

The electric field at the tip of the connecting leader is so high that charge carriers are separated by impact and photoionization around the leader tip. The electric field accelerates the charge carriers, and they move to the tip of the connecting leader. In this way, a current is injected at the tip of the connecting leader, shown in **Figure 1a**. The injected current is given by:

$$i(h) = \int_A \vec{j}\, d\vec{A}, \text{with } \vec{j} = Q''' \cdot b \cdot \vec{E} \tag{1}$$

Q''': Charge density of the charge carriers.
b: Electrical mobility of the charge carriers.
E: Electric field.

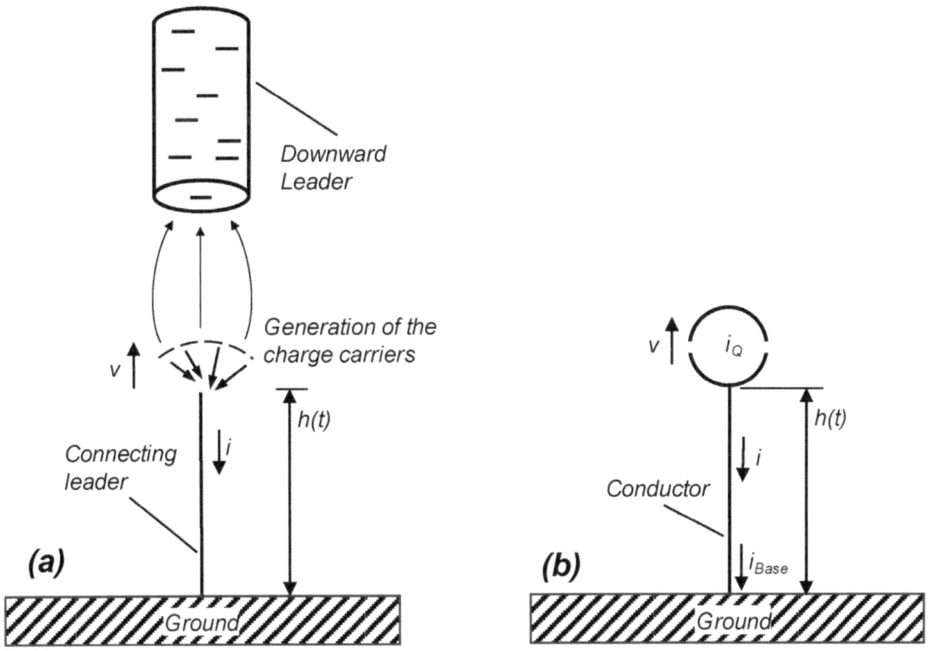

Figure 1.
Assumption for the connecting leader period, showing (a) the physical model and (b) the electrical equivalent circuit.

j: Current density.

h: Height of the upper end of the connecting leader.

In the equivalent circuit, the current injection can be represented by a current source $i_Q = i\ (h)$ located at the tip of the connecting leader in the height h. The current source travels at the connecting leader's upper end, which increases with the velocity v, shown in **Figure 1b**.

A certain time period is needed to separate the charge carriers and for the thermal ionization process at the tip of the connecting leader. For this reason, the traveling velocity (v) is less than the speed of light (c). On the other hand, the lower section of the connecting leader is already ionized, and this section represents a more or less good electrical conductor. Therefore, it is assumed that injected current propagates from the connecting leader tip with the speed of light to the ground.

2.2 Discharge process of downward leader channel

After contacting the upward propagating connecting leader with the downward leader, the negatively-charged shell of the downward leader is discharged, shown in **Figure 2a**. The charge carriers stored in the volume dV are injected into the tip of the lightning channel at the height h during the time interval dt. The current is given by:

$$i(h) = \frac{dQ}{dt} = \int_V \frac{dQ'''}{dt} dV \tag{2}$$

Also, in this case, the current injection can be represented by a current source $i_Q = i(h)$ located at the tip of the increasing return stroke channel. **Figure 2b** shows the electrical equivalent circuit with the current source, which travels at the tip of the increasing return stroke channel towards the thundercloud.

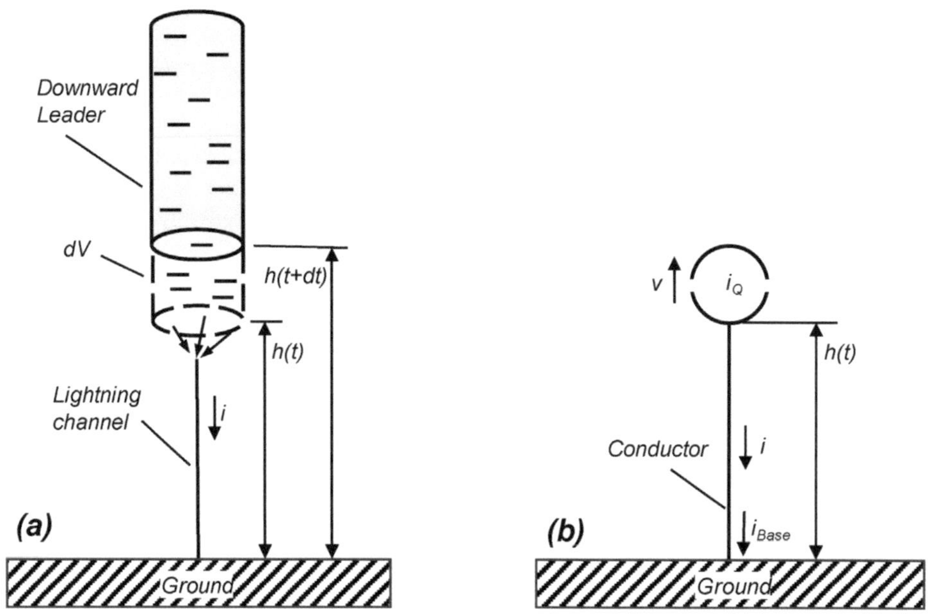

Figure 2.
Assumption for the discharge process of the downward leader channel, showing (a) physical model and (b) electrical equivalent circuit.

A certain time period is needed to collect the charge carriers and the thermal ionization to form a new section of the return stroke. Therefore, in this case, the traveling velocity (v) is less than the speed of light (c).

2.3 Summary

The return stroke process consists of the initial connecting leader process and the subsequent discharge process of the downward leader channel. In the electrical equivalent circuit, both processes can be represented by a current source traveling from the ground in the direction of the thundercloud with the return stroke velocity (v). Therefore, in the TCS model, it is not necessary to distinguish between both processes.

3. Current on the return stroke channel

Figure 3 shows the basic assumptions of the TCS model: The return stroke channel is perpendicular to the earth's surface, and it increases in the z-direction with constant return stroke velocity (v). The return stroke channel is considered as an (ideal) transmission line where the current pulses propagate with the speed of light (c). The ground is taken into account by an ideal-conducting plane.

The current (i_Q) is injected at the top of the return stroke channel from the current source, which begins to travel at t = 0 from ground level. When injected current arrives at the ground, a fraction of the current is reflected depending on the ground resistance. The reflected current moves up, and it is reflected again at the end of the lightning channel.

The current reflections are considered by the ground reflection coefficient ρ and the top reflection coefficient R. The current at the altitude z, $i(z, t)$, is composed by

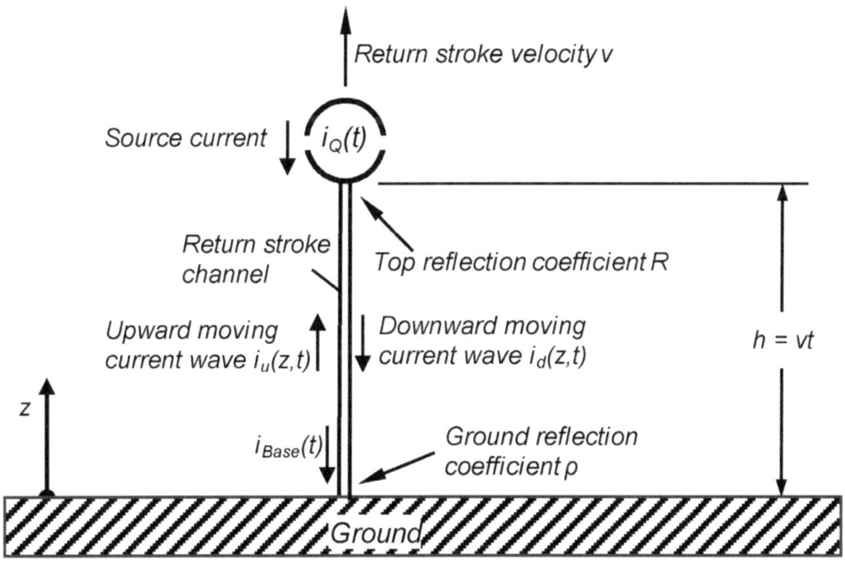

Figure 3
Current distribution on the return stroke channel.

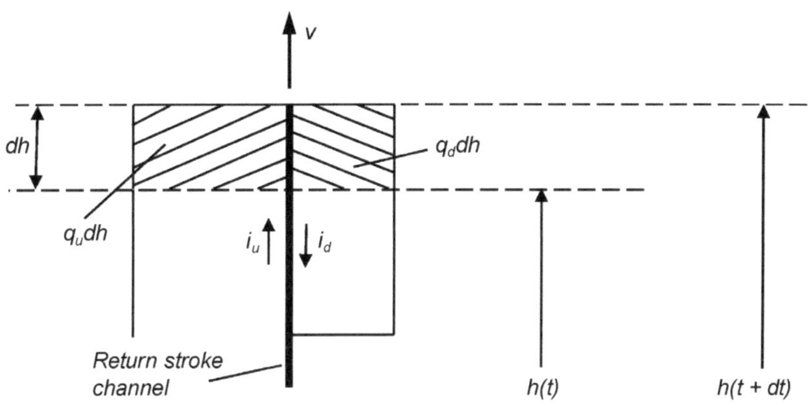

Figure 4.
Current reflection at the upper end of the return stroke channel.

the downward moving current wave $i_d(z, t)$ and the upward-moving current wave $i_u(z, t)$.

3.1 Top reflection coefficient R

Figure 4 shows the upward-moving current wave i_u arrives at the channel top (at the height h). The current wave is reflected at the open end of the lightning channel. For the current refection, the reflection point moves with the return stroke velocity $v = dh/dt$ [13].

The increasing lightning channel creates the new channel segment dh during the time interval dt. The new channel segment is loaded by the charge dQ, composed of the charge of the upward moving current wave (i_u) and the downward moving current wave (i_d). The charge density of the upward moving wave (q_u) and the downward-moving wave (q_d) is given by (c: speed of light):

$$q_u = \frac{i_u}{c} \tag{3}$$

$$q_d = -\frac{i_d}{c} \tag{4}$$

In Eq. (4), the negative sign is due to the current propagation in the opposite direction compared to the coordinate z. The charge dQ is given by:

$$dQ = q_u dh + q_d dh \tag{5}$$

Hence follows:

$$i = \frac{dQ}{dt} = q_u \frac{dh}{dt} + q_d \frac{dh}{dt} = q_u v + q_d v = i_u \frac{v}{c} - i_d \frac{v}{c} \tag{6}$$

The current is given by:

$$i = i_u + i_d \tag{7}$$

With Eqs. (6) and (7), the top reflection coefficient results in:

$$R = \frac{i_d}{i_u} = -\frac{c - v}{c + v} = -A \tag{8}$$

For example, if we assume the return stroke velocity $v = c/3 = 100$ m/μs, the top reflection coefficient is $R = -0.5$, i.e. the half of the current is reflected.

3.2 Source current i_Q and channel base current i_{Base}

The TCS model uses the source current (i_Q) as the input parameter. Because current data are only available from measurements at the striking point, converting the cannel-base current (i_{Base}) into the source current (i_Q) is required. This conversion is deduced in the following.

The channel-base current (i_{Base}) is composed of the downward-moving current wave ($i_{Base/d}$) and the upward-moving current wave ($i_{Base/u}$). With $i_{Base/u} = \rho \cdot i_{Base/d}$, it follows:

$$i_{Base} = i_{Base/u} + i_{Base/d} = (1 + \rho) i_{Base/d} = \frac{1 + \rho}{\rho} i_{Base/u} \tag{9}$$

The reflected current component $\rho \cdot i_{Base/d}(t)$ arrives at the top of the lightning channel after the delay time T. Because at the time $(t + T)$, the distance of propagation cT is equal to the height of the lightning channel $h = v\ (t + T)$, the delay time T can be written as:

$$T = \frac{v}{c - v} t \tag{10}$$

After the reflection at the upper end of the lightning channel, the (reflected) current wave $R\rho \cdot i_{Base/d}(t)$ moves down. The total downward moving current wave $i_d(h, t + T)$ is composed of this reflected current wave and the current from the current source $i_Q(t)$. This current wave arrives at the ground after the delay time T.

At ground level, the downward moving current wave is given by:

$$i_{Base/d}(t + 2T) = i_Q(t + T) + R\rho \cdot i_{Base/d}(t) \tag{11}$$

With Eqs. (8) and (10), it follows:

$$(t + 2T) = \frac{t}{A} \tag{12}$$

$$(t + T) = \frac{t}{Ak} \tag{13}$$

The coefficient k is:

$$k = 1 + \frac{v}{c} \tag{14}$$

Thus, Eq. (11) can be rewritten as:

$$i_{Base/d}\left(\frac{t}{A}\right) - R\rho \cdot i_{Base/d}(t) = i_Q\left(\frac{t}{Ak}\right) \tag{15}$$

Substituting the time t/A by the time t, it further follows:

$$i_{Base/d}(t) - R\rho \cdot i_{Base/d}(At) = i_Q\left(\frac{t}{k}\right) \tag{16}$$

Now, the following iterative procedure is applied to Eq. (16). Between each iteration step, the equation is multiplied by the factor $(R\rho)$ and the time is multiplied by the factor A resulting in the times A^1t, A^2t, A^3t ...:

$$R^1\rho^1 i_{Base/d}(A^1t) - R^2\rho^2 \cdot i_{Base/d}(A^2t) = R^1\rho^1 i_Q\left(\frac{t}{k}A^1\right) \tag{17}$$

$$R^2\rho^2 i_{Base/d}(A^2t) - R^3\rho^3 \cdot i_{Base/d}(A^3t) = R^2\rho^2 i_Q\left(\frac{t}{k}A^2\right) \tag{18}$$

$$R^3\rho^3 i_{Base/d}(A^3t) - R^4\rho^4 \cdot i_{Base/d}(A^4t) = R^3\rho^3 i_Q\left(\frac{t}{k}A^3\right) \tag{19}$$

The addition of the series of equations gives the following formula:

$$i_{Base/d}(t) = \sum_{\nu=0}^{\infty} (R\rho)^{\nu} \cdot i_Q\left(\frac{t}{k}A^{\nu}\right) \tag{20}$$

Substituting the time t by the time kt, Eq. (20) can be rewritten:

$$i_{Base/d}(kt) = i_Q(t) + R\rho \cdot i_Q(At) + \sum_{\nu=2}^{\infty} (R\rho)^{\nu} i_Q(A^{\nu}t) \tag{21}$$

Now, Eq. (21) is multiplied by the factor $(-R\rho)$, and the time t is substituted by the time At. It results:

$$R\rho \cdot i_{Base/d}(Akt) = -R\rho \cdot i_Q(At) - R\rho \cdot \sum_{\nu=1}^{\infty} (R\rho)^{\nu} i_Q(A^{\nu+1}t) \tag{22}$$

and from it

$$-R\rho \cdot i_{Base/d}(Akt) = -R\rho \cdot i_Q(At) - \sum_{\nu=2}^{\infty} (R\rho)^{\nu} i_Q(A^{\nu}t) \tag{23}$$

From adding Eqs. (21) and (23), it follows:

$$i_Q(t) = i_{Base/d}(kt) - R\rho \cdot i_{Base/d}(Akt) \tag{24}$$

With Eqs. (9) and (24), the source current can be rewritten as a function of the channel-base current:

$$i_Q(t) = \frac{1}{1+\rho}\left[i_{Base}(kt) - R\rho \cdot i_{Base}(Akt)\right] \tag{25}$$

With Eqs. (9) and (21), the channel-base current can be rewritten as a function of the source current:

$$i_{Base}(t) = (1+\rho)\sum_{\nu=0}^{\infty}(R\rho)^{\nu} i_Q\left(\frac{t}{k}A^{\nu}\right) \tag{26}$$

Eqs. (25) and (26) are the fundamental equations of the TCS model. They provide the source current $i_Q(t)$ conversion into the channel-base current $i_{Base}(t)$ and vice versa.

3.3 Lightning current along return stroke channel

When the downward propagating current wave, $i_d(z, t)$, starts from the height z, it arrives at the ground after the time delay z/c. With Eq. (9), it follows:

$$i_d(z,t) = i_{Base/d}(t + z/c) = \frac{1}{1+\rho} i_{Base}(t + z/c) \tag{27}$$

In the opposite case, when an upward-moving current starts at the time (t-z/c) from ground level (z = 0), it arrives at time t at the height z. With Eq. (9), it follows:

$$i_u(z,t) = i_{Base/u}(t - z/c) = \frac{\rho}{1+\rho} i_{Base}(t - z/c) \tag{28}$$

From adding Eqs. (27) and (28), the total current results:

$$i(z,t) = \frac{1}{1+\rho}\left[\, i_{Base}\left(t + \frac{z}{c}\right) + \rho \cdot i_{Base}\left(t - \frac{z}{c}\right)\right] \tag{29}$$

At the upper end of the return stroke channel (z = h), the current is given by:

$$i(h,t) = \frac{1}{1+\rho}\left[\, i_{Base}\left(t + \frac{h}{c}\right) + \rho \cdot i_{Base}\left(t - \frac{h}{c}\right)\right] \tag{30}$$

With the channel height h = vt, it follows:

$$i(h,t) = \frac{1}{1+\rho}\left[\, i_{Base}(kt) + \rho \cdot i_{Base}(Akt)\right] \tag{31}$$

3.4 Special case of no ground reflections

The reflections at the ground are often ignored, and the ground reflection coefficient is set to ρ = 0. In this case, the relation between the current along the lightning channel and the channel-base current can be simplified. From Eq. (29), it results for $z \leq h$:

$$i(z,t) = i_{Base}\left(t + \frac{z}{c}\right) \tag{32}$$

From Eq. (25), the relation between the source current (i_Q) and the channel-base current (i_{Base}) follows to:

$$i_Q(t) = i_{Base}(kt) \tag{33}$$

4. Electric and magnetic fields

Figure 5 shows the situation when an electric and magnetic field component is emitted from the infinitesimal small element dz located at the lightning channel in the height z. The field emitted from the lightning channel arrives after the time interval r/c at the point X in the distance s. Thus, the retarded time t_x is introduced to consider the propagation time:

$$t_x = t - \frac{r}{c} = t - \frac{\sqrt{s^2 + z^2}}{c} \tag{34}$$

For an observer in point X, the apparent height of the lightning channel is $h_x = vt_x$ (v = const.). Using $z = h_x$ in Eq. (34), the apparent height follows to [14]:

$$h_x(t) = \frac{v}{c^2 - v^2}\left(c^2 t - \sqrt{v^2c^2t^2 + c^2s^2 - v^2s^2}\right) \tag{35}$$

The differentiation by time provides the apparent return stroke velocity [14]:

$$v_x(t) = \frac{dh_x(t)}{dt} = \frac{v}{1 - \frac{v^2}{c^2}}\left(1 - \frac{v^2 t}{\sqrt{v^2c^2t^2 + c^2s^2 - v^2s^2}}\right) \tag{36}$$

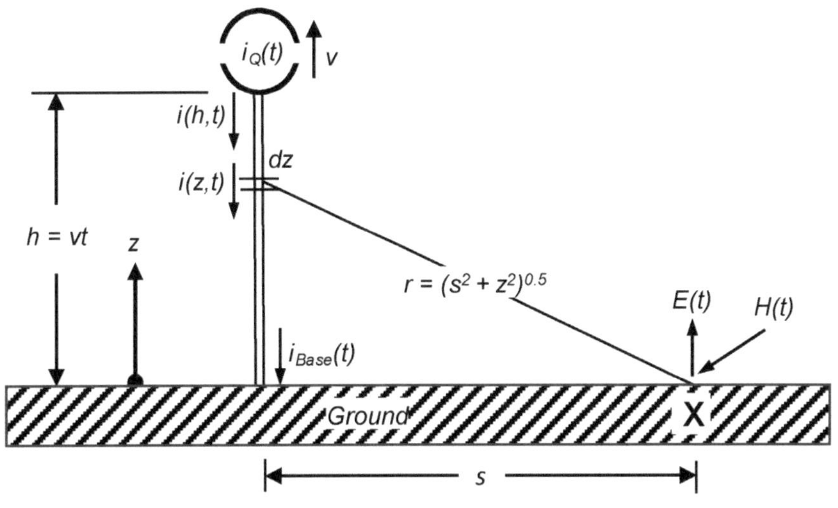

Figure 5.
Field emission from the return stroke channel.

The electric field is perpendicular to the earth's surface. With the apparent height h_x and the retarded time t_x, the vertical electric field in the distance s is given by [14, 15]:

$$E(t) = E_Q(t) + E_i(t) + E_{di}(t) + \Delta E(t) \tag{37}$$

The first term (E_Q) represents the electrostatic field, the second term (E_i) the intermediate field, and the third term (E_{di}) the radiation field. The terms are given by (ε_0 = 8.854 F/m):

$$E_Q(t) = \frac{1}{2\pi\varepsilon_0} \int_0^{h_x} \frac{s^2 - 2z^2}{r^5} \left(\int_{t_{x/0}}^{t_x} i(z,\tau) d\tau \right) dz \tag{38}$$

$$E_i(t) = \frac{1}{2\pi\varepsilon_0} \int_0^{h_x} \frac{s^2 - 2z^2}{cr^4} i(z, t_x) dz \tag{39}$$

$$E_{di}(t) = \frac{1}{2\pi\varepsilon_0} \int_0^{h_x} \frac{s^2}{c^2 r^3} \frac{\partial i(z, t_x)}{\partial t} dz \tag{40}$$

In Eq. (38), the lower constant of integration $t_{x/0}$ denotes the retarded time when the tip of the lightning channel arrived at the height z. The magnetic field is horizontal to the earth's surface. The horizontal magnetic field is given by [14, 15]:

$$H(t) = H_i(t) + H_{di}(t) + \Delta H(t) \tag{41}$$

The first term (H_i) represents the induction field, and the second term (H_{di}) the radiation field. The terms are given by:

$$H_i(t) = \frac{1}{2\pi} \int_0^{h_x} \frac{s}{r^3} i(z, t_x) dz \tag{42}$$

$$H_{di}(t) = \frac{1}{2\pi} \int_0^{h_x} \frac{s}{cr^2} \frac{\partial i(z, t_x)}{\partial t} dz \tag{43}$$

From **Figure 4**, it can be seen that the current changes abruptly at the upper end of the lightning channel. The abrupt current change represents a discontinuity at the open end of a transmission line. The discontinuity moves upwards with the return stroke velocity $v = dh/dt$, i.e., the current $i(h, t)$ is turned on at the channel segment dh during the time interval dt. This is taken into account by the following term:

$$\frac{\partial i(h,t)}{\partial t} dh = \frac{\partial h}{\partial t} di = -v \cdot di \tag{44}$$

The negative value indicates that the propagation of the current is in the opposite direction compared to the coordinate z. The abrupt current change is responsible for the additional far field terms $\Delta E(t)$ and $\Delta H(t)$ in Eq. (37) and in Eq. (41). These terms are often referred to in the literature as turn-on terms [16].

Of course, for an observer in the distance s, the real height has to be substituted by the apparent height h_x and the real return stroke velocity by the apparent return stroke velocity v_x.

The turn-on terms are finally given by [14]:

$$\Delta E(t) = \frac{1}{2\pi\varepsilon_0} \int_{h_{x,-}}^{h_{x,+}} \frac{s^2}{c^2 r^3} \frac{\partial i(z, t_x)}{\partial t} dh_x$$
$$= > \Delta E(t) = \frac{1}{2\pi\varepsilon_0} \frac{s^2}{c^2 \left(\sqrt{s^2 + h_x^2}\right)^3} v_x \cdot i(h_x, t_x) \tag{45}$$

$$\Delta H(t) = \frac{1}{2\pi} \int_{h_{x,-}}^{h_{x,+}} \frac{s}{c r^2} \frac{\partial i(z, t_x)}{\partial t} dh_x$$
$$= > \Delta H(t) = \frac{1}{2\pi} \frac{s}{c\left(s^2 + h_x^2\right)} v_x \cdot i(h_x, t_x) \tag{46}$$

5. Examples

The waveform of the electric and magnetic fields is well-known from measurements at various distances. According to the distance, the field is usually classified into three groups: near field, intermediate field, and far field. The near field distance range is up to several kilometers, the intermediate field distance range is from several kilometers up to several tens of kilometers, and the far field distance range is from several tens of kilometers up to several hundreds of kilometers.

The basic features are as follows [17]:

- The electric and magnetic field exhibits an Initial Peak at distances of more than several hundred meters.
- The near electric field exhibits a Ramp (up to the maximum) after the Initial Peak,
- The near magnetic field exhibits a Hump after the Initial Peak,
- The electric and magnetic far field exhibits a Zero Crossing after the Initial Peak.

The following shows, by using two examples, that the TCS model reproduces these basic features.

The first example analyses the influence of the ground reflection on the current and on the electric and magnetic field for a typical negative first return stroke at a near distance. The second example presents the electric and magnetic field for a typical subsequent return stroke at a near, intermediate, and far distance. In both examples, the return stroke velocity is chosen to $v = c/3 = 100\ m/\mu s$. The following predefined source current is used (For example, see [12]):

$$i_Q(t) = \frac{i_{Q/max}}{\eta} \cdot \frac{\left(\frac{t}{\tau_1}\right)^m}{1 + \left(\frac{t}{\tau_1}\right)^m} \cdot e^{t/\tau_2} \tag{47}$$

The coefficient η denotes the correction factor for the current maximum. The coefficients τ_1 and τ_1 are the front and decay time parameters of the current waveform.

5.1 Influence of ground reflection

In Eq. (47), the current parameters are chosen to $i_{Q/max}$ = 30 kA, m = 5, τ_1 = 1.26 μs and τ_2 = 56.3 μs. The maximum current steepness (di_Q/dt_{max}) is about 32 $kA/\mu s$. These values are typical for a negative first return stroke [18, 19].

The characteristic impedance of the lightning channel is about 1000 Ω [20]. The grounding resistance of poorly grounded buildings is often in the same order of magnitude. In this case, the ground reflection can be neglected. On the other hand, the current ground reflections cannot be ignored when well-grounded structures have much lower resistances. For instance, the well-grounded Peissenberg tower has a ground reflection coefficient of about ρ = 0.7 [21].

In the following, the two cases are analyzed, i.e., the ground reflection coefficient is set to ρ = 0 and ρ = 0.7. **Figure 6a** and **b** show the influence of the ground reflection on the channel-base currents (i_{Base}) (at the striking point). For ρ = 0, the peak current is identical with the peak value of the source current, $i_{Q/max}$ = 30 kA. For ρ = 0.7, the peak current is 43.5 kA, equating to an increase of 45%. **Figure 6c** shows, that the maximum current steepness also got higher by 65.6%, from 24.1 $kA/\mu s$ for ρ = 0 to 39.9 $kA/\mu s$ for ρ = 0.7. The 10–90% rise time decreased accordingly, from $T_{10\text{–}90\%} \approx 1.4$ μs for ρ = 0 to $T_{10\text{–}90\%} \approx 1.1$ μs for ρ = 0.7.

Figures 7 and **8** show the corresponding electric and magnetic fields in a distance of 3 km. The measured Initial Peak of the electric field (E_{max}) and magnetic field (H_{max}) is successfully reproduced, but it is more pronounced for ρ = 0.7, i.e., the initial field peak is about 44% higher for ρ = 0.7 compared to ρ = 0.

The electric field exhibits the Ramp (**Figure 7a**), and the magnetic field exhibits the Hump (**Figure 8a**), known from measurements. The Ramp and Hump are more pronounced for ρ = 0.7 compared to ρ = 0. The different steepness of the Ramp is due to the current reflections, but the final value of the electric field is the same (not shown here) because the total charge transfer is unaltered.

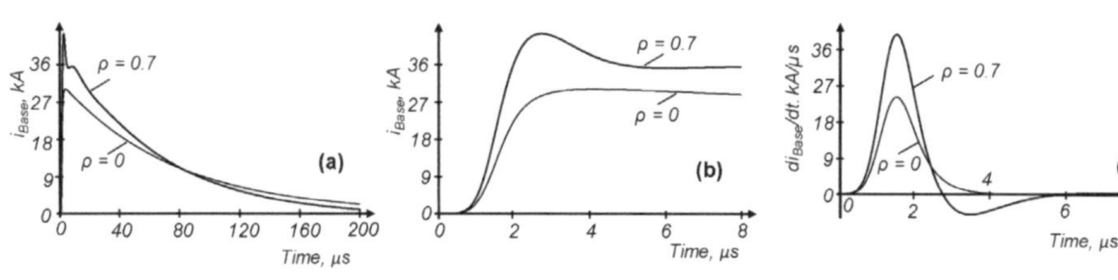

Figure 6.
Channel-base current as a function of the ground reflection factor ρ, showing (a) the total current, (b) the current front, and (c) the current derivative.

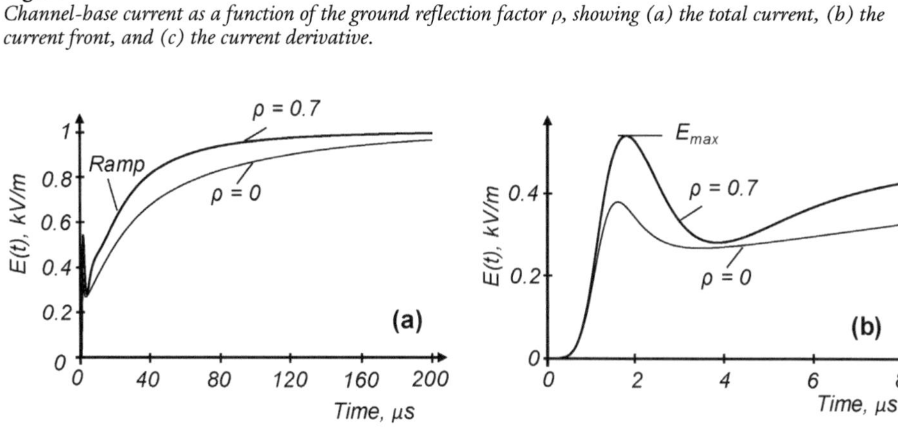

Figure 7.
Electric field in 3 km *distance, showing (a) total field and (b) field front.*

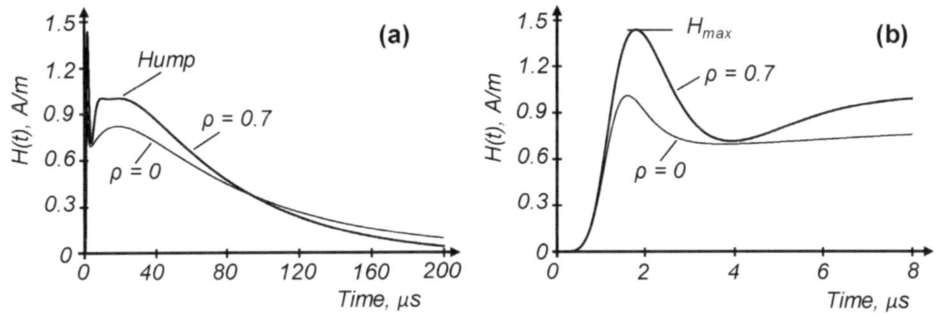

Figure 8.
Magnetic field in 3 km distance, showing (a) total field and (b) field front.

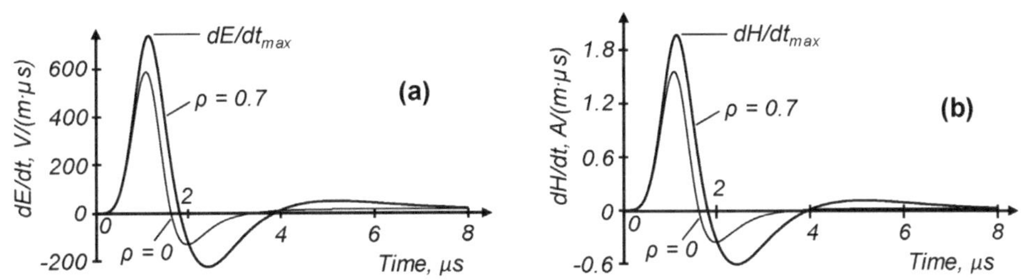

Figure 9
Derivative of (a) Electric field and (b) magnetic field in 3 km *distance.*

Figure 9 shows that the influence of the current reflections on the field derivative is comparably low. For ρ = 0.7, the maximum of the electric and magnetic field derivative (dE/dt_{max}, dH/dt_{max}) is about 26% higher, and the full width at half maximum (*FWHM*) is higher by less than 20%.

5.2 Electric and magnetic field at near, intermediate, and far distance

In Eq. (47), the current parameters are chosen to $i_{Q/max}$ = 10 kA, m = 4, τ_1 = 0.7 μs and τ_2 = 30 μs. Ground reflection is ignored (ρ = 0) and the return stroke velocity is chosen to 100 $m/\mu s$. For the channel-base current (i_{Base}) (at the striking point), the peak value is 10 kA, the 10–90% rise time is 0.94 μs and the time to half value is 31 μs. These values are typical for subsequent return strokes [18, 19, 22].

Figure 10 shows the electric and magnetic fields at 1 km (near field), 10 km (intermediate field), and 100 km (far field). It can be seen that the main characteristics of the electric and magnetic fields are reproduced with the TCS model, i.e., the Initial Peak of the electric and magnetic field, the Ramp of the near electric field, the Hump of the near magnetic field, and the Zero Crossing of the electric and magnetic far field.

At far distances, the electric and magnetic field is approximately given by the radiation term (E_{di}, H_{di}) according to Eqs. (40) and (43). In this case, the electromagnetic field (E_{far}, H_{far}) has a behavior like a plane wave in free space, given by the following formula (Compare [6]):

$$\frac{E_{far}}{H_{far}} = c \cdot \mu_0 = \Gamma_0 \tag{48}$$

μ_0 = 4π ·10^{-7} H/m: Permeability of free space.
Γ_0 = π ·120 Ω ≈ 377 Ω: Impedance of free space.

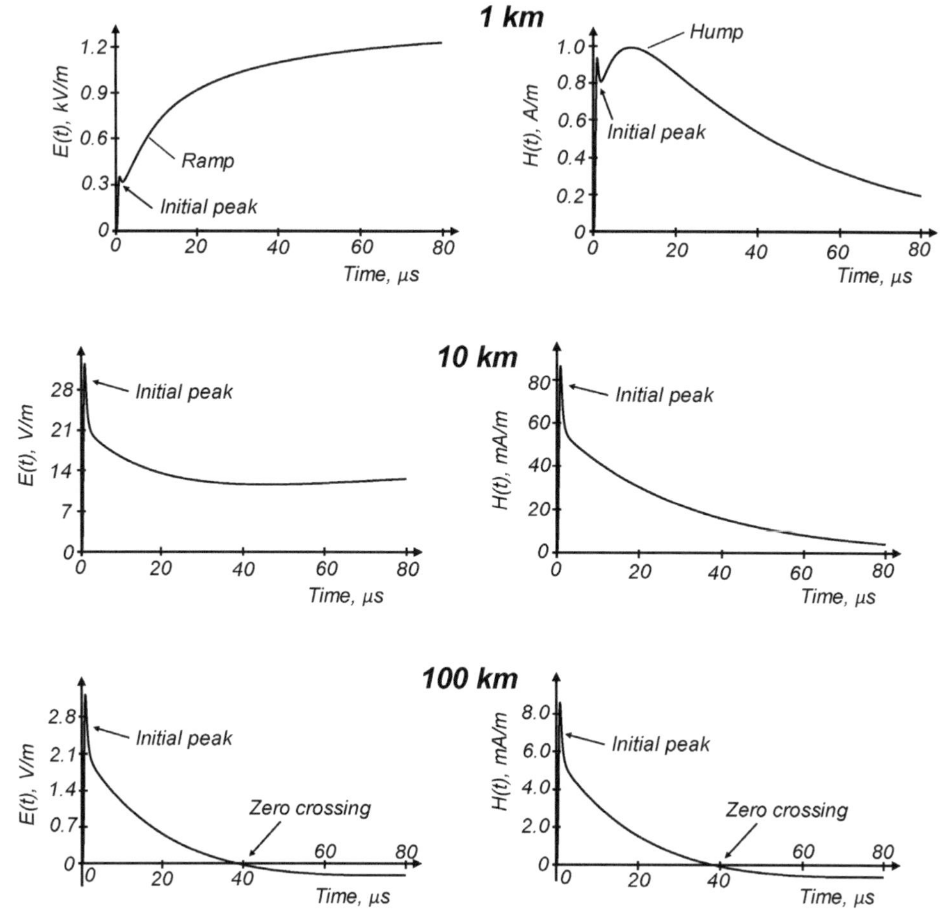

Figure 10.
Electric field, E(t)*, and magnetic field,* H(t)*, of a subsequent return stroke in the distances of 1, 10, and 100* km*. For the channel-base current (at the striking point), the peak current, 10* kA*, rise time, 0.94* μs*, and time to half value, 31* μs.

As shown in Eq. (48), the electric and the magnetic far fields are linked together by the impedance of free space. Therefore, the waveform of the electric and magnetic fields is the same at far distances, as shown in **Figure 10**.

At 100 *km*, the Initial Peak of the electric field is 3.2 *V/m*, which agrees very well with measured data that varies between 2.7 and 5.0 *V/m* [5].

5.3 Summery

The examples show that the main features of the measured electric and magnetic fields are reproduced with the TCS model. These are the Ramp of the near electric field, Hump of the near magnetic field, Zero Crossing of the far distant electric and magnetic fields, and Initial Peak of the electric and magnetic fields for near, intermediate, and far distances.

Author details

Fridolin Heidler
University of the Federal Armed Forces Munich, Neubiberg, Germany

*Address all correspondence to: fridolin.heidler@unibw.de

References

[1] IEC 62305-1. Protection against lightning-Part 1: General principles. 2nd ed. Geneva: IEC; 2010.

[2] IEC 62305-3. Protection against lightning-Part 3: Physical damages and life hazard. 2nd ed. Geneva: IEC; 2010.

[3] Habiger E. Electromagnetic compatibility. 1st ed. Heidelberg: Hueting;1992.

[4] IEC 62305-4. Protection against lightning-Part 4: Electrical and electronic systems within structures. 2nd ed. Geneva: IEC; 2010.

[5] Rakov V, Uman M. Lightning: Physics and Effects. 1st ed. New York: Cambridge Univ. Press; 2003.

[6] Uman M, McLain D, Krider E. The electromagnetic radiation from a finite antenna. Am. J. Phys. 1975; 43:33-38.

[7] Uman M, McLain D. Magnetic field of lightning return stroke. J. Geophysical Res. 1969; 74:6899-6910.

[8] Bruce C, Golde R. "The Lightning Discharge," J. Institution of Electrical Engineers. 1941; 88:6:487-505.

[9] Nucci C, Mazzetti C, Rachidi F, Ianoz M. On lightning return stroke models for LEMP calculations. In: Proceedings of the 19th Intern. Conf. on Lightning Protection (ICLP '88); 25-29 April; Graz, Austria; 1988: p. 463-470.

[10] Rakov V, Dulzon A. A modified transmission line model for lightning return stroke field calculation. In: Proceedings of the 9th Intern. Zurich Symposium on Electromagnetic Compatibility (EMC '91); 12-14 March1991; Zurich, Switzerland; 1991: p. 229-234.

[11] Heidler F. Traveling current source model for LEMP calculation. In: Proceedings of the 6th Intern. Symposium on Electromagnetic Compatibility (EMC '85); 5-7 March 1985; Zurich, Switzerland; 1985: p.157-162.

[12] Heidler F. Review and extension of the TCS model to consider the current reflections at ground and at the upper end of the lightning channel. IEEE Trans. on EMC. 2019; 61:3:644-652. DOI: 10.1109/TEMC.2018.2890342

[13] Heidler F, Hopf C. Influence of the lightning channel termination on the lightning current and lightning electromagnetic impulse. In: Proceedings of the 16th International Aerospace and Ground Conference on Lightning and Static Electricity (ICOLSE '94); 24-27 May 1994; Mannheim, Germany; 1994: p. 65-74.

[14] Heidler F, Hopf C. Lightning current and lightning electromagnetic impulse considering current reflection at the earth's surface. In: Proceedings of the 22nd International Conference on Lightning Protection (ICLP'94); Budapest, Hungary;1994: report R 4-05.

[15] Rakov V, Uman A. Review and evaluation of lightning return stroke models including some aspects of their application. IEEE-Trans. on EMC. 1998; 40:4,403-426. DOI: 10.1109/15.736202.

[16] Rachidi F. The quandary of direct measurement and indirect estimation of lightning current parameters. In: Proceedings of the 27th International Conference on Lightning Protection (ICLP '04); Avignon, France; 2004: p. 36-51.

[17] Lin Y, Uman M, Tiller J, Brantley R, Beasley W, Weidman C. Characterization of lightning return stroke electric and magnetic fields from simultaneous two-station measurements. Journal of Geophysical Research. 1979; 84:C10:6307-6314.

[18] Berger K, Anderson R, Kroeninger H. Parameters of lightning flashes. Electra. 1975; 41:23-37.

[19] Anderson R, Eriksson A. Lightning parameters for engineering application. Electra. 1980; 69:65-102.

[20] Trapp N. Measurement of the current waveform and the current parameters with automatically operating lightning current measuring stations. PhD. Munich: Technical University Munich; 1985; translation from German.

[21] Fuchs F. On the transient behavior of the telecommunication tower at the mountain Hoher Peissenberg. In: Proceedings of the 27th International Conference on Lightning Protection (ICLP `98); Birmingham, UK; 1998: p. 36-41.

[22] Heidler F, Paul Ch. Some return stroke characteristics of negative lightning flashes recorded at the Peissenberg Tower. IEEE-Trans. on EMC. 2017; 59:5,1490-1497. DOI: 10.1109/TEMC.2017.2688587

Chapter 13

Development of Generic Radiating Model for Rectangular Capacitors: Magnetic Near Fields Analysis and Modeling

Walid Labiedh, Bessem Zitouna, Mohamed Tlig and Jaleleddine Ben Hadj Slama

Abstract

This chapter deals with modeling the radiation from rectangular film capacitors as a power electronics component. The rectangular film capacitors are sources of electromagnetic radiation, where its characterization is crucial for electronic circuits EMC. Our study presents the analyses and modeling of the magnetic near field radiated by the plastic and the polyester capacitors. An electromagnetic inverse method is combined with an optimization method based on genetic algorithms to create a radiating equivalent model. A very good agreement is observed between the magnetic near field cartography measured above the studied structure and calculated using the developed model parameters. Finally, a generic radiating model is proposed for various types of rectangular film capacitors. The generic model is validated using the measurements on a rectangular capacitor. The obtained equivalent model can calculate the magnetic field at any near field zone and far field around the capacitors. Circuit designers can use the field distribution to optimize the placement of the capacitors on the printed circuit board to reduce their coupling and potential interaction with other equipment in the vicinity of the system.

Keywords: Electromagnetic compatibility (EMC), Film capacitors, electromagnetic radiation, near field, inverse problems, generic model, analysis and modeling

1. Introduction

In power electronics, the switching frequencies are increasingly high to reduce these systems' weight, volume, and cost. This rise in frequency is accompanied by an increase in conducted and radiated electromagnetic disturbances. It also has significant effects on the behavior of these components at high frequencies. The characterization of the electromagnetic behavior of the various components of power electronics systems is an important step to control these systems' electromagnetic compatibility that should start from the design phase [1–4].

Many researchers have proposed radiating models for components or systems. In [2], concerning systems, a radiating model of a DC-DC converter was put forward and referenced as TEN 40–2412. The suggested model was composed of a network of four magnetic dipoles. The study in [3] proposed two radiating models

IntechOpen

of a circuit based on a microcontroller. The first model was based on 53 magnetic dipoles, while the second one was based on 517 electric dipoles. In [4], a radiating model of the MOSFET was presented, constituted by a magnetic dipole. Two radiating models of toric self-inductance were suggested in [5]. The first developed model consisted of a large number of dipoles (676 dipoles). It was obtained by using a modeling approach based on the matrix inversion. The second model was composed of a network of 12 dipoles. Previously, researchers have already put forward a radiating model for an inductor [6, 7]. The proposed model by [6] consisted of a network of two simple magnetic dipoles. However, the one suggested by [7] was composed of a large number of magnetic and electric dipoles (520 dipoles). The measurements of the electromagnetic radiation were performed as described in [8].

On the other hand, several methods have been developed to model the radiated emissions of components and electrical systems. The study in [3] suggested two methods. The first one was based on magnetic dipoles, and the second one was based on electric dipoles. The two presented methods required both the field amplitude and phase. The study in [4] put forward an electromagnetic inverse method based on the Genetic Algorithms (GA). The method gave good results when the number of searched parameters is not very significant; otherwise, the method was inefficient considering calculation time and convergence. The same authors developed [2] a new approach based on image processing descriptor PZMI by processing only a scan window of the measured magnetic field cartography. Another approach was developed based on the matrix calculation [6, 7].

In this chapter, we are interested in the radiation of film capacitors. The film capacitors are widely used in the power electronics domain replacing electrolytic capacitors. They particularly permit the improvement of reliability and EMI suppression. Indeed, they are often used in new industrial applications, such as power converters for renewable energy and hybrid automotive systems. The film capacitors are mainly characterized by a high insulation resistance, large conduction currents, and good stability of its capacitance. These components are generally cumbersome, and their radiation certainly affects the overall radiation of the circuits. The electromagnetic field radiated by capacitors creates induced disturbances that may cause a dysfunction to the neighboring circuits. Therefore, the mastery of all the circuit's radiated fields is a requirement for the system's proper functionality. It will be worthwhile to dispose of the radiation model of capacitors.

The literature review of papers related to EMC analysis of discrete capacitors shows few studies dealing with conducted disturbances across these components and their coupling effects on other neighboring components. Particularly, in [9–12], the authors first studied the electromagnetic coupling between the components of an EMC filter. The filter was composed of two capacitors and two coils. Second, they proposed an automatic method to determine the optimal placement of components and the associated tracks' design. However, these studies treat the coupling only in EMC filter applications. They were not interested in the capacitor's radiation. In [12], methods for predicting the magnetic field distribution in the capacitors were presented. Although the existence of a magnetic field may seem anecdotal in capacitors whose functionality depends on the electric field, the magnetic field is precisely the cause of imperfections of the capacitors: the equivalent series inductance, the resonant frequencies, the associated losses, and the currents induced in metalized capacitors [13].

The present chapter aims to propose a radiation model for film capacitors, which has never been proposed before. In Section 2, the experimental techniques used to characterize the near field around the capacitors are presented. In Section 3, the magnetic near field above the capacitors is analyzed. The modeling method is explained in Section 4. In the following sections, the measured fields are used to develop the radiating models of the studied capacitors. Finally, a generic model for all rectangular capacitors is presented.

2. Experimental techniques and characterization method

2.1 Excitation circuit

To allow characterization over a wide frequency band, the component under test (CUT) is excited by a current having a large frequency band similar to those found in power electronics. Therefore, It is proposed to install the component under test in a converter circuit supplied by a voltage equal to 30 V and has a duty cycle of 0.5 (**Figure 1**). The output current of the circuit is 0.5 A. The switching frequency is 50 kHz.

2.2 Measurements methods

In our study, we utilize only one component of the field radiated by the system under test. Indeed, the use of a single component in a large number of points is quite sufficient to identify the equivalent source by guaranteeing the uniqueness of the solution [14, 15]. Furthermore, the developed radiating model is capable of modeling the electric field as well as the magnetic field. Actually, based on [16–18], we can calculate from the magnetic field, the electric ones, and vice versa.

Since the method can be applied to any magnetic or electric field component, we have used the vertical component Hz. We may also utilize the other tangential components (Hx or Hy). For measuring the magnetic near field, we place a magnetic probe above the device under test. It is a manually made probe consisting of a 1.6 mm radius circular loop connected to the central conductor on one side and the external shield of a coaxial cable on the other side. To capture the various components of the H-field, it is necessary to place the normal of the collinear loop to the desired component.

In our study, we use two methods for measuring the magnetic field around the component. The first method performs the near field measurements in the frequency domain, and the second one is performed in the time domain.

2.2.1 Frequency measurements

The method of frequency measurements is based on a spectrum analyzer. In fact, to measure the magnetic field radiated of the under-test device, we use the near-field measurement bench presented in **Figure 2**.

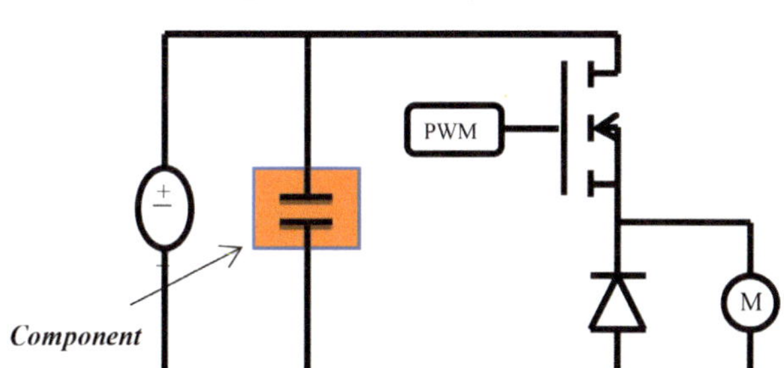

Figure 1.
Chopper circuit where component under test is characterized.

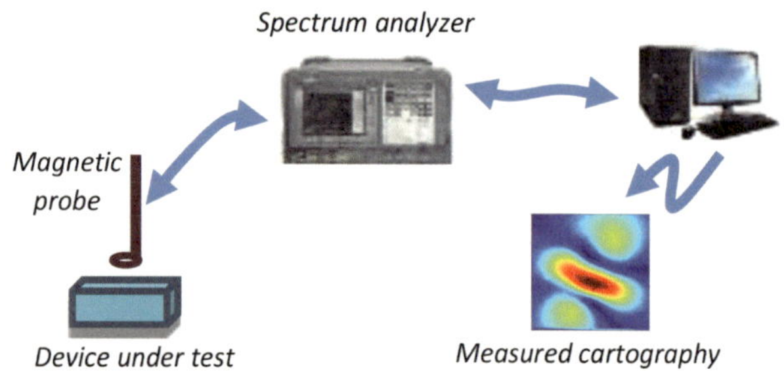

Figure 2.
Frequency measurement bench.

2.2.2 Temporal measurements

The temporal measurements method is based on the utilization of an oscilloscope. This measurement method permits us to have a radiated magnetic field by the device under test at any moment. After measuring the temporal voltage across the probe, a Fast Fourier Transform (FFT) calculation is performed. **Figure 3** presents the adopted methodology.

According to Lenz-Faraday law, for a simple fixed circular conductor loop, having a very small radius R compared to the wavelength, diving in a magnetic field B(t) oriented along the z-axis, the potential difference induced in the loop by the magnetic field is given by the following equations:

In the time domain

$$Vm\ (t) = \oint \frac{\partial B}{\partial t} dS \tag{1}$$

In the frequency domain

$$Vm\ (f) = j\ \omega\ B\ S \tag{2}$$

where $B = \mu_0 \times H$

$$\text{So } Vm\ (f) = j\ \omega \mu_0\ H\ S \tag{3}$$

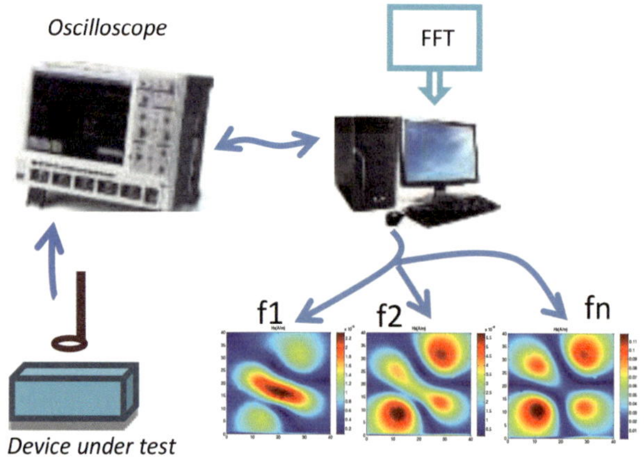

Figure 3.
Methodology using temporal bench.

where $\mu_0 = 4 \times \pi \times 10^{-7}$, is the magnetic permeability $S = \pi \times r^2$ is the surface of the probe, f is the radiation frequency, and $\omega = 2\pi f$.

The magnetic field is assumed to be constant over the entire area of the probe. This is especially true when the probe is very small compared to the capacitor's size. Therefore, the magnetic field measured at the center of the probe is calculated by the following equation:

$$H(f) = Vm(f)/2 \times (2 \times \pi \times \mu_0 \times f \times 2 \times S) \quad (4)$$

where Vm (*f*) is the component at the frequency *f* of the FFT of the voltage measured at the terminals of the probe. Finally, we can extract the near-field cartographies for each radiating frequency.

2.3 Probe calibration

Before using a magnetic field probe, it is necessary to calibrate it. In [5, 6], the authors suggested a method to validate the accuracy of their magnetic field probes. They measured the radiated magnetic field around a simple circuit with a conductive wire above a ground plane. The values of the measured field are compared to those calculated theoretically. Thus, the comparison of the results enabled the validation of the used probes.

Similarly, in this work, to calibrate the magnetic probe, we have used a radiating circuit whose radiation is known theoretically. We compare the measured radiated magnetic field to that calculated by the numerical electromagnetic tool NEC [19] based on the moment's method. The radiating circuit is a rectangular loop of 5 cm length and 3 cm width excited by a sinusoidal voltage of 10 V amplitude at a frequency of 5 MHz (**Figure 4**).

The simulations and measurements are made at the points located in a horizontal plane at 3 mm above the radiating loop and having dimensions of 5×3 cm^2. The calculation step is 2.5 mm along the X-axis and 1.5 mm along the Y-axis (441 measurement points).

Figures 5–7 gives the magnetic field components Hz, Hx, and Hy along the X- and Y-axis, respectively. It shows a good agreement between the measured and calculated curves of the Hz component. In order to have a good signal-to-noise ratio when measuring Hx and Hy components, we chose measurement lines located at the edge of the emitting loop. Consequently, in some probe measurements, we notice a difference between measurements and simulations due to the edge effects and the coupling effects between the probe and the radiating loop. These phenomena are not significant on the Hz component and therefore do not affect the construction of the model.

Figure 4.
Loop used for probe calibration.

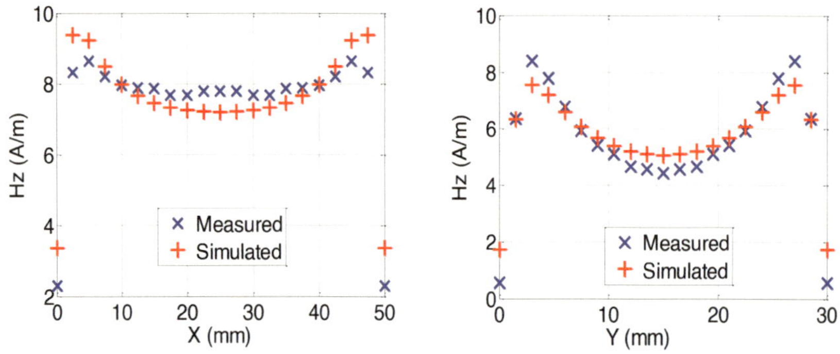

Figure 5.
Hz following X- and Y-axes respectively for Y = 4.5 mm and X = 25 mm.

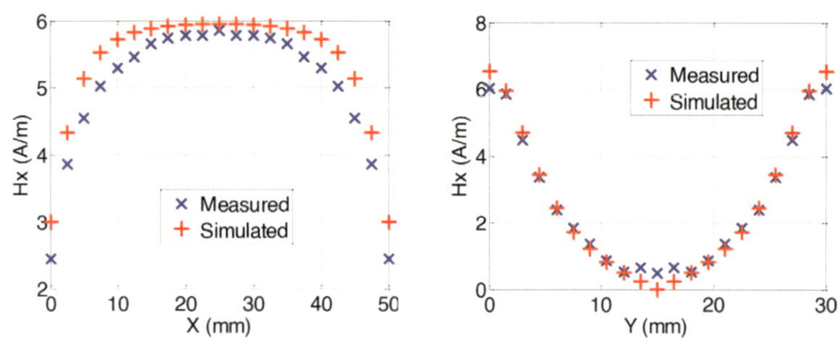

Figure 6.
Hx following X- and Y-axes respectively for Y = 1.5 mm and X = 25 mm.

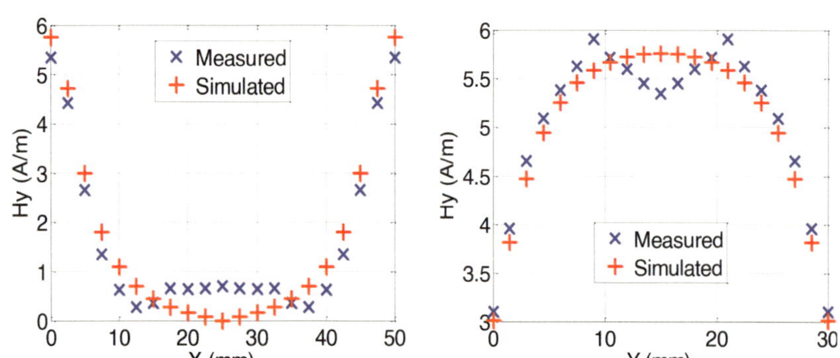

Figure 7.
Hy following X- and Y-axes respectively for Y = 1.5 mm and X = 25 mm.

In order to examine the electric field rejection capability of our magnetic probe, we have placed it over a transmitting electric antenna where the amplitude of the electric field is important. The observed voltage across the probe is so low that we cannot dissociate it from the noise. Hence we deduce that our shielded probe rejects very well the electric field and measures the magnetic field with great accuracy.

3. Measurements and analysis of polypropylene capacitor radiation

Measurements are performed in the time domain using a high-bandwidth oscilloscope. We have measured the vertical component Hz of the magnetic near-field radiated above a plastic polypropylene capacitor of a 2.2 uF capacitance and a maximum voltage of 100 V (the capacitor size is 15 mm × 6 mm × 26.5 mm). To avoid the interferences caused by the capacitor legs, we have folded the legs of the latter. The probe is located at a distance of 2 mm above the capacitor in a plane whose dimensions are 4 cm x 2 cm. The displacements are done with a measuring step of 2 mm on the X-axis and 1 mm on the Y-axis. Thus, we have 400 measurement points. **Figure 8** shows the measurement plane and the component under test. It gives a particular point (3, 5) position in the measurement plane.

Figure 9 shows the temporal voltage measured at the terminals of the magnetic probe once it is situated at the point (3, 5): the third position along the X-axis and the fifth position along the Y-axis. When analyzing the measured magnetic field signal, the existence of switching transients is noticed during turning on and off the power component utilized in the chopper.

We present in **Figure 10** the spectrum of radiated magnetic field Hz above the capacitor in the point (3,5). **Figure 10** shows the spectrum analysis of the whole temporal signal with both switching phases (closing and opening of the MOSFET in the excitation circuit where our studied component is characterized). That measured above the capacitor at the point of the coordinates (3,5) presented in **Figure 9**. By analyzing the measured temporal signal spectrum, four principal harmonics are observed, which have the following frequencies: 3 MHz, 11 MHz, 20 MHz, and 40 MHz. The resonances that appear above the capacitor are principally due to the

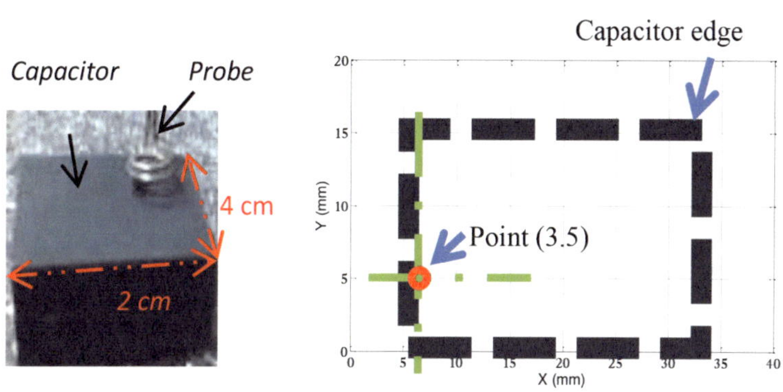

Figure 8.
Measurement plane over component under test.

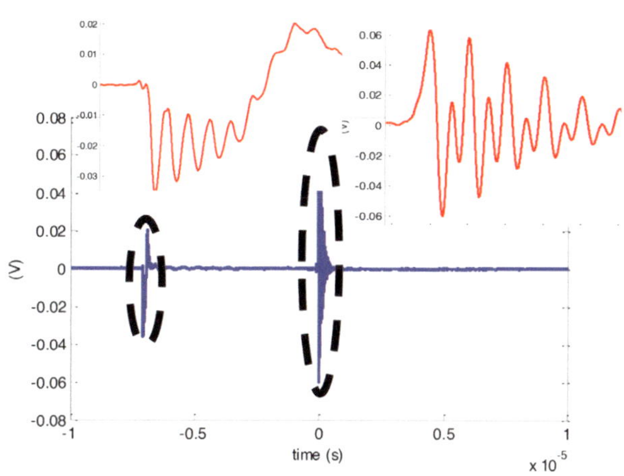

Figure 9.
Measured voltage at the terminals of the probe above capacitor under test.

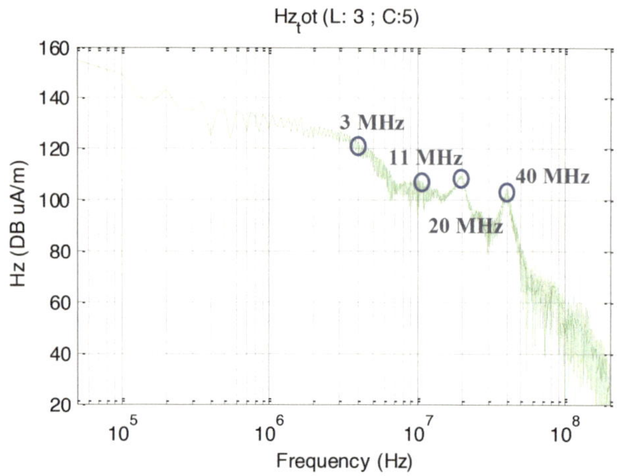

Figure 10.
Spectrum of radiated magnetic field Hz at the terminals of the probe above the studied capacitor at point (3.5).

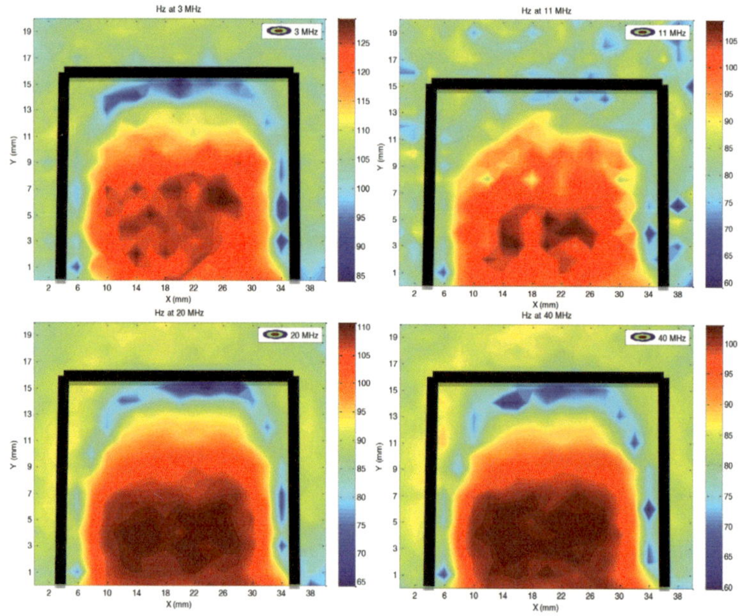

Figure 11.
Cartographies of the near field for 3 MHz, 11 MHz, 20 MHz, and 40 MHz frequencies.

operating current flowing into the chopper circuit. **Figure 11** presents the cartographies of the field above the capacitor at these different frequencies.

According to the previous cartographies, the distribution of the radiated magnetic field above the capacitor is over the entire component surface, as illustrated in **Figure 12**.

4. Modeling methodology

To develop a radiating model that enables producing a near field equal to that is radiated by the system under test, we use the flow diagram given in **Figure 13**. The modeling methodology is based on a hybrid approach consisting of the coupling between the electromagnetic inverse method and the Method of Moments (MoM) [20].

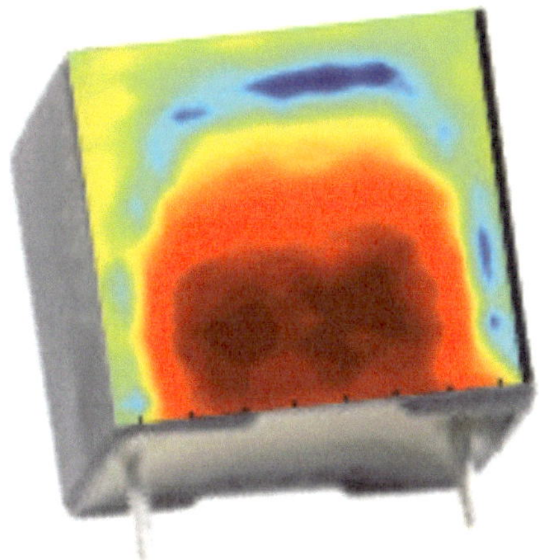

Figure 12.
Distribution of magnetic field above capacitor for f = 20 MHz.

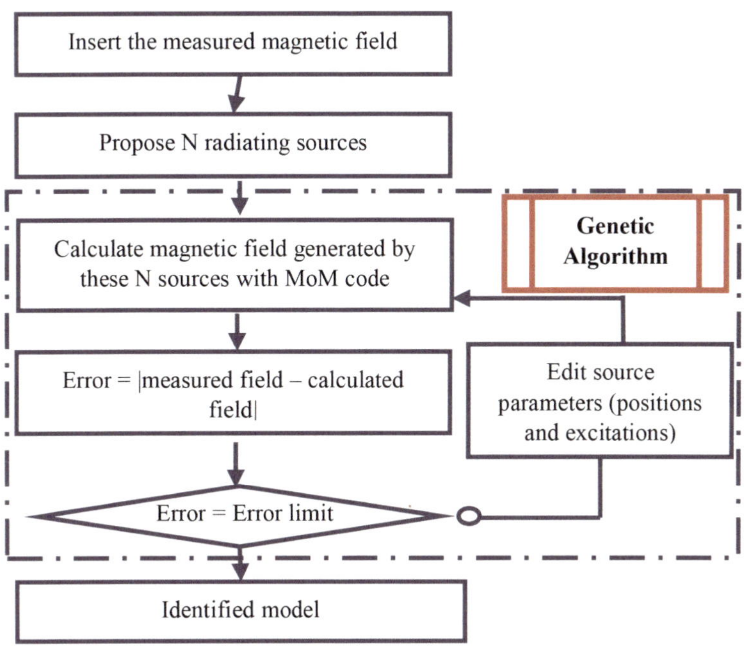

Figure 13.
Flow chart of the modeling approach.

This approach starts by introducing the electromagnetic near field measured around the device under test and then defines the optimization parameters by the Genetic Algorithm (GA) that has been presented in [21, 22]. After, the search is launched to find the parameters of the equivalent radiating structure by comparing the measured cartography to the one calculated numerically by the MoM and using the obtained parameters.

The magnetic field generated by the N sources is calculated using NEC2D software core, which is based on the MoM.

As the capacitor cartographies gather at the radiation of a rectangular loop, the search for models has been launched to find a rectangular current loop using the method based on the coupling between the inverse method and the MoM. Each

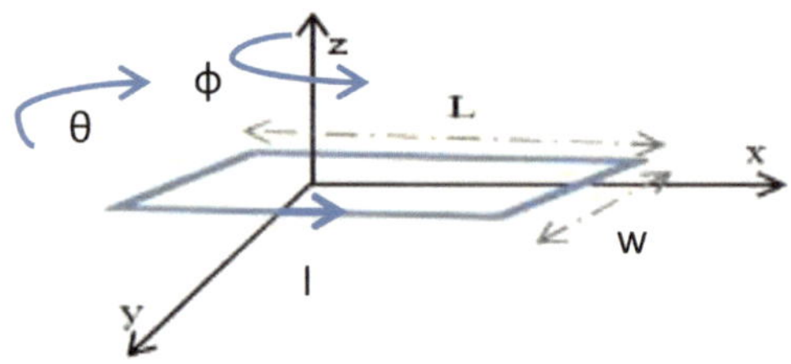

Figure 14.
Rectangular current loop parameters.

rectangular loop is characterized by the coordinates of its center (X0, Y0, Z0), its width (w), its length (L), its orientations (φ, θ), the radius (r) of the loop wire, and the current flowing through it (**Figure 14**).

5. Model for the polypropylene capacitor

5.1 Model research

After presenting the magnetic field cartography radiated by the capacitor in Section 3, we apply our approach to find a model that gives the same radiation as measured over the capacitor. To do this, we used the cartography of the vertical component Hz of the magnetic field at a frequency of 20 MHz. We chose the cartography of 20 MHz because it was the clearest one. **Figure 15** shows the position of the identified loop on the capacitor cartography. The modeling results for the obtained model are shown in **Table 1**.

Figure 16 presents the measured and estimated cartographies of the magnetic field vertical component Hz. A reasonable agreement between the two cartographies is observed.

5.2 Model validation

Further validation of the model is performed on other faces of the capacitor or other components as proposed in [4, 23]. Hence, we measured the magnetic field cartography in the literal face and compared it to the one that has been calculated using the obtained model.

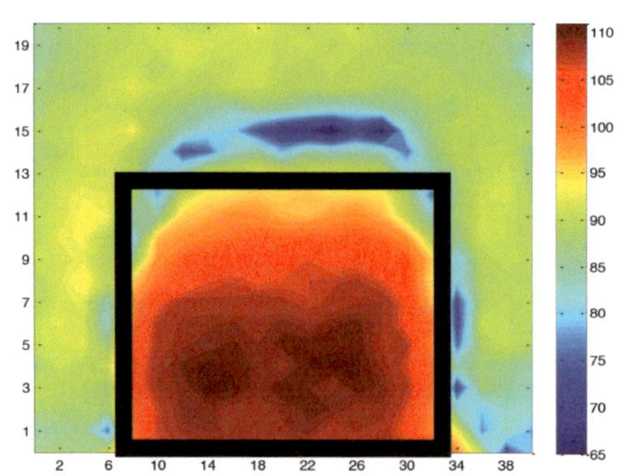

Figure 15.
Obtained loop position.

Center coordinates (X_0, Y_0, Z_0) (mm)		3.6; 19; −2.9
Loop dimensions	Width (mm)	12
	Length (mm)	23
Orientations	φ (rd)	0
	θ (rd)	0
Conductor radius R (mm)		1
Voltage excitation (V)		0.035

Table 1.
Parameters of the equivalent model of the polypropylene capacitor.

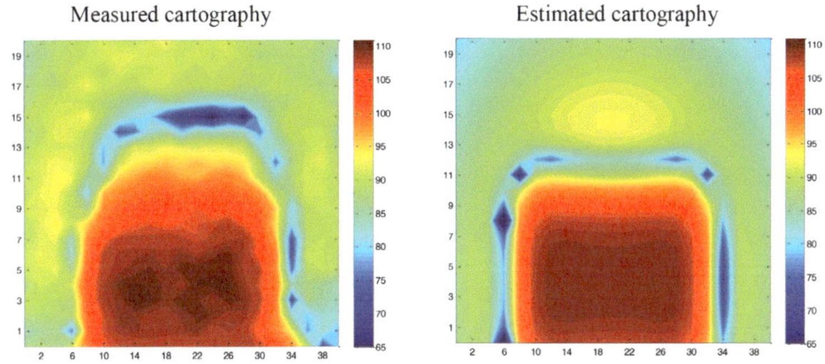

Figure 16.
Measured and estimated cartographies.

Accordingly, the capacitor is placed vertically, and measurements of the cartography are carried out on the other side of the capacitor. **Figure 17** depicts the layout of the measurements.

The validation of the model is made at 20 MHz and a measurement height of 3 mm. **Figure 18** shows the measured and calculated magnetic field cartography above the vertically placed capacitor.

According to the previous results, the magnetic field is reconstructed by the model shows good resembling with the measured one. These results confirm the existence of an equivalent rectangular current loop in the capacitor. To explain the obtained results, we present in **Figure 19** the internal structure of the capacitor. Film capacitors are made out of two pieces of plastic film covered with metallic electrodes wound into a cylindrical-shaped winding, with terminals attached, and then encapsulated.

The analysis of the internal structure of the film capacitor confirms the obtained equivalent model. Indeed, the surface of the capacitor appears as a rectangular loop.

The internal coupling phenomenon certainly causes this current distribution at the capacitor contour in the component. This current distribution in the capacitor film is consistent with that introduced in [12, 13].

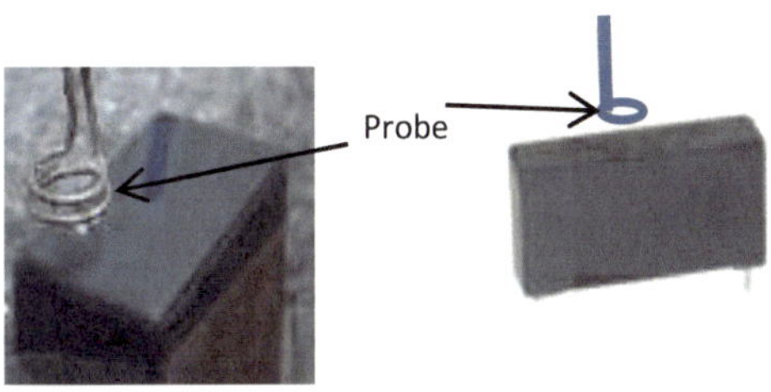

Figure 17.
Measurement of the magnetic field over a vertically placed capacitor.

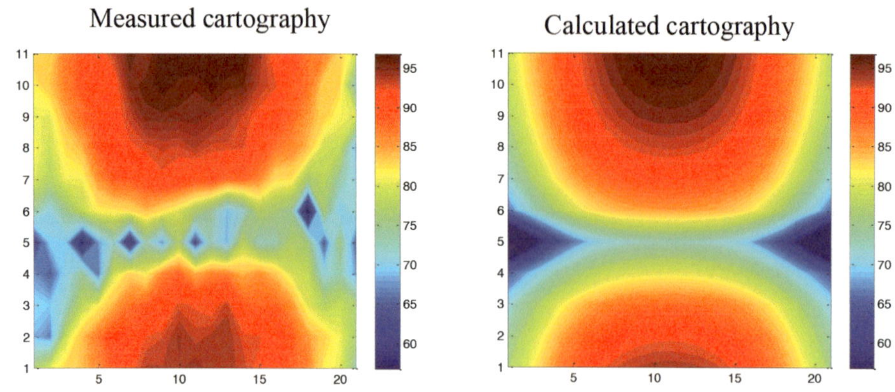

Figure 18.
Cartographies of the measured and calculated field above vertically placed capacitor.

Figure 19.
Film capacitor's internal structure.

6. Model for the metalized polyester capacitor

6.1 Measurement of magnetic near-field radiated by polyester capacitors

Based on the frequency measurement method, we measured the vertical component Hz of the magnetic field radiated above a metalized polyester capacitor of a 1 μF capacitance and having a maximum voltage of 100 V (the capacitor size 14 mm x 7 mm x 18 mm). In order to have only the magnetic field radiated by the capacitor, the capacitor legs are bent. The probe was located at a 3 mm distance from the capacitor. The magnetic field measurements were performed in a plane whose dimensions were 30 mm x 20 mm with a measuring step of 1.5 mm on the X-axis and 1 mm on the Y-axis. **Figure 20** shows the measurement plane and the component under test.

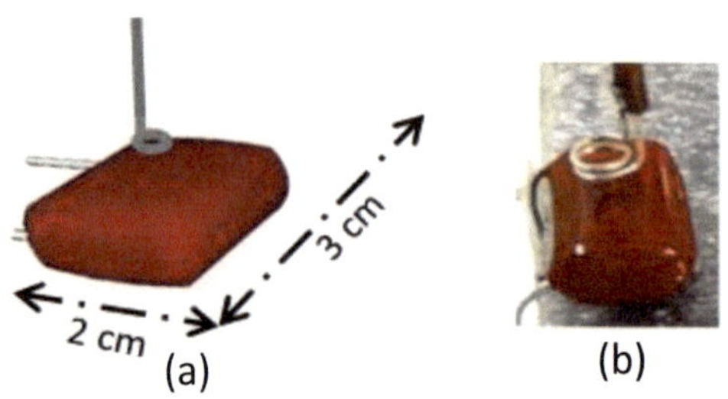

Figure 20.
(a) Measurement plan; (b) device under test.

After making Fast Fourier transformation (FFT) of the measured signals above the capacitor, we notice that the frequency of 26 MHz has maximum amplitude. We present in the following **Figure 21** the cartographies of the Hz and Hy components measured at the frequency of 26 MHz.

6.2 Radiated-field modeling

As for the polypropylene capacitor, the radiation of the polyester capacitor looks like that of a rectangular loop. The same method exposed previously is used to the cartography of the measured magnetic field Hz at the frequency of 26 MHz. The modeling result is a rectangular current loop whose parameters are given in **Table 2**. **Figure 22** exposes the cartographies of the magnetic field Hz measured and estimated by the obtained model above the capacitor. A good agreement between the two cartographies is observed.

6.3 Model validation

To validate the obtained model better, we examined the cartographies of the other components of the magnetic field above the capacitor. We specifically compared the measured component, Hy of the magnetic field, calculated using the obtained model (**Figure 23**).

The previous results show that the magnetic field reconstructed by the model resembles, to a great extent, the measured magnetic field.

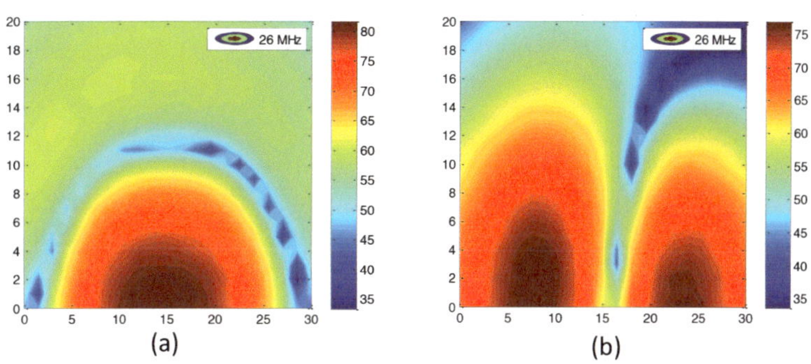

Figure 21.
Cartographies of the field above the capacitor: (a) Hz and (b) Hy.

Center coordinates (X_0, Y_0, Z_0) (mm)		1; 15; −3.3
Loop dimensions	Width (mm)	12
	Length (mm)	16
Orientations	φ (rd)	0
	θ (rd)	−0.2
Conductor radius R (mm)		1
Voltage excitation (V)		0.0017

Table 2.
Parameters of the equivalent model of the metallized polyester capacitor.

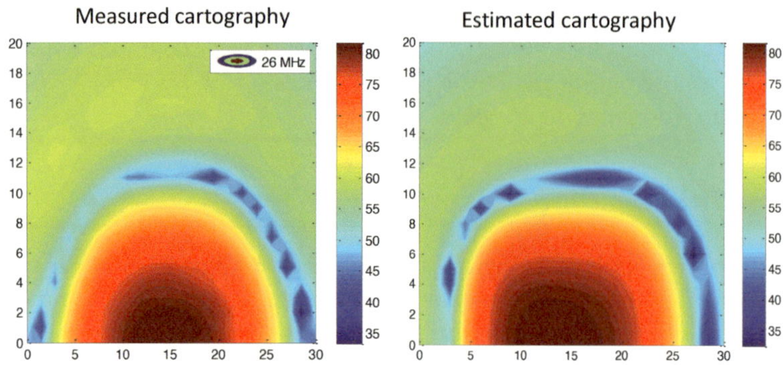

Figure 22.
Measured and estimated cartographies of polyester capacitor.

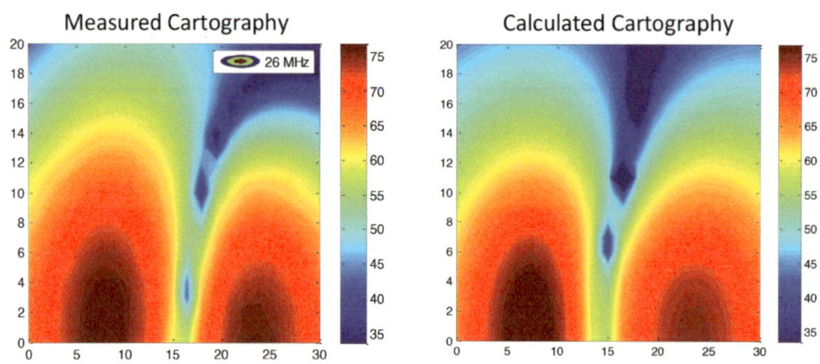

Figure 23.
Cartographies of measured and estimated Hy component of the magnetic field above the capacitor.

7. Capacitor generic model

The examination of the results obtained previously allows us to conclude that the equivalent radiating model for the different types of rectangular film capacitors is a rectangular loop whose parameters depend on the geometry of the capacitor as illustrated in **Figure 24** and explained in **Table 3**.

We suggest in **Table 3** formulas in order to determine by using capacitor dimensions the size of the source rectangular loop of the generic model.

To validate the proposed generic model, we applied it to another rectangular capacitor. Therefore, we measured the magnetic field above a third rectangular

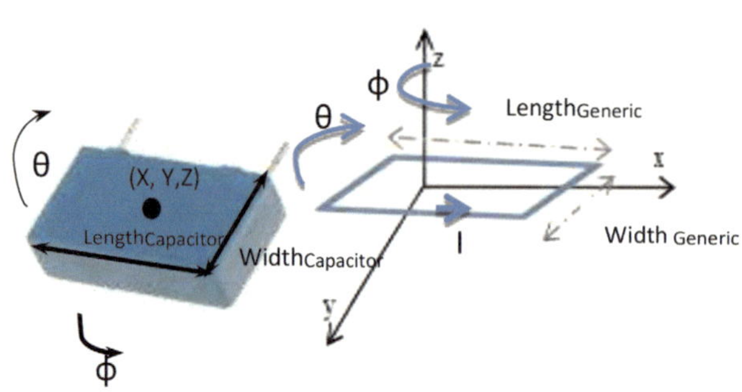

Figure 24.
Capacitor parameters and the analogy with the equivalent loop.

The center coordinates	The Gravity center of the capacitor
Loop dimensions	WidthGeneric ≅ 0.8 × Widthcapacitor LengthGeneric ≅ 0.86 × Lengthcapacitor
Orientations	$\varphi = 0$ θ depends on the position of the capacitor under test relative to the x-axis ≅ 0.
Voltage excitation	Dependence on the field amplitude

Table 3.
Size of the source rectangular loop of the generic model.

capacitor having 2.2 μF capacitance and a maximum voltage of 250 V (the capacitor size is 25 mm x 14 mm x 41.5 mm).

We performed measurements of the near field cartography above the studied capacitor, and then we applied the method based on the coupling of the electromagnetic inverse method with the MoM to find the equivalent model that will be compared to the one obtained by the generic model formulas.

According to **Table 4**, we notice that the calculated parameters by the formulas of the generic model are very close to those obtained by the inverse electromagnetic method.

We observe from the previous table that the searching model using the generic model is seven times faster than when being used without the generic model. Indeed, in the generic model, we search only one parameter (the excitation voltage), whiles the others parameters are calculated using the formulas given in **Table 3**.

Figure 25 shows Hy and Hz's components of the magnetic near-field measured and estimated by the two obtained models above the capacitor under test at a frequency of 27 MHz.

According to the previous results, the magnetic field reconstructed using the proposed generic model shows an agreement with the measured magnetic field. In fact, in our modeling approach, we are interested in the high amplitudes of the magnetic field (red parts in the cartographies) because they have the most important magnetic field radiated by the component under test. We notice a good agreement between the measured and calculated fields by the generic model is observed.

Figure 26 presents a comparison between the two models for cuts of Hz and Hy components along the x-axis. It shows a reasonable agreement between the measured magnetic field and the obtained one with the two models.

		Without generic model	Generic model
The center coordinates (X_0, Y_0, Z_0) (mm)		2; 26; −7.3	2.3,25,-7
Loop dimensions	Width (mm)	21	20
	Length (mm)	37	35.69
Orientations	φ (rd)	0	0
	θ (rd)	0	0
Conductor radius R (mm)		1	1
Voltage excitation (V)		0.0007	0.00055
Modeling time (min)		62	8

Table 4.
Parameters of rectangular loop obtained by two methods.

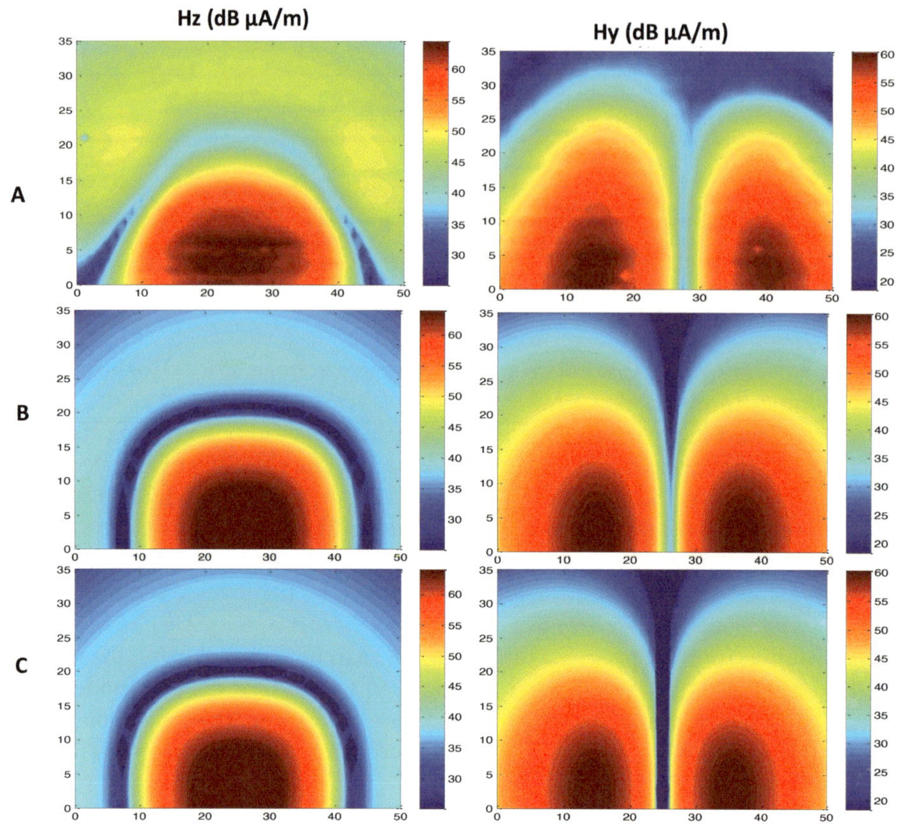

Figure 25.
Cartographies of magnetic field: (A) measured and (B) estimated using the inverse method and (C) using generic model (at 27 MHz).

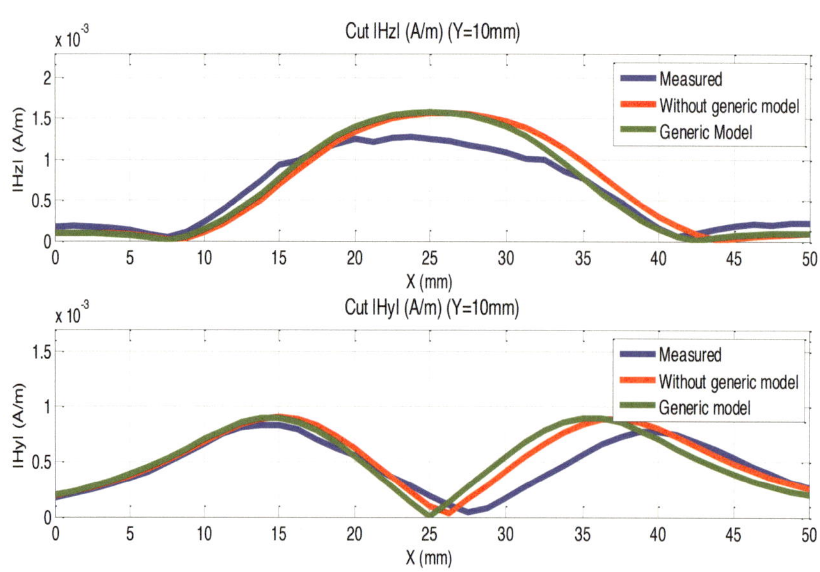

Figure 26.
Comparison of the two models for cuts of Hz and Hy components along the x-axis.

DOI: http://dx.doi.org/10.5772/intechopen.98894

The examination of all the obtained results enables us to conclude that the suggested generic model is a slight model that permits, in a very short time, the calculation of the radiation of the different types of rectangular capacitors.

8. Conclusion

Film capacitors are among the larger components on power electronics boards. Therefore, they cause considerable radiation above all the circuits. In this paper, we have presented the characterization measurements of the magnetic field radiated by different film capacitors used in power electronics. We have then utilized the inverse electromagnetic method based on coupling between the Genetic algorithms and the MoM to find models of equivalent radiating sources of these capacitors. A generic radiating model for different rectangular film capacitors has been finally put forward and validated using measurements on another capacitor.

This equivalent model is special; hence, its radiation is much more complicated than the one usually used by the electromagnetic inverse problem (elementary magnetic and electric dipoles).

The proposed radiating model reduces the number of parameters to be determined by the GA method and consequently reduces the convergence time of the inverse method. Circuit designers can use it to optimize the placement of capacitors on the printed circuit board to reduce their coupling with other equipment located in the vicinity of this system.

Author details

Walid Labiedh, Bessem Zitouna*, Mohamed Tlig and Jaleleddine Ben Hadj Slama
LATIS, Laboratory of Advanced Technology and Intelligent Systems, ENISo, National Engineering School of Sousse, University of Sousse, Tunisia

*Address all correspondence to: bessem.zitouna@yahoo.fr

References

[1] Jaleleddine BEN HADJ SLAMA and Walid LABIEDH, "Library of EMC Models for Passive Components and Printed Circuit Board," Electrotechnical Conference (MELECON), 16th IEEE Mediterranean, pp. 325-330, 25-28 March 2012.

[2] Saidi, S.; Ben Hadj Slama, J., "A Near-Field Technique Based on PZMI, GA, and ANN: Application to Power Electronics Systems," IEEE Transactions on Electromagnetic Compatibility, VOL. 56, no. 4, pp. 784-791, Aug. 2014.

[3] Y. Vives-Gilabert, C. Arcambal, A. Louis, F. De Daran, P. Eudeline, B. Mazari, "Modeling Magnetic Radiations of Electronic Circuits Using Near-Field Scanning Method," IEEE Transactions on Electromagnetic Compatibility, VOL. 49, no.2, pp.391-400, May 2007.

[4] Sofiene Saidi and Jaleleddine Ben Hadj Slama, "Analysis and modeling of power MOSFET radiation," Progress In Electromagnetics Research M, VOL. 31, pp 247-262, 2013.

[5] Yolanda Vives-Gilabert, Christian Arcambal, Anne Louis, Philippe Eudeline, and Belahcene Mazari, "Modeling Magnetic Emissions Combining Image Processing and an Optimization Algorithm" IEEE Transactions on Electromagnetic Compatibility, VOL. 51, NO. 4, pp 909-918, NOVEMBER 2009.

[6] Levy, P.-E.; Gautier, C.; Costa, F.; Revol, B.; Labarre, C., "Accurate Modeling of Radiated Electromagnetic Field by a Coil With a Toroidal Ferromagnetic Core," IEEE Transactions on Electromagnetic Compatibility, VOL. 55, no. 5, pp. 825-833, Oct. 2013.

[7] Shall, H.; Riah, Z.; Kadi, M., "A 3-D Near-Field Modeling Approach for Electromagnetic Interference Prediction," IEEE Transactions on Electromagnetic Compatibility, VOL. 56, no. 1, pp.102-112, Feb. 2014.

[8] Baudry, D.; Arcambal, C.; Louis, A.; Mazari, B.; Eudeline, P., "Applications of the Near-Field Techniques in EMC Investigations," IEEE Transactions on Electromagnetic Compatibility, VOL. 49, no. 3, pp. 485-493, Aug. 2007.

[9] De-Oliveira, T.; Schanen, J.-L.; Guichon, J.-M.; Gerbaud, L., "Optimal Stray Magnetic Couplings for EMC Filters," IEEE Transactions on Industry Applications, VOL. 49, no. 4, pp. 1619-1627, July-Aug. 2013.

[10] De Oliveira, T.; Guichon, J.-M.; Schanen, J.-L.; Gerbaud, L., "PEEC-models for EMC filter layout optimization," 2010 6th International Conference on Integrated Power Electronics Systems (CIPS), VOL., no, pp. 1-6, 16-18 March 2010.

[11] Henglin Chen; Zhaoming Qian, "Modeling and Characterization of Parasitic Inductive Coupling Effects on Differential-Mode EMI Performance of a Boost Converter," IEEE Transactions on Electromagnetic Compatibility, VOL. 53, no. 4, pp. 1072-1080, Nov. 2011.

[12] E. L. Nativel, T. Talbert, T. Martire, C. Joubert, N. Daude and P. Falgayrettes, "Near-field electromagnetic tomography applied to current density reconstruction in metallized capacitors," IEEE Transactions on Power Electronics, VOL. 20, no. 1, pp. 11-16, Jan. 2005.

[13] C. Joubert, A. Béroual, G. Rojat, "Magnetic Field and Current Distribution in Metallized Capacitors," Journal of Applied Physics, VOL. 76, No. 9, pp. 5288-52931 November 1994.

[14] Jun Fan, "Near-field scanning for EM emission characterization," IEEE

Electromagnetic Compatibility Magazine, VOL. 4, no. 3, pp. 67-73, 3rd Quarter 2015.

[15] X. Gao, J. Fan, Y. Zhang, H. Kajbaf and D. Pommerenke, "Far-Field Prediction Using Only Magnetic Near-Field Scanning for EMI Test," IEEE Transactions on Electromagnetic Compatibility, VOL. 56, no. 6, pp. 1335-1343, Dec. 2014.

[16] Y. Liu, B. Ravelo, and A. K. Jastrzebski, "Calculation of time-domain near field Ex,y,z(t) from Hx,y(t) with PWS and FFT Transforms," in Proc. Int. Symp. Electromagn. Compat., Roma, Italy, pp. 1–6, 17–20 September 2012.

[17] B. Ravelo, "E-field extraction from H-near-field in time-domain by using PWS method," Progress Electromagn. Res. B, VOL. 25, pp. 171–189, 2010.

[18] Y. Liu, B. Ravelo, A. K. Jastrzebski, and J. B. Hadj Slama, "Calculation of the time domain z-component of the EM-near-field from the x- and y-components," in Proc. 41st Eur. Microw. Conf., pp. 317–320, 10 - 13 October 2011.

[19] G. J. Burke and A. J. Poggio, "Numerical Electromagnetics Code (NEC) - Method of Moments," NOSC TD 116 Part I, January 1981.

[20] J. Ben Hadj Slama, S. Saidi, "Coupling the Electromagnetic Inverse Problem Based on Genetic Algorithms with Moment's Method for EMC of Circuits," 15th IEEE Mediterranean Electrotechnical Conference MELECON'10, Malta, pp. 709-714, 26-28 April 2010.

[21] Sofiene Saidi and Jaleleddine Ben Hadj Slama, "Effect of Genetic Algorithm parameters on Convergence of the Electromagnetic Inverse Method," 8th International Multi-Conference on Systems, Signals & Devices, pp. 1-5, 22-25 March 2011.

[22] Zitouna, B., Ben Hadj Slama, J., "Enhancement of time-domain electromagnetic inverse method for modeling circuits radiations," IEEE Trans.Electromagn. Compat., VOL. 58, (2), pp. 534–542, 2016.

[23] Bessem Zitouna and J. Ben Hadj Slama, "Time domain inverse method based on the near field technique to solve electromagnetic interference problems: application to an AC/DC flyback converter," IET Power Electronics, VOL. 11, no. 13, pp. 2133–2139, Jun. 2018.

Chapter 14

Efficient Simulation of Airborne SAR Raw Data in Case of Motion Errors

Yuhua Guo, Shichao Jin and Yuhong Guo

Abstract

In the simulation of SAR raw data, it is well-known that the frequency-domain algorithm is more efficient than a time-domain algorithm, making it is more suitable for extended scene simulation. However, the frequency-domain algorithm is perhaps better suited for ideal linear motion and requires some degrees of approximations to take the nonlinear motion effects. This chapter presents an efficient simulation approach based on hybrid time and frequency-domain algorithms under certain assumptions. The algorithm has high efficiency and is suitable for the simulation of extended scenes, which demands highly computational resources. The computational complexity of the proposed algorithm is analyzed, followed by numerical results to demonstrate the effectiveness and efficiency of the proposed approach.

Keywords: synthetic aperture radar (SAR), trajectory deviations, antenna pointing errors, raw data simulation, Fourier transform (FT)

1. Introduction

There are two major approaches to the SAR raw echo data simulation: time-domain and frequency-domain methods. For airborne SAR, the sensor platform trajectory, due to nonideal factors such as atmospheric turbulence, deviates from the linear motion state in the synthetic aperture time, which seriously affects the imaging quality if no proper compensation is made [1]. The time-domain algorithm can accurately reflect the influence of various error factors, but the pitfall is that the computational efficiency is low [2–4]. On the other hand, the frequency-domain algorithm, based on the two-dimensional frequency-domain expression of the original echo signal, uses a fast FT to realize simulation [5, 6]. This kind of algorithm delivers high efficiency to a high degree. However, it is only for simulation under ideal flight conditions.

The fast simulation of SAR imaging signals has a long history of research. The main idea of the existing methods is to realize the SAR raw data simulation in the frequency-domain with the high efficiency of the fast FT [7]. In recent years, Franceschetti's team in Italy has conducted a systematic study on the fast simulation algorithm of SAR raw data. A simulation method based on the two-dimensional frequency domain is proposed, which assumes slow platform motion with a narrow antenna beam [8]. Then, the algorithm is modified by introducing a one-dimensional azimuth Fourier transform and range time-domain integration [9].

Reference [10] improved the traditional 2-D frequency-domain method and extended it to the squint mode. However, these two algorithms only consider the case of the tracking error or the case of the squint mode and do not consider the case of the squint with the trajectory deviations. The time-frequency hybrid simulation methods were proposed to improve SAR raw echo data processing speed [11–20]. The original raw data is usually acquired in the squint situation due to the instability of the attitude. Furthermore, the change of antenna attitude will cause the change of beam direction, which leads to the degradation of image quality. The fast echo simulation algorithm for antenna attitude change has been proposed on the premise that the amplitude of linear antenna error is far less than the beam width [21–23]. For sinusoidal antenna azimuth attitude jitter, it will cause paired echoes. Similarly, sinusoidal antenna range attitude jitter will also cause paired echoes.

In this work, we present the fast echo simulation algorithm of trajectory offset and attitude jitter error. We first attempt to extend the Fourier domain approach to account for trajectory deviations at a non-zero squint angle. Then we consider the Fourier domain approach to account for trajectory deviations and antenna beam pointing errors, given rising more problematic for airborne SAR systems. The rest of the chapter is organized as follows: The algorithm for trajectory deviations is presented in Section 2. In Section 3, SAR signal simulations are carried out considering the trajectory deviations and antenna attitude variations. Finally, conclusions are drawn to close the chapter.

2. Raw echo algorithm of squint airborne strip SAR with trajectory offset error

In this section, an algorithm of raw echo with trajectory error in squint is proposed. The approach adopts the Acquisition Doppler (AD) [24] geometry instead of the standard cylindrical reference system (see **Figure 1**).

2.1 Raw echo algorithm of squint airborne strip SAR with trajectory offset error

The echo generation algorithm with trajectory error in squint is deduced by rotating the trajectory error along the line of sight. The motion error in the cone coordinate system is shown in **Figure 2**. The transformation relationship between a cylindrical coordinate system and a conical coordinate system is as follows:

$$r_0 = r\cos\phi, x_0 = x + r\sin\phi_d \tag{1}$$

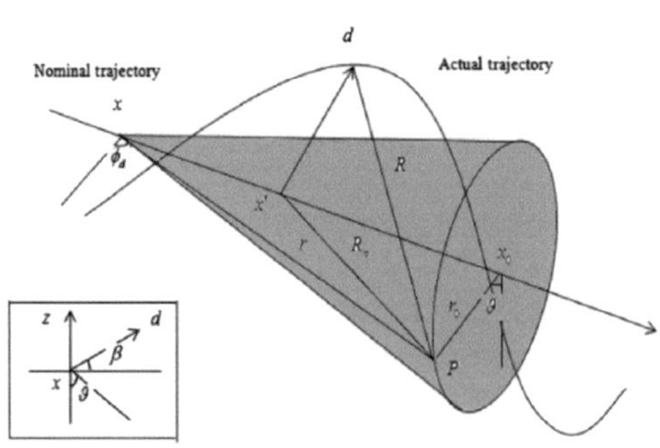

Figure 1.
Geometric diagram of SAR with squint and trajectory offset.

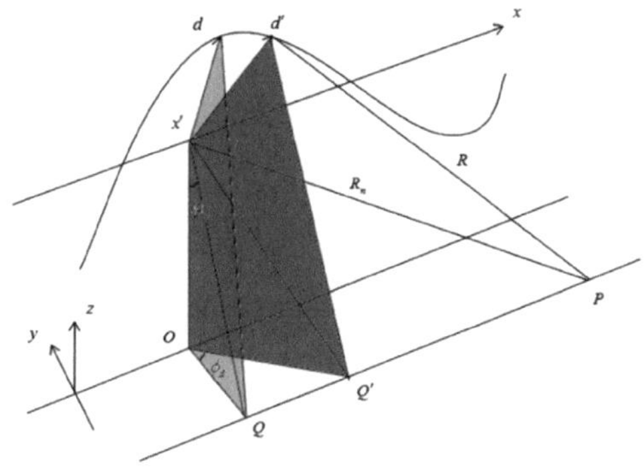

Figure 2.
AD processing geometry.

where x are r the coordinates of the point target in the cone coordinate system, x_0 and r_0 are the coordinates of the point target in the cylinder coordinate system. d is the trajectory error, β is related to the horizontal and vertical components of the trajectory error. ϕ_d is the squint angle. R_n and R are the ideal distance and the distance from the carrier to the target, respectively. In the cone coordinate system, R_n and R are transformed into:

$$R_n = \sqrt{{r_0}^2 + (x' - x_0)^2} = \sqrt{(r\cos\phi)^2 + (x' - x - r\sin\phi_d)^2} \tag{2}$$

$$R = \sqrt{(r\cos\phi_d + \delta\tilde{r}_{xr})^2 + (x' - x - r\sin\phi_d)^2} \tag{3}$$

where [25].

$$\begin{aligned}\delta\tilde{r}(x', x, r, \phi_d) \approx d(x')\{&\sin\beta(x')\cos\phi_d\cos\vartheta(x,r) - \cos\beta(x')\sin\vartheta(x,r)\\ &+\frac{1}{2}(\sin\phi_d)^2[\cos\vartheta(x,r)]^2\sin\beta(x')\cos\phi_d\cos\vartheta(x,r)\\ &+\frac{1}{2}\cos\beta(x')\sin\vartheta(x,r)(\sin\phi_d)^2[\cos\vartheta(x,r)]^2\Big\}\end{aligned} \tag{4}$$

Then the trajectory deviations $\delta R(x', x, r, \phi_d)$ can be decomposed into the following terms:

$$\delta R(x', x, r, \phi_d) \approx \delta\tilde{r}(x', \phi_d) + \tilde{\psi}(x', r, \phi_d) + \tilde{\varphi}(x', x, r, \phi_d) \tag{5}$$

where $\delta\tilde{r}(x', \phi_d)$ is the track deviation caused by the deviation of radar track from the ideal track when the radar beam points to the center of the scene, which is only related to the azimuth position of radars and is the main component of track deviation, $\tilde{\psi}(x', r, \phi_d)$ is the range to space variability of track deviation, and $\tilde{\varphi}(x', x, r, \phi_d)$ accounts for the influence of radar platform position, azimuth and range variations.

Suppose that the radar transmits the following Linear frequency modulation (LFM) signal. After heterodyne, the echo signal obtained is

$$\begin{aligned}h(x', r') = &\iint dxdr\gamma(x,r)rect\left[\frac{r' - R}{D}\right]\omega^2\left[x' - x - r\frac{\sin(\phi - \phi_d)}{\cos\phi_d}\right]\\ &\times\exp\left[-j\frac{4\pi}{\lambda}R - j\alpha_r(r' - R)^2\right]\end{aligned} \tag{6}$$

where r' is the sampling coordinate of the range signal, λ is the wavelength, $\gamma(x, r)$ is the reflection function, $rect[\cdot]$ represents the rectangular envelope of the pulse. After applying the range FT of $h(x', r')$, we obtain

$$\begin{aligned} H(x',\eta) &= rect\left[\frac{\eta}{2\alpha_r D}\right]\cdot \exp\left[j\frac{\eta^2}{4\alpha_r}\right]\times \iint dxdr\gamma(x,r) \\ &\times \exp\left[-j\left(\eta+\frac{4\pi}{\lambda}\right)\cdot(r+\Delta R+\delta R)\right]\cdot \omega^2\left(x'-x-r\frac{\sin(\phi-\phi_d)}{\cos\phi_d}\right) \end{aligned} \tag{7}$$

Furthermore, by separating the factor $\exp\left[-j\left(\eta+\frac{4\pi}{\lambda}\right)\delta r(x',\phi_d)\right]$, we obtain

$$H(x',\eta) = \exp\left[-j\left(\eta+\frac{4\pi}{\lambda}\right)\delta r(x',\phi_d)\right]\tilde{H}(x',\eta) \tag{8}$$

where

$$\begin{aligned} \tilde{H}(x',\eta) &= rect\left[\frac{\eta}{2\alpha_r D}\right]\cdot \exp\left[j\frac{\eta^2}{4\alpha_r}\right]\times \iint dxdr\gamma(x,r) \\ &\times \exp\left[-j\left(\eta+\frac{4\pi}{\lambda}\right)\cdot\left(r+\Delta R(x',x,r,\phi_d)+\tilde{\psi}(x',r,\phi_d)+\tilde{\varphi}(x',x,r,\phi_d)\right)\right] \\ &\times \omega^2\left(x'-x-r\frac{\sin(\phi-\phi_d)}{\cos\phi_d}\right) \end{aligned} \tag{9}$$

If the following conditions are satisfied:

$$|\tilde{\varphi}(x',x,r,\phi_d)| \ll \frac{\lambda}{4\pi} \tag{10}$$

$$|\tilde{\psi}(x',x,r,\phi_d)| \ll \frac{f}{\Delta f}\frac{\lambda}{2\pi} \tag{11}$$

we can obtain the azimuth FT of $\tilde{H}(x',\eta)$ as follows:

$$\begin{aligned} \tilde{H}(\xi,\eta) &= \int dr \exp\left[-j\overline{\eta}r\right]\int drG(\xi-l,\eta,r)\Gamma(\xi-l,r)Q_\eta(l,\eta,r) \\ &= \int dr \exp\left[-j\overline{\eta}r\right]\{Q_\eta(\xi,\eta,r)\otimes_\xi[G(\xi,\eta,r)\Gamma(\xi,r)]\} \end{aligned} \tag{12}$$

where

$$Q_\eta(l,\eta,r) = FT_{x'}\{\exp\left[-j\overline{\eta}\tilde{\psi}(x',r,\phi_d)\right]\} \tag{13}$$

with $\overline{\eta} = \eta + \frac{4\pi}{\lambda}$, After evaluating $\tilde{H}(\xi,\eta)$, it can be direct get $\tilde{H}(x',\eta)$ by using the inverse FT of $\tilde{H}(\xi,\eta)$ along the azimuth direction. Then, using Eq. (8) and the range inverse FT, the original echo signal can be computed. The flowchart of the algorithm is shown in **Figure 3**. The validity of the range needs to satisfy Eqs. (10) and (11).

2.2 Computational complexity of the algorithm

Let us now consider the computational complexity of the algorithm in **Figure 3**. Suppose the data size is $N_a \times N_r$, with N_a corresponding to the number of azimuth

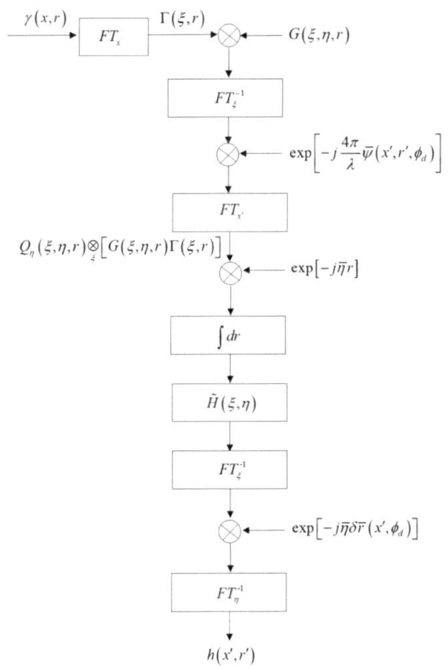

Figure 3.
Raw data simulation process of strip squint SAR with trajectory error.

sampling points, and N_r to the number of range sampling points. It follows that the amount of computation required by the algorithm is

$$\begin{aligned} N &\approx N_a N_{r2}(3+\log_2 N_a) + N_a N_r\left(1+\log_2 N_a + \frac{1}{2}\log_2 N_r\right) \\ &\approx N_a N_r^2(3+\log_2 N_a) \end{aligned} \tag{14}$$

Recalled that the required computation of time domain algorithm is as follows

$$\tilde{N} \approx N_a N_r N_{tp} N_{sa} \tag{15}$$

where N_{tp} is the size of the emitted pulse and N_{sa} is the synthetic aperture, N_{tp} is usually equal to the size of N_{r}. The ratio of Eqs. (14) and (15) takes the form

$$\frac{\tilde{N}}{N} \approx \frac{N_{sa}}{3+\log_2 N_a} \tag{16}$$

It is evident that compared with the time domain simulation method, the proposed method has higher computational efficiency.

2.3 Algorithm simulation verification

In what follows, we use the SAR system parameters listed in **Table 1** to verify the SAR raw data simulation algorithm of non-ideal track under squint condition. **Figure 4** shows the track deviation between the actual track and the ideal track. For a point target located at the near range $r = (4840/\cos 4^o)m$, the phase difference between the simulated and the exact raw signal, along the range direction and the azimuth direction, is displayed in **Figure 5(a)** and **(b)**, respectively. We see that the

Nominal height	**4000 m**	**Range pixel dimension**	**3 m**
Midrange coordinate	5140 m	Chirp bandwidth	45 MHz
Wavelength	3.14 cm	Chirp duration	5 μs
Pulse Repetition Frequency	400 Hz	Velocity	100 m/s
Sampling Frequency	50 MHz	Number of azimuth samples of the raw data	972
Azimuth pixel dimension	25 cm	Number of range samples of the raw data	416

Table 1.
Simulation parameters.

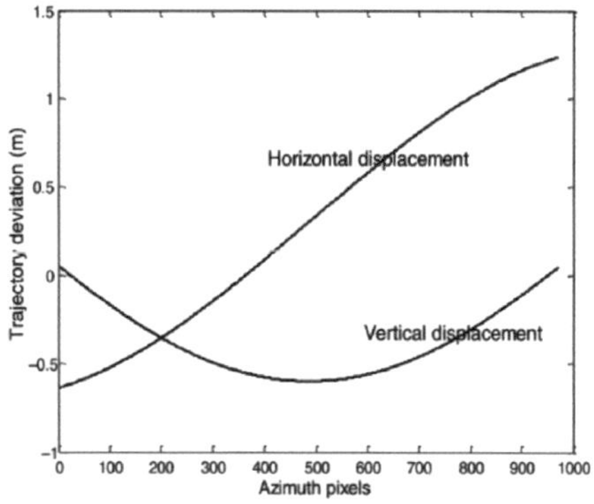

Figure 4.
Sensor displacements.

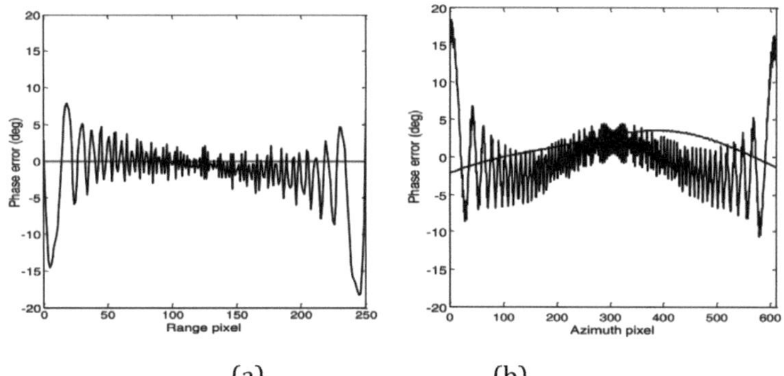

Figure 5.
(a) Range cut, (b) azimuth cut, the point target is located in the close range position ($r = (4840/\cos 4^o)$ m) of the scene.

phase error is much smaller than $\pi/10$, which in practice is negligible. By comparing the "depurated" phase error and the expected phase error $(4\pi/\lambda)\tilde{\varphi}(x', x, r, \phi_d)$ (see **Figure 6**), we may conclude that the azimuth "depurated" phase error is given rise by the approximation of Eq. (10). Moreover, the point target impulse response for midrange at squint angle 15° after the motion compensation are shown in **Figure 7 (a)** and **(b)**. In addition, the figure of merits by point target analysis for midrange shown in **Figure 7(a)** and **(b)** between the proposed approach and the time-domain one is given in **Table 2**. The proposed raw data generator has high precision at least as that by the time domain algorithm.

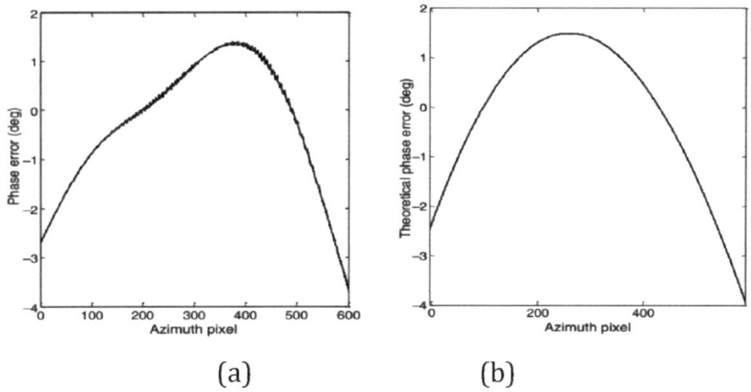

Figure 6.
(a) The phase error (smooth line) after removing the stationary phase influence. (b) the phase error along the azimuth direction ($r = (4840/\cos\phi_d)m$) according to the formula $|\overline{\varphi}(x',x,r,\phi_d)| << \frac{4\pi}{\lambda}$.

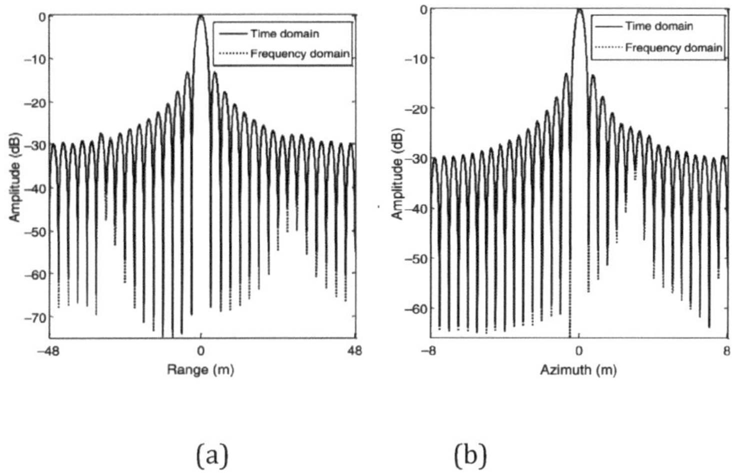

Figure 7.
Point target imaging results after motion error compensation. (a) Range cut and (b) azimuth cut.

Simulation method	**Azimuth direction**			**Range direction**		
	IRW (m)	**PSLR (dB)**	**ISLR (dB)**	**IRW (m)**	**PSLR (dB)**	**ISLR (dB)**
Time domain simulation algorithm	0.2344	−13.1035	−9.6665	2.8125	−13.2115	−9.6759
Simulation algorithm of squint with trajectory offset	0.2344	−13.2322	−9.6862	2.8125	−13.2160	−9.6871

Table 2.
Point target evaluation results.

Finally, the extended scene is considered. The final results are given in **Figure 8**, where **Figure 8(a)** shows the image without motion compensation, **Figure 8(b)** displays the results by a two-step MOCO algorithm, and **Figure 8(c)** is the image after the motion compensation [25]. Concerning the computational efficiency, it took only about 2 min for the simulation run, as shown in **Figure 8(c)**, on an 8-GHz Intel Core i5 personal computer.

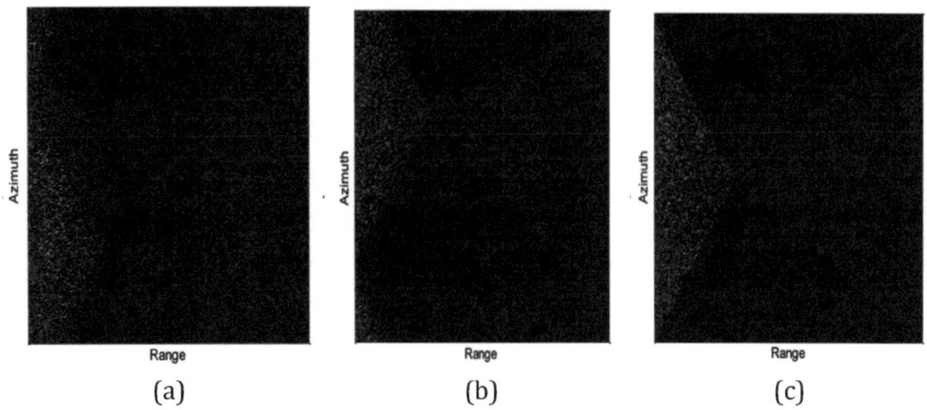

Figure 8.
Comparison of imaging results: (a) No motion compensation, (b) second order motion error compensation, (c) the compensation method in ref. [25] is used.

3. Raw echo algorithm of airborne stripmap SAR with trajectory error and attitude jitter

Due to the air disturbance or the flight instability of the SAR platform, the platform trajectory deviation and attitude error are inevitably introduced into the sensor parameters. This section considers the raw echo algorithm of airborne stripmap SAR with trajectory error and attitude jitter.

3.1 Raw echo algorithm of airborne stripmap SAR with trajectory error and attitude jitter

Firstly, the simulation algorithm of antenna beam pointing error is briefly introduced. In Ref. [22], a two-dimensional frequency-domain echo simulation algorithm is proposed to reflect the beam pointing error. The difference between the raw data echo, including antenna beam pointing error, and ideal echo is the azimuth envelope. When the amplitude of the antenna beam jitter is less than the beam width, the azimuth amplitude weighting function can be approximated by the Taylor expansion. From reference [22], When the beam pointing is sinusoidal jitter, a pair of echoes is erroneously produced.

Suppose the received echo is as follows:

$$\begin{aligned} h(x',r') = & \iint dxdr\gamma(x,r) \cdot \exp\left\{-j\frac{4\pi}{\lambda}R - j\frac{4\pi}{\lambda}\frac{\Delta f/f}{c\tau}(r'-R)^2\right\} \\ & \times rect\left[\frac{r'-R}{c\tau/2}\right] \cdot W^2\left[\frac{x'-x-\delta_x}{X}\right] \end{aligned} \tag{17}$$

where $\delta_x = (r + \delta\tilde{r}_{xr}(x',x,r,\phi_d))\tilde{\delta}_{az}(x')$ is the antenna pointing error in azimuth direction. When the amplitude of antenna beam jitter is less than the beam width, the azimuth amplitude weighting function is approximated by the second-order Taylor expansion:

$$\begin{aligned} W^2\left(\frac{x'-x-\delta_x}{X}\right) \approx & W^2\left(\frac{x'-x}{X}\right) - W^{2(1)}\left(\frac{x'-x}{X}\right)(r+\delta r_{xr}(x',x,r))\delta_{az}(x') \\ & + \frac{1}{2}W^{2(2)}\left(\frac{x'-x}{X}\right)(r+\delta r_{xr}(x',x,r))^2\delta_{az}^2(x') \end{aligned} \tag{18}$$

where r is the distance coordinate between the platform and the target point, and the original echo is decomposed into three parts:

$$h(x', r') \approx h_0(x', r') + h_1(x', r') + h_2(x', r') \tag{19}$$

where $h_0(x', r')$ is equivalent to the original echo with trajectory deviations, which can be calculated by the traditional frequency-domain algorithm. The last two terms of Eq. (19) contain trajectory offset error and antenna attitude jitter error. $h_0(x', r')$ can be estimated by echo simulation algorithm, which only contains trajectory offset error. Therefore, the main task is to evaluate the two terms of $h_1(x', r')$ and $h_2(x', r')$.

In order to expand the range of distance effectiveness, we assume that

$$\bar{f}_1(x, x', \eta, r) \approx (r + \delta r_r(x', r)) \cdot \gamma(x, r) \cdot \exp\left\{-j\frac{4\pi}{\lambda}\psi(x', r)\right\} \tag{20}$$

which satisfies the following condition: $\left|\left(\eta + \frac{4\pi}{\lambda}\right)\varphi(x', x, r) + \eta\psi(x', r)\right| \ll 1$. The azimuth Fourier transform of $\bar{f}_1$ can be expressed as: $\bar{F}_1(\chi, l, \eta, r) = \Gamma(\chi, r) Q_1(l, r)$. where $\Gamma(\cdot)$ is the azimuth FT of $\gamma(x, r)$.

$$Q_1(l, r) = FT_{x'}\left\{(r + \delta r_r(x', r)) \cdot \exp\left[-j\frac{4\pi}{\lambda}\psi(x', r)\right]\right\} \tag{21}$$

Then $\bar{H}(\xi, \eta) = \int \exp\left[-j\bar{\eta}r\right]\left\{Q_1(\xi, r) \underset{\xi}{\otimes} \left[G(\xi, \eta, r) \cdot \Gamma(\xi, r)\right]\right\}dr$.

The system function $G(\cdot)$ of SAR is decomposed into two parts

$$G(\xi, \eta, r) = G_{A1}(\xi, r) \cdot G_{B1}(\xi, \eta, r) \tag{22}$$

where the term.

$G_{A1}(\xi, r) = \exp\left[-j\left(\sqrt{\left(\frac{4\pi}{\lambda}\right)^2 - \xi^2} - \frac{4\pi}{\lambda}\right)r\right]$ describes the azimuth frequency modulation, including the change of focusing depth, and

$$G_{B1}(\xi, \eta, r) = rect\left[\frac{\eta}{2bc\tau/2}\right] \cdot \left(\frac{PRF}{v}\right) \cdot \left[W^2\left(\frac{\xi - \Delta\xi}{2\Omega_r}\right) - W^2\left(\frac{\xi}{2\Omega_r}\right)\right] \exp\left\{j\frac{\eta^2}{4b}\right\}$$
$$\times \exp\left\{-j\left(\sqrt{\bar{\eta}^2 - \xi^2} - \bar{\eta}\right)r\right\} \cdot \exp\left\{-j\left(\sqrt{\left(\frac{4\pi}{\lambda}\right)^2 - \xi^2} - \frac{4\pi}{\lambda}\right)r\right\} \tag{23}$$

the migration effect of range element and the error of antenna beam jitter. Then, we have

$$\tilde{h}_1(\xi, \eta) \approx \int \exp\{-j\bar{\eta}r\} \cdot G_{B1}(\xi, \eta, r) \cdot \left\{Q_1(\xi, r) \otimes_{\xi} \left[G_A(\xi, r) \cdot \Gamma(\xi, r)\right]\right\}dr \tag{24}$$

Similarly, let

$$\bar{f}_2(x, x', \eta, r) \approx \frac{1}{2}(r + \delta r_r(x', r))^2 \cdot \gamma(x, r) \cdot \exp\left\{-j\frac{4\pi}{\lambda}\psi(x', r)\right\} \tag{25}$$

and

$$Q_2(\xi,r) = \underset{x'}{FT}\left\{\frac{1}{2}(r+\delta r_r(x',r))^2 \cdot \exp\left[-j\frac{4\pi}{\lambda}\psi(x',r)\right]\right\} \tag{26}$$

$$\begin{aligned} G_{B2}(\xi,\eta,r) = rect\left[\frac{\eta}{2bc\tau/2}\right] \cdot \left(\frac{PRF}{v}\right)^2 \cdot \left[W^2\left(\frac{\xi}{2\Omega_x}\right) - 2W^2\left(\frac{\xi-\Delta\xi}{2\Omega_x}\right)\right. \\ \left. + W^2\left(\frac{\xi-2\Delta\xi}{2\Omega_x}\right)\right] \times \exp\left\{j\frac{\eta^2}{4b}\right\} \cdot \exp\left\{-j\left(\sqrt{\overline{\eta}^2-\xi^2}-\overline{\eta}\right)r\right\} \\ \times \exp\left\{j\left(\sqrt{\left(\frac{4\pi}{\lambda}\right)^2-\xi^2}-\frac{4\pi}{\lambda}\right)r\right\} \end{aligned} \tag{27}$$

we yield

$$\tilde{h}_2(\xi,\eta) \approx \int \exp\left[-j\overline{\eta}r\right] \cdot G_{B2}(\xi,\eta,r) \times \left\{Q_2(\xi,r) \otimes_\xi [G_A(\xi,r)\Gamma(\xi,r)]\right\}dr \tag{28}$$

From Eqs. (24) and (28), we obtain the efficient computation of $\tilde{h}_1(\xi,\eta)$ and $\tilde{h}_2(\xi,\eta)$. **Figure 9** summaries the computation flow chart of the procedures outlined above. Finally, $h(x',r')$ can be evaluated from **Figure 10**. The effectiveness of the algorithm for antenna jitter is $\delta_{az}(x') \ll \frac{X}{r+\delta r_r(x',r)} < \frac{X}{r} = \frac{\lambda}{L}$.

Note that the trajectory deviations should satisfy the following condition:

$$\left|\left(\eta+\frac{4\pi}{\lambda}\right)\varphi(x',x,r) + \eta\psi(x',r)\right| \ll 1 \tag{29}$$

which ensures that the range of distance effectiveness of the algorithm.

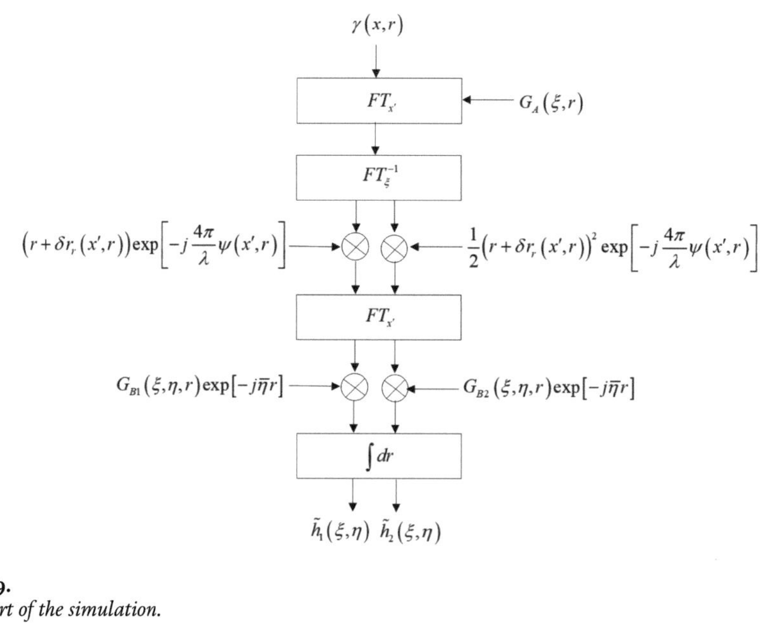

Figure 9.
Flowchart of the simulation.

3.1.1 Computational complexity

The complexity of algorithms given in Section 3.1 increases with the order of antenna pattern decomposition. Suppose N_a and N_r are the azimuth and range pixel

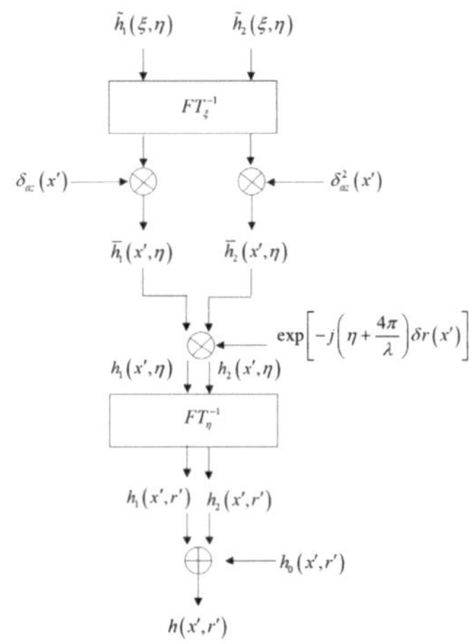

Figure 10.
Horizontal and vertical components of medium trajectory offset error.

size of the original signal, $N_a N_r$ is the computational complexity of $\gamma(\mathrm{x}, \mathrm{r})$. Let $\tilde{N}$ and N be the computational complexity of 3.1 part and medium trajectory offset error algorithm, then $\tilde{N} = (n+1)N$, where $N = N_x N_r^2$. Now we compare the computational complexity of part 3.1 and the exact time domain algorithm. It is known that the computational complexity of time domain is $\hat{N} = N_x N_r N_{tp} N_{sa}$, where N_{tp} and N_{sa} are the sizes of the emitted pulse and the synthetic aperture. Generally, n is less than N_{sa} and N_{tp} are the same size with N_r. Then the algorithm given in this part is more efficient than the time-domain algorithm. The calculation time-saving rate is

$$\frac{\hat{N}}{\tilde{N}} \approx \frac{N_{\mathrm{sa}}}{n+1} \tag{30}$$

Therefore, the algorithm given in Section 3.1 can simulate the extended scene in a reasonable time.

3.1.2 Algorithm verification

In this part, we will give some simulation results to verify the proposed algorithm. The simulation parameters are selected from the X-band SAR data in Ref.s [8, 9], and the main parameters are shown in **Table 3**. Note that the precise time domain simulation can be obtained from Eq. (17). In order to simplify, the algorithm given in part 3.1 is called algorithm A, and the time-domain algorithm is

Nominal height	4000 m	Range pixel dimension	3 m
Midrange coordinate	5140 m	Chirp bandwidth	45 MHz
Wavelength	3.14 cm	Chirp duration	5 μs
Pulse Repetition Frequency	400 Hz	Azimuth antenna dimension	1 m
Sampling Frequency	50 MHz	Number of azimuth samples in the raw data	1941
Azimuth pixel dimension	25 cm	Number of range samples in the raw signals	830

Table 3.
Sensor simulation parameters.

called algorithm B. Let a single point target be located in the middle of the scene (r = 5140 m), and the horizontal and vertical components are shown in **Figure 11**. Let the antenna pointing error be as follows:

$$\delta_a(x') = \delta_m \sin\left(\frac{2\pi}{vT_b}x' + \varphi_0\right) \tag{31}$$

where δ_m is the amplitude, T_b is the period of antenna jitter and φ_0 is the initial phase. Assumptions $\delta_m = \frac{1}{20}\frac{\lambda}{L}$, $T_b = \frac{1}{10}T_s$ and $\varphi_0 = 0$. **Figure 11** shows the horizontal and vertical components of the trajectory offset, and the antenna pointing error is Eq. (31). Now we consider the near range point (r = 4840 m) in the scene. The amplitude and phase comparison results of algorithms A and B are shown in **Figures 12** and **13**, respectively. Similar conclusion can be drawn that the amplitude error and phase error of algorithm A and B are very small, which verifies the effectiveness of algorithm A. **Table 4** shows the imaging quality evaluation results of algorithms A and B at near range (r = 4840 m), center range (r = 5140 m) and far range (r = 5440 m), including impulse response width (IRW), peak sidelobe ratio (PSLR) and integral sidelobe ratio (ISLR).

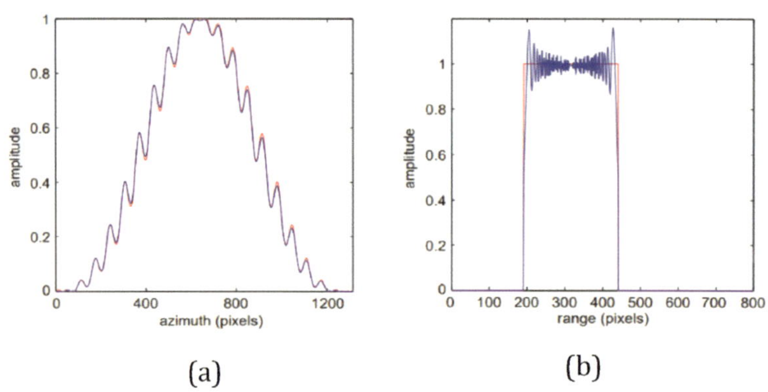

Figure 11.
Amplitude comparison results of algorithm A and B (the blue line represents algorithm a and the red line represents algorithm B) (a) azimuth cut. (b) Range cut. The point target is located in the scene, and the close range R = 4840 meters.

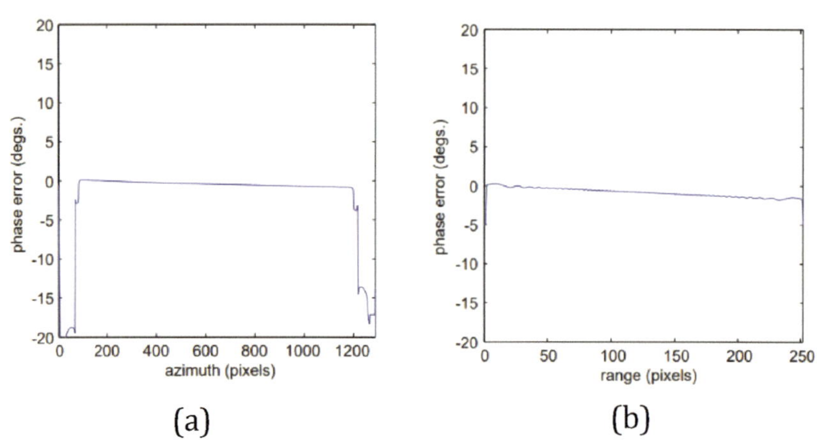

Figure 12.
Phase comparison results of algorithm A and algorithm B. (a) Azimuth cut. (b) Range cut. The point target is located in the scene, and the close range R = 4840 meters.

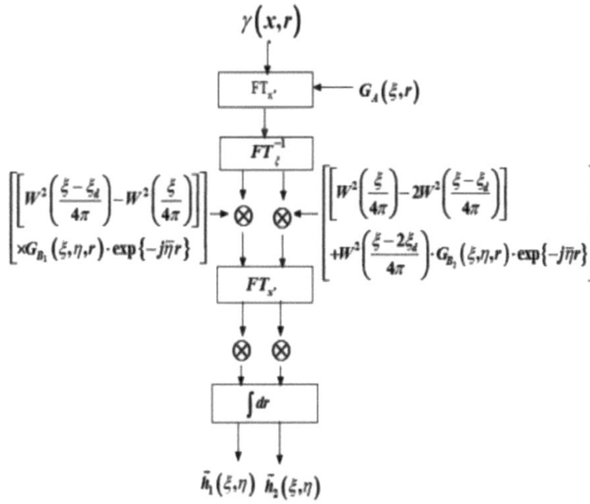

Figure 13.
Flowchart of medium trajectory error and attitude jitter at squint angle.

Algorithm	r (m)	Azimuth direction			Range direction		
		IRW (dB)	PSLR (dB)	ISLR (dB)	IRW (dB)	PSLR (dB)	ISLR (dB)
A	4840	0.2266	−5.0346	−3.9095	−2.7188	−6.8351	−5.6503
	5140	0.2227	−3.4025	−2.2344	2.8125	−12.3214	−9.5441
	5440	0.2188	−2.4523	−1.2622	2.7656	−8.5	−7.2288
B	4840	0.2266	−5.3933	−4.2801	2.7656	−7.1473	−6.0101
	5140	0.2227	−3.3611	−2.1945	2.7656	−12.8208	−9.6497
	5440	0.2188	−2.4399	−1.2496	2.7656	−8.0237	−6.8483

Table 4.
The focus performance of the point target response of algorithm A and B.

3.2 Raw echo algorithm of airborne stripmap SAR with trajectory error and attitude jitter under squint conditions

This section presents the simulation algorithm of airborne SAR raw echo with trajectory offset and attitude jitter error under squint conditions.

3.2.1 Simulation algorithm

Suppose the received echo is as follows:

$$h(x', r') = \iint dxdr\gamma(x, r) \cdot \exp\left\{-j\frac{4\pi}{\lambda}R - j\frac{4\pi}{\lambda}\frac{\Delta f/f}{c\tau}(r' - R)^2\right\} \\ \times rect\left[\frac{r' - R}{c\tau/2}\right] \cdot W^2\left[\frac{x' - x - \delta_x}{X}\right] \tag{32}$$

where $\delta_x = (r + \delta\overline{r}_{xr}(x', x, r, \phi_d))\tilde{\delta}_{az}(x')$ is the antenna pointing error in azimuth direction. From [25] we have

$$\delta\overline{r}_{xr}(x',x,r,\phi_d) \approx d(x')\big(\sin\beta(x')\cos\phi_d\cos\vartheta(x,r) - \cos\beta(x')\sin\vartheta(x,r) + \frac{1}{2}(\sin\phi_d)^2(\cos\vartheta(x,r))^2\sin\beta(x')\cos\phi_d\cos\vartheta(x,r) + \frac{1}{2}\cos\beta(x') \times \sin\vartheta(x,r)(\sin\phi_d)^2(\cos\vartheta(x,r))^2\big) \tag{33}$$

Eq. (32) is a signal model for the exact time-domain simulation and can be used as the criterion to judge the validity of the algorithm proposed below.

When the beam pointing error is less than the beam width, the antenna pattern can be approximated as follows:

$$W^2\left(\frac{x'-x-\delta_{x'}}{X}\right) \approx W^2\left(\frac{x'-x}{X}\right) - W^{2(1)}\left(\frac{x'-x}{X}\right)(r+\delta\overline{r}_{xr}(x',x,r,\phi_d))\delta_{az}(x') + \frac{1}{2}W^{2(2)}\left(\frac{x'-x}{X}\right)(r+\delta\overline{r}_{xr}(x',x,r,\phi_d))^2\delta_{az}{}^2(x') \tag{34}$$

where $W^{2(n)}(\cdot)$ is the nth derivative of the antenna pattern. Accordingly, the echo can be decomposed as follows:

$$h(x',r') \approx \hat{h}_0(x',r') + \hat{h}_1(x',r') + \hat{h}_2(x',r') \tag{35}$$

The above equation, $\hat{h}_0(x',r')$ only includes trajectory deviations, the latter two terms including trajectory offset and attitude jitter.

Assuming that the range Fourier transform of $\hat{h}_1(x',r')$ is $\hat{h}_1(x',\eta)$, then

$$\hat{h}_1(x',\eta) = \exp\left[-j\left(\eta+\frac{4\pi}{\lambda}\right)\delta\overline{r}(x',\phi_d)\right]\overline{\overline{h}}_1(x',\eta) \tag{36}$$

The term $\tilde{\delta}_{az}(x')$ can be extracted from the integral, that is

$$\overline{\overline{h}}_1(x',\eta) = \overline{\delta}_{az}(x')\tilde{\hat{h}}_1(x',\eta) \tag{37}$$

$$\hat{h}_1(x',\eta) = rect\left[\frac{\eta}{bc\tau}\right]\exp\left\{j\frac{\eta^2}{4b}\right\}\iint dxdr\gamma(x,r)\exp\left\{-j\left(\eta+\frac{4\pi}{\lambda}\right) \times (r+\Delta R(x'-x,r)+\overline{\psi}(x',r,\phi_d)+\overline{\varphi}(x',x,r,\phi_d))\right\} \times [-(r+\delta\overline{r}_{xr}(x',x,r,\phi_d))]W^{2(1)}\left(\frac{x'-x}{X}\right) \tag{38}$$

The azimuth Fourier transform of $\overline{\overline{\mathrm{h}}}_1(x',\eta)$ is

$$\overline{\overline{\mathrm{h}}}_1(\xi,\eta) = FT\left[\tilde{\delta}_{az}(x')\right]\otimes\tilde{\hat{h}}_1(\xi,\eta) \tag{39}$$

where

$$\tilde{\hat{h}}_1(\xi,\eta) = \int dr\exp\left[-j\overline{\eta}r\right]\int dl\tilde{G}_1(\xi-l,\eta,r)\tilde{\hat{F}}_1(\xi-l,l,\eta,r) \tag{40}$$

$\tilde{\hat{F}}_1(\cdot)$ is the Fourier transform of $\tilde{\hat{f}}_1(\cdot)$, and

$$\tilde{\hat{f}}_1(x,x',\eta,r) = -[r+\delta\overline{r}_{xr}(x',x,r,\phi_d)]\gamma(x,r) \times \exp\left\{-j\left(\eta+\frac{4\pi}{\lambda}\right)[\overline{\psi}(x',r,\phi_d)+\overline{\varphi}(x',x,r,\phi_d)]\right\} \tag{41}$$

Similarly, a two-dimensional Fourier transform $\hat{h}_2(x', r')$ can be obtained. By performing a range FT of $h_2(x', r')$ and separating the factor $\exp\{-j\eta\overline{\delta r}(x', \phi_d)\}$, we get $\overline{\hat{h}}_2(x', \eta)$. Then, after azimuth FT of $\overline{\hat{h}}_2(x', \eta)$, we obtain: $\overline{\hat{h}}_2(\xi, \eta) = FT_{x'}\left[\tilde{\delta}^2_{az}(x')\right] \otimes \tilde{\hat{h}}_2(\xi, \eta)$, where

$$\tilde{\hat{h}}_2(\xi, \eta) = \int dr \exp\left[-j\overline{\eta}r\right] \int dl \tilde{\hat{G}}_2(\xi - l, \eta, r)\tilde{\hat{F}}_2(\xi - l, l, \eta, r) \tag{42}$$

where $\tilde{\hat{F}}_2(\cdot)$ is the two-dimensional FT of $\tilde{\hat{f}}_2(\cdot)$. The difference between $\tilde{\hat{f}}_1(\cdot)$ and $\tilde{\hat{f}}_2(\cdot)$ is that the former contains a factor $[r + \overline{\delta r}_{xr}(x', x, r, \phi_d)]$ while the latter contains a factor $\frac{1}{2}[r + \overline{\delta r}_{xr}(x', x, r, \phi_d)]^2$. If the two-dimensional spectrum of $\tilde{\hat{h}}_1(\xi, \eta)$ and $\tilde{\hat{h}}_2(\xi, \eta)$ is obtained from Eqs. (40) and (42), then the estimation of $h(x', r')$ can be obtained from **Figure 14**. However, due to the coupling of azimuth and distance between $\tilde{\hat{f}}_1(\cdot)$ and $\tilde{\hat{f}}_2(\cdot)$, it is necessary to decouple them. Therefore, we adopt the following approach:

The amplitude approximation $\tilde{\hat{f}}_1(\cdot)$ and $\tilde{\hat{f}}_2(\cdot)$ satisfy the following conditions: $\overline{\delta r}_{xr}(x', x, r, \phi_d) \approx \overline{\delta r}(x', r, \phi_d)$. Phase approximation needs to meet the following requirements:

$$\tilde{\hat{f}}_1(\cdot) \approx (r + \overline{\delta r}_{xr}(x', r, \phi_d))\gamma(x, r) \exp\left\{-j\frac{4\pi}{\lambda}\overline{\psi}(x', r, \phi_d)\right\} \tag{43}$$

$$\tilde{f}_2(x, x', \eta, r) \approx \frac{1}{2}(r + \overline{\delta r}_r(x', r, \phi_d))^2\gamma(x, r) \times \exp\left[-j\frac{4\pi}{\lambda}\overline{\psi}(x', r, \phi_d)\right] \tag{44}$$

$$|\overline{\varphi}(x', x, r, \phi_d)| < < \frac{\lambda}{4\pi} \tag{45}$$

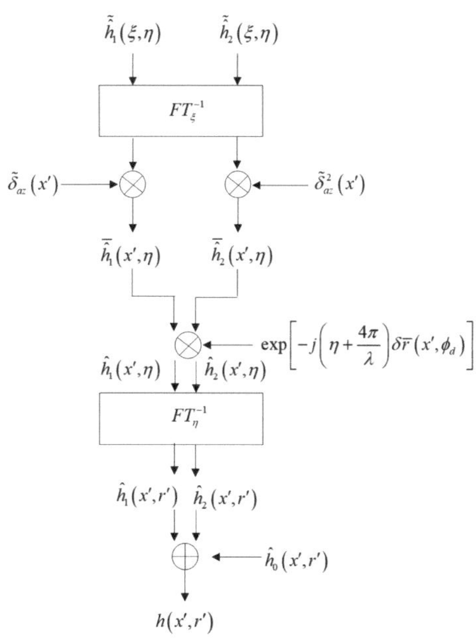

Figure 14.
SAR raw data echo simulation diagram.

$$|\overline{\psi}(x', r, \phi_d)| < < \frac{f}{\Delta f}\frac{\lambda}{2\pi} \tag{46}$$

then

$$\begin{aligned}\tilde{\hat{h}}_1(\xi,\eta) &= \int \exp(-j\overline{\eta}r) \int dl G_1(\xi - l, \eta, r)\Gamma(\xi - l, r) Q(l, \eta, r) \\ &= \int dr \exp(-j\overline{\eta}r)\{Q(\xi,\eta,r) \otimes_{\xi} [G_1(\xi,\eta,r)\Gamma(\xi,r)]\}\end{aligned} \tag{47}$$

where

$$Q(l, \eta, r) = FT_{x'}\{\exp[-j\overline{\eta}\overline{\psi}(x', r, \phi_d)]\} \tag{48}$$

$$\begin{aligned}G_1(\xi,\eta,r) &\approx rect\left(\frac{\eta}{bc\tau}\right)\left[W^2\left(\frac{\xi - \xi_d - \Delta\xi(r)}{2a}\right) - W^2\left(\frac{\xi - \xi_d}{2a}\right)\right] \\ &\times \exp\left(j\frac{\eta^2}{4b}\right)\exp\left[-j\left(\sqrt{\overline{\eta}^2 - \xi^2}\cos\phi_d - \overline{\eta}\right)r - j\xi r \sin\phi_d\right]\end{aligned} \tag{49}$$

Similarly, we can get an estimate of $\tilde{\hat{h}}_2(\xi,\eta)$, with the flow chart shown in **Figure 15**.

In the simulation, the attitude change needs to meet the following condition:

$$\delta_{az}(x') \ll \frac{X}{r + \delta\overline{r}_r(x', r, \phi_d)} < \frac{X}{r} = \frac{\lambda}{L} \tag{50}$$

Combined with conditions (45) and (46), we see that the algorithm is suitable for squint with medium trajectory offset error and antenna beam pointing error.

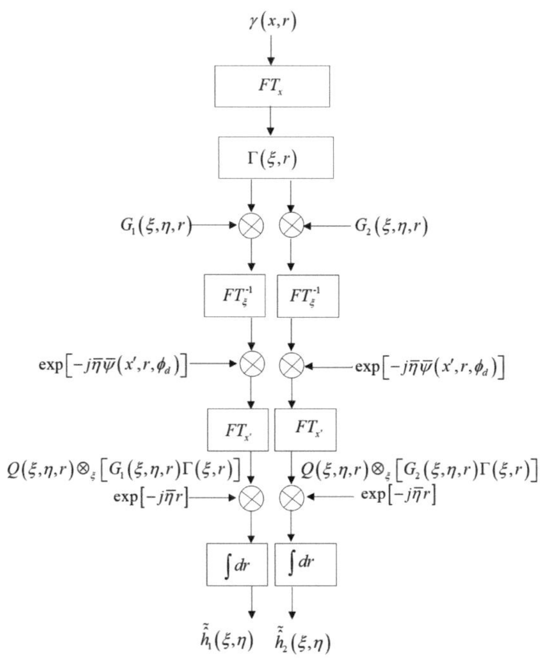

Figure 15.
The flowchart of medium trajectory error and attitude jitter.

The computational efficiency of the algorithm is analyzed below. Obviously, the computational complexity of the algorithm increases with the expansion order of the antenna pattern. Let N be the computational complexity with

$$N \approx nN_aN_r^2(3 + \log_2 N_a) \tag{51}$$

and $\tilde{N}$ be the computational complexity of the time domain method. It turns out that the following ratio is obtained

$$\frac{\tilde{N}}{N} \approx \frac{N_{sa}}{n(3 + \log_2 N_a)} \tag{52}$$

We observe from the above equation that the proposed algorithm has higher computational efficiency than the time-domain algorithm for the same order of antenna pattern expansion n.

3.2.2 Simulation results

In this section, the proposed algorithm is verified by comparing with the simulation results of time-domain algorithm. The trajectory offset error is shown in **Figure 16** and the antenna pointing error is given by Eq. (31). **Figure 17** shows the phase comparison results of azimuth and range directions, where **Figure 17(a)** is

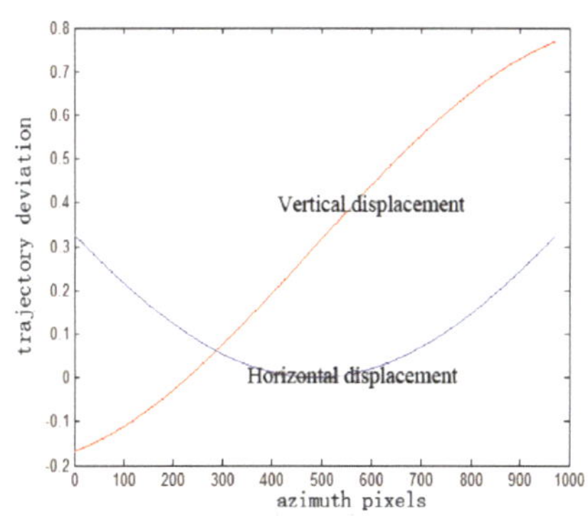

Figure 16.
Moderate trajectory deviations [m] vs. the azimuth pixels.

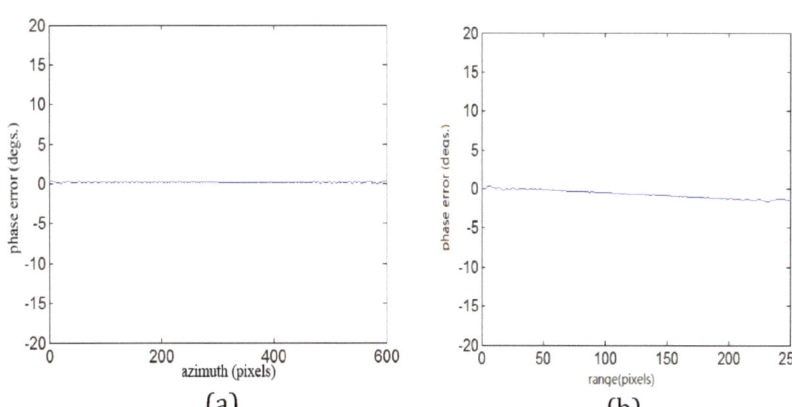

Figure 17.
SAR raw data phase comparison of the algorithm proposed in Section 3.2 and the time domain algorithm.

the azimuth cut and **Figure 17(b)** is the range cut. It can be seen that the phase errors of the proposed algorithm and the time-domain algorithm are both minimal, proving the effectiveness of the proposed algorithm.

4. Conclusion

In this chapter, we present a fast echo simulation algorithm of airborne striped SAR with trajectory error in squint condition, the fast echo simulation algorithm of airborne stripmap SAR with antenna attitude jitter and trajectory error, and the fast echo simulation algorithm of airborne stripmap SAR with attitude jitter and trajectory error in squint condition. From the comparison between the proposed hybrid-domain echo algorithm and the time-domain algorithm, it can be seen that the phase error slice is within a reasonable range, but the hybrid algorithm runs faster than the time-domain simulation by using FT, which is conducive to the simulation of large scenes.

Acknowledgements

The author would like to thank Professor Kunshan Chen for his support in reviewing. This work is supported by Innovation Funds of China Aerospace Science and Technology (No. Y-Y-Y-FZLDTX-20).

Author details

Yuhua Guo[1,2]*, Shichao Jin[1,2] and Yuhong Guo[3]

1 State Key Laboratory of Space-Ground Integrated Information Technology, Beijing, China

2 Beijing Institute of Satellite Information Engineering, Beijing, China

3 Department of Mathematics, Taiyuan Normal University, Jinzhong, China

*Address all correspondence to: yuhuaguo321@163.com

References

[1] Fornaro G. Trajectory deviations in airborne SAR: Analysis and compensation. IEEE Transactions on Aerospace and Electronic Systems. 1999; **35**:997-1009. DOI: 10.1109/7.784069.

[2] Mori A, De Vita F. A time-domain raw signal simulator for interferometric SAR. IEEE Transactions on Geoscience and Remote Sensing. 2004; 42: 1811-1817, DOI:10.1109/ TGRS.2004.832242.

[3] Dogan O, Kartal M. Time domain SAR raw data simulation of distributed. targets. 2010:1-11. DOI: 10.1155/2010/ 784815.

[4] Zhang F, Hu C, Li W, Hu W, Li HC. Accelerating time-domain SAR raw data simulation for large areas using multi-GPUs. IEEE Journal of selected topics in applied earth observations and remote sensing. 2014;7:3956-3965. DOI: 10.1109/JSTARS.2014.2330333.

[5] Franceschetti G. Schirinzi G. a SAR processor based on two dimensional FFT code. IEEE Transactions on Aerospace and Electronic Systems. 1990;**26**:356-365. DOI: 10.1109/7.53462.

[6] Franceschetti G, Miliaccio M, Riccio D, Schirinzi GSARAS. A SAR raw signal simulator. IEEE Transactions on Geoscience and Remote Sensing. 1992; **30**:110-123. DOI: 10.1109/36.124221.

[7] Khwaja AS, Ferro-Famil L, Pottier E. Efficient stripmap SAR raw data generation taking into account sensor trajectory deviations. IEEE Transactions on Geoscience and Remote Sensing Letters. 2011;**8**:794-798. DOI: 10.1109/ LGRS.2011.2111411

[8] Franceschetti G, Iodice A, Perna S. Riccio D. SAR sensor trajectory deviations: Fourier domain formulation and extended scene simulation of raw data. IEEE Transactions on Geoscience and Remote Sensing. 2006;**9**:2323-2334. DOI: 10.1109/TGRS.2006.873206.

[9] Franceschetti G, Iodice A, Perna S, Ricco D. Efficient simulation of airborne SAR raw data of extended scenes. IEEE Transactions on Geoscience and Remote Sensing. 2006;**10**:2851-2860. DOI: 10.1109/TGRS.2006.875786.

[10] Diao GJ, Xu XJ. Fast algorithm for SAR raw data simulation with large squint angles. Journal of Electronics and Information Technology. 2011;**33**: 684-689. DOI: 10.3724/SP. J.1146.2010.00498

[11] Franceschetti G, Guida R, Iodice A, Riccio D. Ruello G. Efficient Simulation of hybrid stripmap/spotlight SAR raw signals from extended scenes. 2004;**42**: 2385-2396. DOI: 10.1109/ TGRS.2004.834763

[12] Xu W, Deng YK. Efficient Spaceborne TOPS mode SAR raw signal simulation of extended scenes. Journal of Electronics and Information Technology. 2012;**34**:187-193. DOI: 10.3724/SP.J.1146.2011.00216.

[13] Boerner E, Balss U, Eineder M. Simulation of distributed spotlight raw data from X-SAR/STRM stripmap data. In: International Geoscience and Remote Sensing Symposium. 2004;**3**:1770-1773

[14] Xu W, Huang PP, Deng YK. Efficie nt sliding spotlight SAR raw signal simulation of extended scenes. EURASIP Journal of Advances in Signal Processing. 2011;**52**:1-7. DOI: 10.1186/ 1687-6180-2011-52.

[15] Cimmino S, Giorgio F, Iodice A, Riccio D, Ruello G. Efficient spot-light SAR raw signal simulation of extended scenes. IEEE Transactions on Geoscience and Remote Sensing. 2003; **41**:2329-2337. DOI: 10.1109/ TGRS.2003.815239

[16] Domenico D, Gerardo DM, Antonio I, Daniele R, Giuseppe R. Efficient simulation of extended-scene SAR raw signals with any acquisition mode. In: International Geoscience and Remote Sensing Symposium. 2018. pp. 6715-6718

[17] Shahrezaei IH, Kazerooni M. Fallah M. a complex target terrain SAR raw data generation and evaluation based on inversed equalized hybrid-domain algorithm processing. Waves in Random and Complex Media. 2017;**27**: 47-66. DOI: 10.1080/ 17455030.2016.1198062.

[18] Sofiani R, Heidar H. Kazerooni M. An efficient raw data simulation algorithm for large complex marine targets and extended sea clutter in spotlight SAR. 2018:1223-1230. DOI: 10.1002/mop.31143

[19] Cimmino S, Franceschetti G, Iodice A, Riccio D, Ruello G. Efficient spotlight SAR raw signal simulation of extended scenes. IEEE Transactions on Geoscience and Remote Sensing. 2003; **41**:2329-2337. DOI: 10.1109/ TGRS.2003.815239

[20] Jia L, Jia X, Xu XJ, He YH. SAR raw data generation for complex airport scenes. Journal of Radars. 2014;**3**: 565-573. DOI: 10.3724/SP. J.1300.2014.14071

[21] Vandewal M, Speck R, Suess H. Efficient SAR raw data generation including low squint angles and platform instabilities. IEEE Transactions on Geoscience and Remote Sensing Letters. 2008;**5**:26-30. DOI: 10.1109/ LGRS.2007.907419.

[22] Tang XQ, Xiang MS. Wei L. SAR raw signal simulation accounting for antenna attitude variations. In: The IEEE International Geoscience and Remote Sensing Symposium (IGARSS). IEEE. 2009. pp. 613-616

[23] Dogan O, Kartal M. Efficient stripmap-mode SAR raw data simulation including platform angular deviations. IEEE Transactions on Geoscience and Remote Sensing Letters. 2011;**8**:784-788. DOI: 10.1109/ LGRS.2011.2112633

[24] Fornaro G, Sansosti E, Lanari R, Tesauro M. Role of processing geometry in SAR raw data processing. IEEE Transactions on Aerospace and Electroic Systems. 2002;**38**:441-454. DOI: 10.1109/TAES.2002.1008978

[25] Guo YH, Yang XY. Zhou. F F. a new motion compensation method for squinted stripmap synthetic aperture radar. Journal of Computational Information Systems. 2013;**9**:3453-3460. DOI: 10.12733/jcis5836.

Chapter 15

Signal Propagation in Soil Medium: A Two Dimensional Finite Element Procedure

Frank Kataka Banaseka, Kofi Sarpong Adu-Manu, Godfred Yaw Koi-Akrofi and Selasie Aformaley Brown

Abstract

A two-Dimensional Finite Element Method of electromagnetic (EM) wave propagation through the soil is presented in this chapter. The chapter employs a boundary value problem (BVP) to solve the Helmholtz time-harmonic electromagnetic model. An infinitely large dielectric object of an arbitrary cross-section is considered for scattering from a dielectric medium and illuminated by an incident wave. Since the domain extends to infinity, an artificial boundary, a perfectly matched layer (PML) is used to truncate the computational domain. The incident field, the scattered field, and the total field in terms of the z-component are expressed for the transverse magnetic (TM) and transverse electric (TE) modes. The radar cross-section (RCS), as a function of several other parameters, such as operating frequency, polarization, illumination angle, observation angle, geometry, and material properties of the medium, is computed to describe how a scatterer reflects an electromagnetic wave in a given direction. Simulation results obtained from MATLAB for the scattered field, the total field, and the radar cross-section are presented for three soil types – sand, loam, and clay.

Keywords: electromagnetic signal in soil, finite element method, radar cross-section, underground wireless communication

1. Introduction

Signal propagation underground and in soil medium constitute the backbone of the Internet of underground things (IoUT), which power many applications such as precision agriculture, border monitoring for intrusion detection, pipeline monitoring, etc. [1–3]. In the recent past, the study of signal propagation in soil medium for underground wireless communication has focused mainly on empirical techniques [4–8]. The most commonly used modeling techniques for implementing IoUT include electromagnetic waves, magnetic induction, and acoustic waves [9, 10]. Electromagnetic field analysis is performed in this chapter using numerical modeling with the finite element method to examine the signal strength in the soil medium.

The process of numerical modeling of how electromagnetic fields propagate and interact with physical objects and the environment is usually referred to as computational electromagnetics (CEM), numerical electromagnetics. The primary motivation of this process is to develop efficient approximations to Maxwell's equations through numerical schemes for cases where closed-form analytical solutions of Maxwell's equation cannot be obtained due to the complexity of geometries, material parameters, and boundary conditions. Therefore, several real-life problems that are not analytically computable, such as electromagnetic scattering, antenna radiation, electromagnetic wave propagation, electromagnetic compatibility, etc., can effectively be solved by numerical techniques. The mathematical model of the electromagnetic problem is usually obtained in terms of partial differential equations, integral equations, or integro-differential equations derived from Maxwell's equations and a set of a priori constraints of the problem such as boundary and initial conditions material parameters and geometry. The problem is ideally defined on an infinite-dimensional function space. Numerical methods apply a discretization to the continuum to reduce infinite degrees of freedom to a finite degree of freedom. In other words, the solution of an infinite dimension-dimensional problem is projected into a finite-dimensional space. Hence, the solution to the problem becomes amenable on a digital computer. The main philosophy in most of the numerical methods is to apply the divide-and-conquer strategy. The idea is to divide an intractable continuous problem into smaller pieces (divide), express the solution over each small piece (conquer), and then combine the piece-wise solutions to obtain a global solution. In this chapter, FEM will be applied to the two-dimensional boundary value problem in EM wave propagation through the soil to evaluate the signal strength of the wave propagation in soil. This evaluation will be based on the incidence angle of the transmitted wave. The radar cross-section of the scatterer will be used to evaluate the direction of the wave.

2. Time-harmonic EM model

When deriving the wave equation for electromagnetic wave propagation through a vacuum or a dielectric medium such as soil or pure water, the free charge density $\rho_f = 0$ and the free current density $J_f = 0$. However, in the case of conductors like seawater or metals, we cannot control the flow of charges and, in general, J_f is certainly not equal to zero. With this, Maxwell's equations for linear media assume the form [11].

$$\begin{cases} (i)\quad \nabla \cdot E = \frac{1}{\varepsilon}\rho_f \qquad (iii)\quad \nabla \text{x} E = -\frac{\partial B}{\partial t} \\ (ii)\quad \nabla \cdot B = 0 \qquad (iv)\quad \nabla \text{x} B = \mu\sigma E + \mu\varepsilon\frac{\partial E}{\partial t} \end{cases} \tag{1}$$

Apply the curl to (iii) and (iv), and we obtain modified wave equations for the electric field E and the magnetic field B:

$$\nabla^2 u = \mu\varepsilon\frac{\partial^2 u}{\partial t^2} + \mu\sigma\frac{\partial u}{\partial t} \tag{2}$$

where u represents the scalar component of the electric or magnetic field, ε is the permittivity of the medium $\mu = \mu_r\mu_0$, is the magnetic permeability (μ_r is the relative permeability of the soil $\mu_r = 1$, for non-magnetic soil, $\mu_0 = 4\pi\text{x}10^{-7}N/A^2$ is the magnetic constant in vacuum), and σ is the electric conductivity of the soil medium. Eq. (2) admits the plane wave solution

$$u(x,t) = ue^{i(kx-\omega t)} \tag{3}$$

where $\omega = 2\pi f$ is the angular frequency, t is the time, and k is the complex wavenumber or propagation constant, which can be derived from Eq. (3). The boundary value problem (BVP) used to solve the time-harmonic electromagnetic problem in 2-D, can be expressed in its generic form as

$$-\frac{\partial}{\partial x}\left(p_x \frac{\partial u}{\partial t}\right) - \frac{\partial}{\partial y}\left(p_y \frac{\partial u}{\partial y}\right) + qu = f \qquad for\ (x,y) \in \Omega \tag{4}$$

$$u = u_0 \quad \text{on } \Gamma_{\text{D}} : \text{ Dirichlet Boundary Condition} \tag{5}$$

$$\left(p_x \frac{\partial u}{\partial t} \hat{a}_x + p_y \frac{\partial u}{\partial y} \hat{a}_y\right).\ \hat{n} = \beta \qquad \text{on } \Gamma_{\text{N}} : \text{ Neumann Boundary Condition} \tag{6}$$

$$\left(p_x \frac{\partial u}{\partial t} \hat{a}_x + p_y \frac{\partial u}{\partial y} \hat{a}_y\right).\ \hat{n} + \alpha u = \beta \qquad \text{on } \Gamma_{\text{M}} : \text{ Mixed Boundary Condition} \tag{7}$$

where $u(x,y)$ is the unknown function to be determined, and $p_x(x,y)\ p_y(x,y)$ $q(x,y) f(x,y)$ are given functions. $u_0\ \alpha$, and β are given functions in boundary conditions (BCs); $\Gamma_D\ \Gamma_N$ and Γ_M refers to boundaries where Dirichlet, Neumann, and mixed BCs are imposed, respectively; $\hat{n} = \hat{a}_x n_x + \hat{a}_y n_y$ is the unit vector normal to the boundary in the outward direction.

The weak form is used to provide the finite element solution. The weak form is written as [12–14].

$$\iint_{\Omega} \left[p_x \frac{\partial u}{\partial x}\frac{\partial \psi}{\partial x} + p_y \frac{\partial u}{\partial y}\frac{\partial \psi}{\partial y}\right] ds - \oint_{\Gamma} \psi \left[p_x \frac{\partial u}{\partial y} n_x + p_y \frac{\partial u}{\partial y} n_y\right] dl + \iint_{\Omega} q\psi u\ ds - \iint_{\Omega} \psi f\ ds = 0 \tag{8}$$

where $\psi(x,\ y)$ is the weight function.

The weak form is applied in each element domain, and element matrices are formed by expressing the unknown function as a weighted sum of nodal shape functions. The sum of line integrals of two neighboring elements cancels out while combining the element matrices. Therefore, the line integral can be omitted for interior elements and should be considered only for elements adjacent to the boundary. Due to its special form, the line integral makes easier the imposition of mixed types of BCs. The mesh generation will be discussed first, and the shape functions will be given, and then the finite element solution of the time-harmonic problem of electromagnetic scattering in a dielectric medium (soil) will be presented.

3. The Helmholtz time harmonic model

In 2-D, the geometry and boundary conditions do not vary along an axis (say the z-axis). Hence fields can be represented as a superposition of fields of two orthogonal polarizations using the linearity property. Any field can be decomposed into transverse magnetic (TM) and transverse electric (TE) parts for the z-variable. In the TM case (horizontal polarization), only the z-component of the electric field ($\mathbf{E} = \hat{a}_z E_z(x,y)$) and the x-y-components magnetic field ($\mathbf{H} = \hat{a}_x H_x(x,y) + \hat{a}_y H_y(x,y)$) exist. However, in the TE case (vertical polarization), which is a dual of TM, only the z-component of the magnetic field ($\mathbf{H} = \hat{a}_z H_z(x,y)$) and the x-y-components electric field ($\mathbf{E} = \hat{a}_x E_x(x,y) + \hat{a}_y E_y(x,y)$) exist.

The material tensors in this case are defined as

$$\boldsymbol{\varepsilon}_{rc} = \begin{bmatrix} \varepsilon_{rc}^{xx} & \varepsilon_{rc}^{xy} & 0 \\ \varepsilon_{rc}^{yx} & \varepsilon_{rc}^{yy} & 0 \\ 0 & 0 & \varepsilon_{rc}^{zz} \end{bmatrix} \quad with \quad (\boldsymbol{\varepsilon}_{rc})_{sub} = \begin{pmatrix} \varepsilon_{rc}^{xx} & \varepsilon_{rc}^{xy} \\ \varepsilon_{rc}^{yx} & \varepsilon_{rc}^{yy} \end{pmatrix}, \tag{9}$$

$$\boldsymbol{\mu}_r = \begin{bmatrix} \mu_r^{xx} & \mu_r^{xy} & 0 \\ \mu_r^{yx} & \mu_c^{yy} & 0 \\ 0 & 0 & \mu_r^{zz} \end{bmatrix} \quad with \quad (\boldsymbol{\mu}_r)_{sub} = \begin{pmatrix} \mu_r^{xx} & \mu_r^{xy} \\ \mu_r^{yx} & \mu_r^{yy} \end{pmatrix}, \tag{10}$$

The generalized homogeneous Helmholtz equation in TM and TE case can be written, respectively, in the following forms [15].

$$\nabla \cdot \left(\boldsymbol{\Lambda}_\mu \cdot \nabla E_z\right) + k_0^2 \varepsilon_{rc}^{zz} E_z = 0, \tag{11}$$

$$\nabla \cdot \left(\boldsymbol{\Lambda}_\varepsilon \cdot \nabla H_z\right) + k_0^2 \mu_r^{zz} H_z = 0, \tag{12}$$

Where

$$\boldsymbol{\Lambda}_\mu = \frac{1}{\left|(\boldsymbol{\mu}_r)_{sub}^T\right|} (\boldsymbol{\mu}_r)_{sub}^T, \tag{13}$$

$$\boldsymbol{\Lambda}_\varepsilon = \frac{1}{\left|(\boldsymbol{\varepsilon}_{rc})_{sub}^T\right|} (\boldsymbol{\varepsilon}_{rc})_{sub}^T, \tag{14}$$

Where the superscript T indicates the transpose and $|.|$ refers to the determinant of the corresponding submatrix; k_0 is the free-space wavenumber; $\varepsilon_{rc} = \varepsilon_r - j\sigma/\omega\varepsilon_0$ is the complex relative permittivity ε_r, and μ_r are the relative permittivity and permeability, respectively; and σ is the conductivity. For isotropic mediums, Eq. (8) becomes

$$\nabla \cdot \left(\mu_r^{-1} \nabla E_z\right) + k_0^2 \varepsilon_{rc} E_z = 0, \tag{15}$$

Or

$$\frac{\partial}{\partial x}\left(\frac{1}{\mu_r}\frac{\partial E_z}{\partial x}\right) + \frac{\partial}{\partial x}\left(\frac{1}{\mu_r}\frac{\partial E_z}{\partial y}\right) + k_0^2 \varepsilon_{rc} E_z = 0, \tag{16}$$

The same way Eq. (11) becomes

$$\nabla \cdot \left(\varepsilon_{rc}^{-1} \nabla H_z\right) + k_0^2 \mu_r H_z = 0, \tag{17}$$

Or

$$\frac{\partial}{\partial x}\left(\frac{1}{\varepsilon_{rc}}\frac{\partial H_z}{\partial x}\right) + \frac{\partial}{\partial x}\left(\frac{1}{\varepsilon_{rc}}\frac{\partial H_z}{\partial y}\right) + k_0^2 \mu_r H_z = 0, \tag{18}$$

4. Scattering from a dielectric medium (soil)

In [16], a new wave number model is proposed with the combination of the Peplinski principle and multiple scattering from particles in the soil medium. The new wave number is used in the computation of the path loss. In another recent work, sensitivity analysis of the Ku-band scattering coefficient to soil moisture was performed under single-polarized, dual-polarized, and dual-angular combinations [17]. Similarly, a model of parabolic equations for reflection and refraction in an environment with an obstacle where the area is decomposed into two different domains. The discrete mixed Fourier transformation is used to compute the field strength in the upper subdomain, and the finite difference method is used to calculate the field strength in the lower subdomain [18].

In our case, an infinitely large dielectric object of an arbitrary cross-section is considered and illuminated by an incident wave that is not a function of z. An illustration of the scattering problem for TM mode is shown in **Figure 1** where a general FEM is depicted. The problem is defined in TE mode by replacing the electric field with a magnetic field.

4.1 The perfectly matched layer (PML)

Since the domain extends to infinity, an artificial boundary or layer is used to truncate the computational domain. That is an absorbing boundary condition (ABC) or perfectly matched layer (PML) [15].

Locally-Conformal PML (LC-PML) is a powerful PML method whose implementation is straightforward. It is implemented by replacing the real coordinates ($\boldsymbol{\rho} = (x, y)$) inside the PML region with their corresponding complex coordinates ($\tilde{\boldsymbol{\rho}} = (\tilde{x}, \tilde{y})$) [15]. The computational domain is the union of the PML region, the free-space region, and the scatterer region.

Mesh generation is the process of representing the domain of interest as a collection of elements. The two most commonly used elements in 2-D problems are triangular and quadrilateral. **Figure 2** shows the mesh that is formed by triangular elements for a rectangular domain of the boundary value problem (BVP). The figure was generated in MATLAB.

For a computational domain with such curved boundaries, discretization errors will occur due to the inability to capture the exact geometry of the domain. Triangular elements are preferred due to their simplicity and the possibility of developing algorithms for automatic triangulation for a computational domain such as Delaunay triangulation [19, 20]. Although fewer elements are needed when quadrilateral elements are used, Triangular elements are well-suited for complex geometries and cause fewer numerical dispersion errors. Furthermore, the calculation of element matrices is easier in triangular elements. Discretization error might inevitably occur unless sufficient mesh density is used. This happens regardless of whether triangular or quadrilateral elements are used for meshing. One way to overcome this problem is to refine the mesh. Mesh refinement might be needed

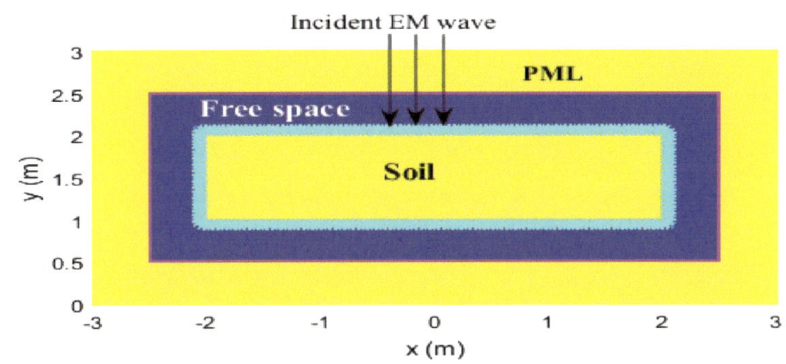

Figure 1.
Electromagnetic scattering in soil: FEM modeling with PML.

especially if the geometry has a curved boundary or some corners, sharp edges, small features, or discontinuities. This might be important especially if linear interpolation functions are used. Another approach that increases accuracy is to use high-order elements at the expense of increased computational load. To achieve this, we use extra nodes within an element and use high-order interpolation functions. **Figure 3** shows a simple mesh with six linear triangular elements having eight nodes. Local node numbers must follow the anticlockwise orientation in all elements to guarantee that the area of each element is obtained as a positive quantity. During the mesh generation, certain data arrays must be created [19].

An element connectivity matrix of size Mx3, where M is the number of elements vectors of node coordinates (x and y), each of size Nx1, where N is the number of nodes representation of the given functions (p_x, p_y, q, and f) in each element (each of which is of size Mx1 Arrays containing special nodes and/or element. The connectivity matrix belonging to **Figure 3** is given in **Table 1**.

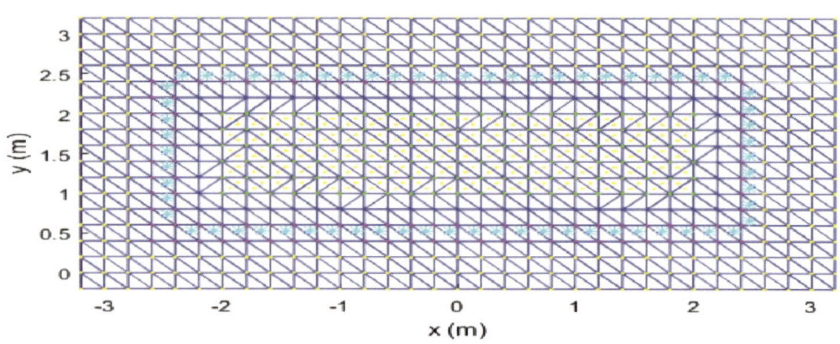

Figure 2.
Mesh formed by triangular elements for the rectangular domain of the BVP.

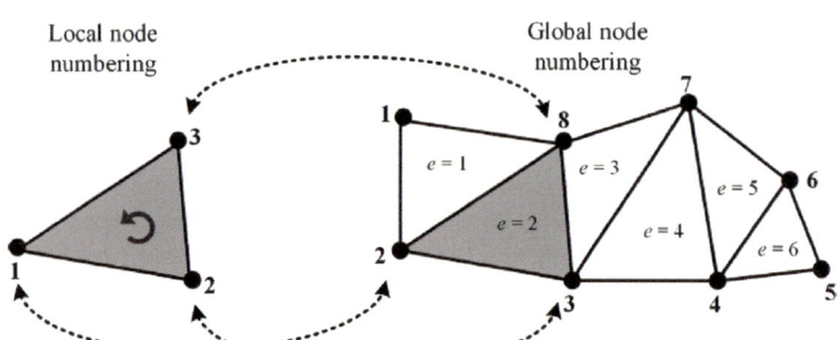

Figure 3.
Mesh of a 2-D domain using linear triangular elements.

Element (e)	Node 1	Node 2	Node 3
1	1	2	8
2	2	3	8
3	3	7	8
4	3	4	7
5	4	6	7
6	4	5	6

Table 1.
Structure of element connectivity for triangular mesh.

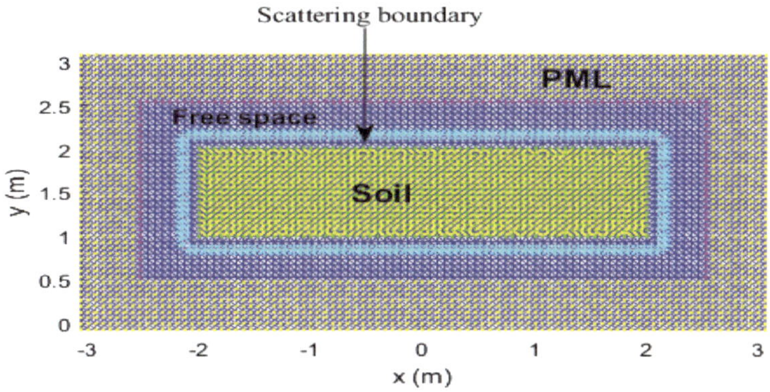

Figure 4.
Triangular mesh generated with $\Delta h = \lambda_0/15$.

The Delaunay function is used to create the connectivity matrix and automatically enumerates the nodes. The triangular mesh is generated for a rectangular domain. The element size (els) = $\Delta h = \lambda_0/5$, $\Delta h = \lambda_0/10$, $\Delta h = \lambda_0/20$ and the like. The higher the denominator, the finer the element. **Figure 4** shows the mesh generated in MATLAB for els = $\Delta h = \lambda_0/15$.

4.2 The scattered field formulation

Fields in the presence of the scatterer can be decomposed into two parts:

- The first one is the incident field which is produced without the scatterer and
- The second one is the scattered field produced by an equal amount of current induced on the scatterer or the surface enclosing the scatterer.

We first assume the TM polarization case, where the incident field $\left(\mathbf{E}^{\text{inc}} = \hat{a}_z E_z^{\text{inc}}\right)$, and the scattered field $\left(\mathbf{E}^{\text{scat}} = \hat{a}_z E_z^{\text{scat}}\right)$, are expressed in terms of the z-component, the z-component of the total field can be expressed as $E_z = E_z^{\text{scat}} + E_z^{\text{inc}}$. For an isotropic case, the total field satisfies the differential Eq. [9].

$$-\frac{\partial}{\partial x}\left(\frac{1}{\mu_r}\frac{\partial E_z^{\text{scat}}}{\partial x}\right) - \frac{\partial}{\partial x}\left(\frac{1}{\mu_r}\frac{\partial E_z^{\text{scat}}}{\partial y}\right) - k_0^2 \varepsilon_{rc} E_z^{\text{scat}} = f(x,y), \tag{19}$$

Where the source term is given by

$$f(x,y) = \frac{\partial}{\partial x}\left(\frac{1}{\mu_r}\frac{\partial E_z^{\text{inc}}}{\partial x}\right) + \frac{\partial}{\partial x}\left(\frac{1}{\mu_r}\frac{\partial E_z^{\text{inc}}}{\partial y}\right) + k_0^2 \varepsilon_{rc} E_z^{\text{inc}}, \tag{20}$$

This differential equation is a special form of Eq. (4a), where $u = E_z^{\text{scat}}(x,y)$, $p_x = p_y = p = \frac{1}{\mu}$ and $q = -k_0^2 \varepsilon_{rc}$.

The incident field can be arbitrary and usually chosen as a uniform plane wave since the incident field sources are sufficiently far away from the object. The incident field with the unit magnitude is given by

$$\mathbf{E}^{\text{inc}} = \hat{a}_z E_z^{\text{inc}} = \hat{a}_z \exp\left[jk\left(x\cos\varphi^{\text{inc}} + y\sin\varphi^{\text{inc}}\right)\right], \tag{21}$$

where φ^{inc} is the angle of incidence for the x-axis in cartesian coordinates.

Similar calculations can be performed in TE polarization mode by replacing the electric field with the magnetic field and permittivity with permeability. Hence, the differential equation in terms of the scattered magnetic field is given by

$$-\frac{\partial}{\partial x}\left(\frac{1}{\varepsilon_{rc}}\frac{\partial H_z^{\text{scat}}}{\partial x}\right)-\frac{\partial}{\partial x}\left(\frac{1}{\varepsilon_{rc}}\frac{\partial H_z^{\text{scat}}}{\partial y}\right)-k_0^2\mu_r H_z^{\text{scat}}=f(x,y), \tag{22}$$

Where the source term is given by

$$f(x,y)=\frac{\partial}{\partial x}\left(\frac{1}{\varepsilon_{rc}}\frac{\partial H_z^{\text{inc}}}{\partial x}\right)+\frac{\partial}{\partial x}\left(\frac{1}{\varepsilon_{rc}}\frac{\partial H_z^{\text{inc}}}{\partial y}\right)+k_0^2\mu_r H_z^{\text{inc}}, \tag{23}$$

This differential equation is a special form of Eq. (4a), where $u=H_z^{\text{scat}}(x,y)$, $p_x=p_y=p=\frac{1}{\varepsilon_{rc}}$ and $q=-k_0^2\mu$.

For dielectric objects, the right-hand side of the matrix equation can be obtained by using the source term $f(x, y)$ in each element. An alternative simpler approach is that the source term in Eq. (20) is just the differential operator applied to the known incident field within the object. The left-hand side of Eq. (19) is the same differential operator being applied to the unknown scattered field, which yields the left-hand side of the matrix equation ($\mathbf{Au}$), where $\mathbf{u}$ refers to the vector of nodal values of the scattered field. Therefore, $\mathbf{Au}=-\mathbf{Au}^{\text{inc}}$ must be satisfied within the scatterer region. After forming the global matrix, as usual, the right-hand side vector can be modified by just multiplying the incident field vector by the global matrix, i.e. $(\mathbf{b}=-\mathbf{Au}^{\text{inc}})$, only for entries $\mathbf{b}$ corresponding to the nodes lying inside the object.

4.3 Radar cross section

The radar cross-section (RCS) of the scatterer is perhaps the most critical parameter that must be evaluated in the post-processing phase of FEM for the electromagnetic scattering problem [9]. RCS is the reflection of the scattering electromagnetic ware at an incident angle in a particular direction. In other words, it is the area capturing that amount of scattered power produced at the receiver in an isotropic medium. This is a density that is equal to that scattered by the actual target. It is a function of several parameters, such as operation frequency, polarization, illumination angle, observation angle, geometry, and material properties of the object. In 2-D, it is mathematically defined as

$$\sigma_{2D}=\lim_{r\to\infty} 2\pi\rho\frac{\left|\mathrm{u}_{\text{far}}^{\text{scat}}\right|^2}{\left|\mathrm{u}^{\text{inc}}\right|^2} \tag{24}$$

Where $\mathrm{u}_{\text{far}}^{\text{scat}}=\frac{1}{\sqrt{\rho}}f(\varphi)$ is the scattered electric or magnetic field at far-zone observed in a given direction, when ρ satisfies the inequality $\rho\gg 2D^2/\lambda$, where D is the largest dimension of the scatterer and λ is the wavelength.

If the incident and observation directions are the same, the RCS is called monostatic or backscatter RCS; otherwise, it is referred to as bistatic RCS. In 2-D, RCS is usually normalized for λ (wavelength) or m (meter). The unit for σ_{2D}/λ is dBw, and for σ_{2D}/m is dBm.

5. Scattering inside the dielectric medium

For each such element, the far-zone field is computed and then superposed over all elements, as follows:

$$E_{\text{far}}^{\text{scat}} = \eta\sqrt{\frac{k}{8\pi\rho}}e^{-j3\pi/4}e^{-jk\rho}\sum_{i=1}^{K} J_z^{(i)} e^{jk\left(x_m^{(i)}cos\varphi+y_m^{(i)}sin\varphi\right)}\Delta l^{(i)}, \tag{25}$$

where K is the number of elements adjacent to the boundary, $\left(x_m^{(i)}, y_m^{(i)}\right)$ the i-th segment midpoint, and $\Delta l^{(i)}$ the i-th segment length.

Since FEM is formulated in terms of the scattered electric field, computation of the derivatives of the scattered field requires additional effort. This can simply be achieved by using the weighted sum of derivatives of shape functions in terms of nodal scattered fields.

5.1 Scattering inside the dielectric medium, TM case

For the dielectric object, the TM case, the interior part of the object should be included. Since the magnetic current density becomes nonzero, the integral containing the magnetic current density should be evaluated. The magnetic current density has x- and y-components, which is defined as $\mathrm{M} = \hat{a}_x M_x + \hat{a}_y M_y$. Hence, it shows that $\hat{\rho} \times \mathrm{M} = \left(\cos\varphi M_y - \sin\varphi M_x\right)\hat{a}_z$. The components of the magnetic current density can be determined in terms of the scattered electric field as follows:

$$\mathrm{M} = \mathrm{E} \times \hat{n} = -n_y E_z \hat{a}_x + n_x E_z \hat{a}_y, \tag{26}$$

Finally, the scattered electric field can be obtained as follows:

$$\begin{aligned} E_{\text{far}}^{\text{scat}} &= \sqrt{\frac{k}{8\pi\rho}}e^{-j3\pi/4}e^{-jk\rho}\sum_{i=1}^{K}\left(n_x \cos\varphi + n_y \sin\varphi\right)E_z^{(i)} e^{jk\left(x_m^{(i)}cos\varphi+y_m^{(i)}sin\varphi\right)}\Delta l^{(i)} \\ &+\eta\sqrt{\frac{k}{8\pi\rho}}e^{-j3\pi/4}e^{-jk\rho}\sum_{i=1}^{K} J_z^{(i)} e^{jk\left(x_m^{(i)}cos\varphi+y_m^{(i)}sin\varphi\right)}\Delta l^{(i)} \end{aligned} \tag{27}$$

Here, $E_z^{(i)}$ can be determined as the average of nodal field values connected to the boundary.

5.2 Scattering inside the dielectric medium, TE case

For dielectric object, TE case: The magnetic current density is nonzero and has only z-component, $\mathrm{M} = \hat{a}_z M_z$ because $\hat{\rho} \times (\hat{\rho} \times \mathrm{M}) = -M_z \hat{a}_z$. The scattered magnetic field is determined as:

$$\begin{aligned} H_{\text{far}}^{\text{scat}} &= \sqrt{\frac{k}{8\pi\rho}}e^{-j3\pi/4}e^{-jk\rho}\sum_{i=1}^{K}\left(J_x^{(i)} \sin\varphi - J_y^{(i)} \cos\varphi\right)e^{jk\left(x_m^{(i)}cos\varphi+y_m^{(i)}sin\varphi\right)}\Delta l^{(i)} \\ &+\eta\sqrt{\frac{k}{8\pi\rho}}e^{-j3\pi/4}e^{-jk\rho}\sum_{i=1}^{K} M_z^{(i)} e^{jk\left(x_m^{(i)}cos\varphi+y_m^{(i)}sin\varphi\right)}\Delta l^{(i)} \end{aligned} \tag{28}$$

6. Simulation results and discussion

The operating frequency used in the simulation is 300 MHz, and the free-space wavelength is $\lambda_0 = 1m$. The mesh size is approximatively $\Delta h = \lambda_0/40$. The dielectric medium is a rectangular computational domain, the relative permittivity for sandy soil $\varepsilon_{rc1} = 4 + i0.1$, the relative permittivity for loamy soil $\varepsilon_{rc2} = 12 + i1.8$, and the relative permittivity for clay soil $\varepsilon_{rc3} = 25.3 + i5.7$. The angle of incidence is 0°, that is, the incident plane wave travels in the $-y$ direction. The wavelength inside the dielectric medium (soil) decreases in terms of the wavelength in free-space and relative permittivity, i.e. $\lambda = \lambda_0/\sqrt{\varepsilon_{rc}}$, and hence, mesh size $\Delta h = \lambda_0/40$ resolution corresponds to $\lambda\sqrt{\varepsilon_{rc}}/40$ resolution inside the dielectric. Therefore, smaller elements are used in dielectric media to preserve the level of accuracy. All simulation results are obtained using MATLAB.

Figure 5 shows scattering from a dielectric medium for TE mode, with sandy soil characteristics. In (a) and (b), the scattered and total fields are shown respectively. We observe a minimal electric field scattering and an intense total electric field inside the dielectric medium of sandy soil. This is because this medium is porous and allows for better signal propagation.

Figure 6 shows scattering from a dielectric medium for TM mode, with sandy soil characteristics. In this case, we observe a minimal magnetic field scattering in (a) and an intense total magnetic field in (b) inside the dielectric medium of sandy soil.

Figure 7 shows scattering from a dielectric medium for TE mode, with loamy soil characteristics. In (a) and (b), the scattered and total fields are shown respectively. We observe an intense electric field scattering and a minimal total electric

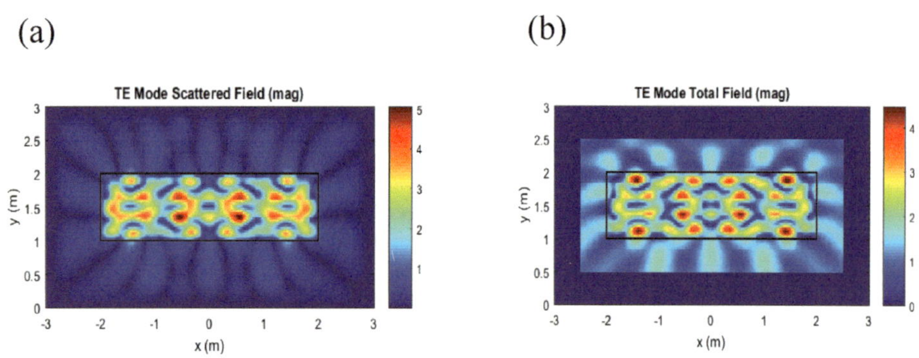

Figure 5.
Scattering from a dielectric medium (Sandy soil): (a) scattered field (TE) (b) Total field (TE).

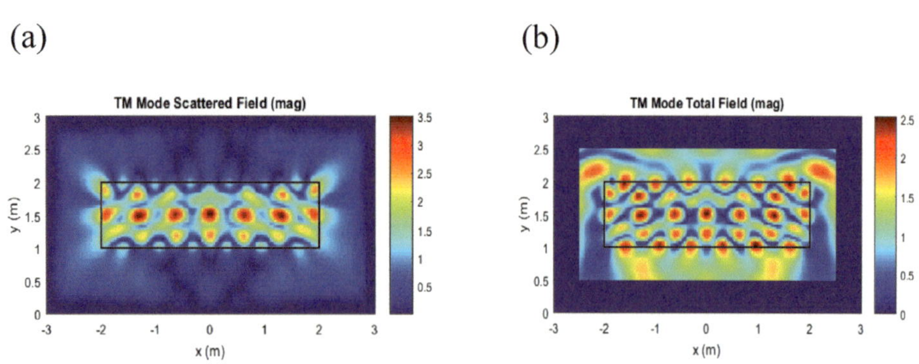

Figure 6.
Scattering from a dielectric medium (Sandy soil): (a) scattered field (TM) (b) Total field (TM).

field inside the dielectric medium of loamy soil. This is because this medium is less porous and presents some challenges in signal propagation.

Similarly, **Figure 8** shows scattering from a dielectric medium for TM mode, with loamy soil characteristics. In this case, we observe a high magnetic field scattering in (a) and a low total magnetic field in (b) inside the dielectric medium of loamy soil.

Figure 9 shows scattering from a dielectric medium for TE mode, with clay soil characteristics. In (a) and (b), the scattered and total fields are shown respectively. We observe very high electric field scattering and a low total electric field inside the dielectric medium of clay soil. This is because this medium is non-porous and presents a very poor signal propagation.

Similarly, **Figure 10** shows scattering from a dielectric medium for TM mode, with clay soil characteristics. In this case, we observe a very high magnetic field

(a) (b)

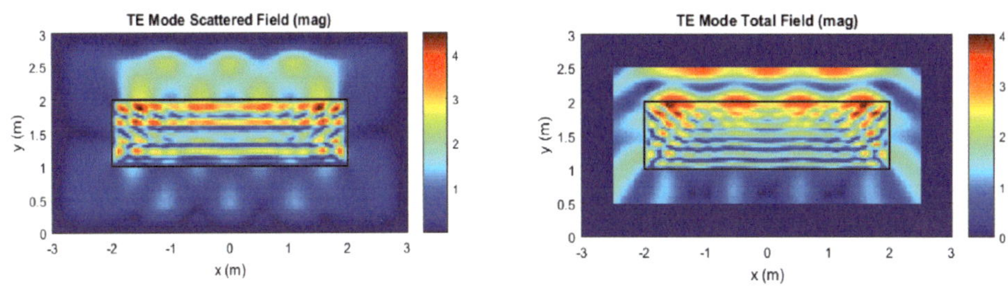

Figure 7.
Scattering from a dielectric medium (loamy soil): (a) scattered field (TE) (b) Total field (TE).

(a) (b)

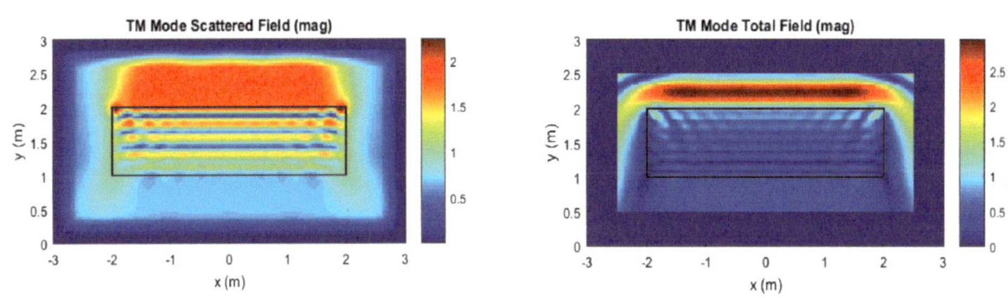

Figure 8.
Scattering from a dielectric medium (loamy soil): (a) scattered field (TM) (b) Total field (TM).

(a) (b)

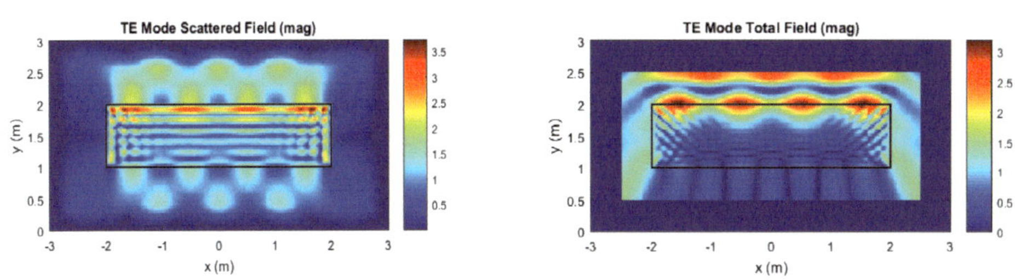

Figure 9.
Scattering from a dielectric medium (clay soil): (a) scattered field (TE) (b) Total field (TE).

scattering in (a) and a low total magnetic field in (b) inside the dielectric medium of clay soil.

Figures 11–13 show the bistatic RCS profiles to describe how scatterers reflect the incident electromagnetic wave in a given direction. This is the area intercepting

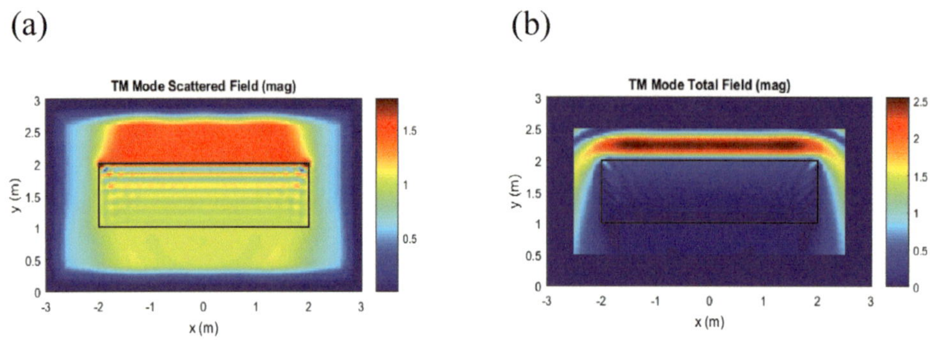

Figure 10.
Scattering from a dielectric medium (clay soil): (a) scattered field (TM) (b) Total field (TM).

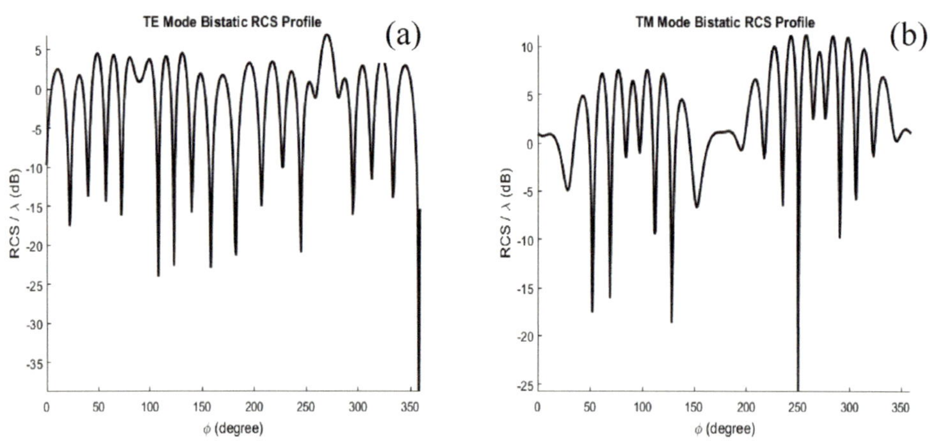

Figure 11.
Radar cross section in Sandy soil for (a) TE mode and (b) TM mode.

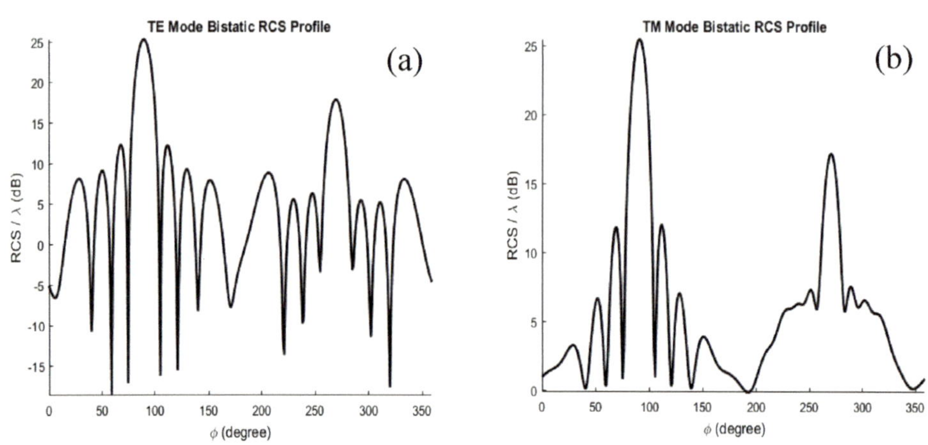

Figure 12.
Radar cross section in loam soil for (a) TE mode and (b) TM mode.

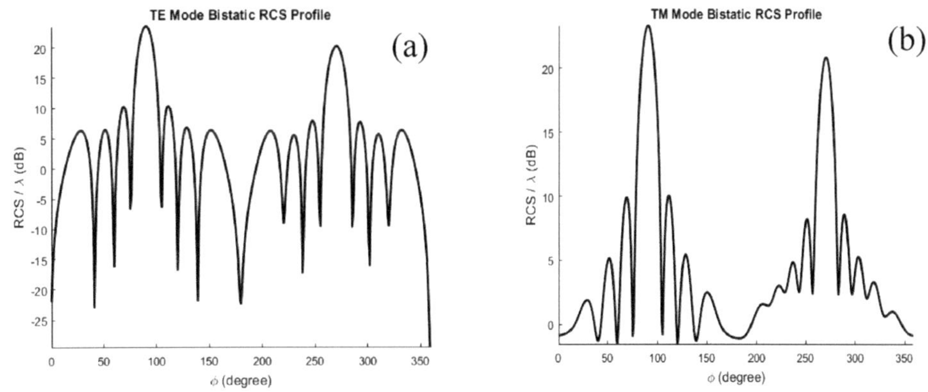

Figure 13.
Radar cross section in clay soil for (a) TE mode and (b) TM mode.

that amount of power which, when scattered in the soil medium, produces at the receiver a density which is equal to that scattered by the actual target. The RCS is a function of several parameters, such as operation frequency, polarization, illumination angle, observation angle, geometry, and properties of the soil medium. It is shown in the TE modes (a) and TM modes (b) for sandy soil, loamy soil, and clay soil, respectively.

7. Conclusion

In a two-Dimensional Finite Element Analysis of EM wave Propagation through the soil, a boundary value problem (BVP) used to solve the time-harmonic electromagnetic problem in 2-D, has been expressed in its generic form. In TM and TE cases, the Helmholtz model has considered an infinitely large dielectric object of an arbitrary cross-section for scattering from a dielectric medium and illuminated by an incident wave. Since the domain extends to infinity, an artificial boundary, an absorbing boundary condition (ABC), or a perfectly matched layer (PML), has been used to truncate the computational domain. The incident field, the scattered field, and the total field in terms of the z-component are expressed for the TM and TE modes. The radar cross-section (RCS), a function of several parameters, such as operation frequency, polarization, illumination angle, observation angle, geometry, and material properties of the medium, has been computed to describe how a scatterer reflects an incident electromagnetic wave in a given direction. Simulation results for the scattered field, the total field, have been presented for soil types, and the radar cross-section for different element refinements have also been presented.

Author details

Frank Kataka Banaseka[1*], Kofi Sarpong Adu-Manu[2], Godfred Yaw Koi-Akrofi[1] and Selasie Aformaley Brown[1]

1 University of Professional Studies, Accra, Ghana

2 University of Ghana, Accra, Ghana

*Address all correspondence to: frank.banaseka@upsamail.edu.gh

References

[1] Salam A, Subsurface MIMO. A Beamforming Design in the Internet of underground things for digital agriculture applications. MDPI Journal of Sensors and Actuators. 2019. DOI: 10.3390/jsan8030041

[2] Vuran MC, Salam A, Wong R, Irmak S. Internet of underground things in precision agriculture: Architecture and technology aspects. Ad Hoc Netw. 2018

[3] Akyildiz IF, Sun Z, Vuran MC. Signal propagation techniques for wireless underground communication networks. Phys. Commun. J. 2009

[4] Jabbar S, Asif Habib M, Minhas AA, Ahmad M, Ashraf R, Khalid S, et al. Analysis of factors affecting energy-aware routing in wireless sensor network. Wirel. Commun. Mob. Comput. 2018

[5] Konda A, Rau A, Stoller MA, Taylor JM, Salam A, Pribil GA, et al. Soft microreactors for the deposition of conductive metallic traces on planar, embossed, and curved surfaces. Adv. Funct. Mater. 2018

[6] Vuran MC, Akyildiz IF. Channel model and analysis for wireless underground sensor networks in soil medium. Phys. Commun. 2010

[7] Sun Z, Akyildiz I. Magnetic induction Communications for Wireless Underground Sensor Networks. IEEE Trans. Antennas Propag. 2010

[8] Engmann F, Adu-Manu K S, Abdulai J-D, Katsriku F. Application of Prediction Approaches in Wireless Sensor Networks. In: Wireless Sensor Networks – Design, Deployment, and Application. Intech Open. 202. DOI: 10.5772/intechopen.94500

[9] Banaseka FK, Katsriku FA, Abdulai J-D, Adu-Manu KS, Engmann F. Signal propagation models in soil medium for the study of wireless underground sensor networks: A review of current trends. In: Hindawi Journal of Wireless Communication and Mobile Computing. 2021

[10] Saeed N, Alouini M-S, Al-Naffouri TY. Toward the internet of underground things: A systematic survey. IEEE Communication Survey and Tutorials. 2019;**21**(4):3443-3466

[11] David JG. Introduction of Electrodynamics. 2nd Ed. ISBN No 0–13–481367-7. Prentice – Hall: Inc; 1989

[12] Arshad K, Katsriku F, Lasebae A. Radiowave VHF propagation modeling in forest using finite elements. In proceedings – 2006 international conference on information and communication technologies, pp. 2146–2149, Damascus. Syria. 2005

[13] Arshad K, Katsriku F. Lasebae a. Modeling obstructions in straight and curved rectangular tunnels by finite element approach. Journal of Electrical Engineering–Bratislava. 2007;**59**(1): 09-13

[14] Arshad K, Katsriku F, Lasebae A. Effects of different parameters on attenuation rates in circular and arch tunnels. Piers Online. 2006;**3**(5):607-611

[15] Ozgun O, Kuzuoglu M. Matlab-based finite element programming in electromagnetic Modeling. CRC Press is an imprint of Taylor and Francis Group Informa. 2019

[16] Banaseka FK, Franklin H, Katsriku F, Abdulai J-D, Ekpezu A, Wiafe I. Soil medium electromagnetic scattering model for the study of wireless underground sensor networks. Hindawi Journal of Wireless Communication and Mobile Computing. 2021

[17] Li T, Hajnsek I, Chen K-S. Sensitivity analysis of Bistatic scattering for soil moisture retrieval. MDPI Remote Sens. 2021;**13**:188. DOI: 10.3390/rs13020188

[18] Li A-Q, Yin C-Y, Wu W. Accurate Prediction of Radio Wave Propagation in an Environment of Dielectric Ground and Obstacles Based on the Principle of Domain Decomposition. Antennas and Propagation: IET Microwave; 2021. DOI: 10.1049/mia2.12130

[19] Delaunay B. Sur la sphere vide. Bulletin de l'Academie des Sciences de l'URSS. Classe des sciences mathematiques et naturelles. Vol. 1934; **6**:789-800

[20] Shewchuk JR. Lecture notes on Delaunay mesh generation. University California at Berkeley. 2012